D1616903

HANDBOOK OF GEOPHYSICAL EXPLORATION

SEISMIC EXPLORATION

VOLUME 24

QUANTITATIVE BOREHOLE
ACOUSTIC METHODS

HANDBOOK OF GEOPHYSICAL EXPLORATION

SEISMIC EXPLORATION

Editors: Klaus Helbig and Sven Treitel

PUBLISHED VOLUMES

1984 – Mathematical Aspects of Seismology. 2nd Enlarged Edition (M. Båth and A.J. Berkhout)*
1984 – Seismic Instrumentation (M. Pieuchot) ISBN 0-08-036944-8
1984 – Seismic Inversion and Deconvolution (a) Classical Methods (E.A. Robinson
1985 – Vertical Seismic Pro.ling (a) Principles. 2nd Enlarged Edition (B.A. Hardage
1987 – Pattern Recognition & Image Processing (F. Aminzadeh)*
1987 – Seismic Stratigraphy (B.A. Hardage)*
1987 – Production Seismology (J.E. White and R.L. Sengbush)*
1989 – Supercomputers in Seismic Exploration (E. Eisner)*
1994 – Seismic Coal Exploration (b) In-Seam Seismics (L. Dresen and H. Rüter)*
1994 – Foundations of Anisotropy for Exploration Seismics (K. Helbig) ISBN 0-08-037224-4
1998 – Physical Properties of Rocks: Fundamentals and Principles of Petrophysics (J.H. Schön) ISBN 0-08-041008-1
1998 – Shallow High-Resolution Reflection Seismics (J. Brouwer and K. Helbig) ISBN 0-08-043197-6
1999 – Seismic Inversion and Deconvolution (b) Dual-Sensor Technology (E.A. Rc ISBN 0-08-043627-7
2000 – Vertical Seismic Profiling: Principles. 3d Updated and Revised Edition (B.A. Hardage) ISBN 0-08-043518-1
2001 – Seismic Signatures and Analysis of Reflection Data in Anisotropic Media (I. Tsvankin) ISBN 0-08-043649-8
2001 – Computational Neural Networks for Geophysical Data Processing (M.M. Pc ISBN 0-08-043986-1
2001 – Wave Fields in Real Media: Wave Propagation in Anisotropic, Anelastic an Porous Media (J.M. Carcione) ISBN 0-08-043929-2
2002 – Multi-Component VSP Analysis for Applied Seismic Anisotropy (C. MacB ISBN 0-08-0424439-2
2002 – Nuclear Magnetic Resonance. Petrophysical and Logging Applications (K.J. Dunn, D.J. Bergman and G.A. LaTorraca) ISBN 0-08-043880-6
2003 – Seismic Amplitude Inversion in Reflection Tomography (Y. Wang) ISBN 0-08-044243-9
2003 – Seismic Waves and Rays in Elastic Media (M.A. Slawinski) ISBN 0-08-04
2003 – Quantitative Borehole Acoustic Methods (X.-M. Tang and A. Cheng) ISBN 0-08-044051-7

* *Book out of print.*

SEISMIC EXPLORATION

Volume 24

QUANTITATIVE BOREHOLE ACOUSTIC METHODS

by

X.M. TANG
Baker Atlas
Houston, TX, U.S.A.

A. CHENG
Consultant
Houston, TX, U.S.A.

2004

ELSEVIER

Amsterdam – Boston – Heidelberg – London – New York – Oxford
Paris – San Diego – San Francisco – Singapore – Sydney – Tokyo

ELSEVIER B.V.
Sara Burgerhartstraat 25
P.O.Box 211, 1000 AE
Amsterdam, The Netherlands

ELSEVIER Inc.
525 B Street, Suite 1900
San Diego, CA 92101-4495
USA

ELSEVIER Ltd
The Boulevard, Langford Lane
Kidlington, Oxford OX5 1GB
UK

ELSEVIER Ltd
84 Theobalds Road
London WC1X 8RR
UK

First edition 2004

Library of Congress Cataloging in Publication Data
A catalog record is available from the Library of Congress.

British Library Cataloguing in Publication Data
A catalogue record is available from the British Library.

ISBN: 0-08-044051-7
ISSN: 0950-1401 (Series)

∞ The paper used in this publication meets the requirements of ANSI/NISO Z39.48-1992 (Permanence of Paper).
Printed in The Netherlands.

THE AUTHORS

Xiao-Ming Tang, Sc.D., is chief staff scientist within the Acoustic Program of the Baker Hughes Houston Technology Center, Houston, Texas.

Dr. Tang received a B.S. degree in 1982 from Beijing University, an M.S. degree from the Institute of Geophysics, State Seismological Bureau, Beijing, China, in 1984, and a Sc. D. degree from the Massachusetts Institute of Technology in 1990, all in Geophysics. He started doing research in borehole acoustics in 1986 at the Earth Resources Laboratory (ERL) of M.I.T. After graduating from M.I.T. in 1990, he worked as research scientist at New England Research and was appointed visiting scientist at ERL, M.I.T. He joined Baker Atlas (then Western Atlas) in 1994 and became project leader for acoustic processing and interpretation development.

Dr. Tang is a member of the Society of Professional Well Log Analysts and the Society of Exploration Geophysicists. He is the recipient of the Best Paper Award of the Society of Professional Well Log Analysts at its 2002 Symposium. He currently serves in the society's Technology Committee.

Dr. Tang has published more than 60 technical publications, mostly on acoustic theory, measurement, and data processing and interpretation. He authored or co-authored twelve patents. His current interests include borehole acoustics, petrophysics, and rock mechanics.

Arthur (Chuen Hon) Cheng is a consultant in borehole geophysics and petrophysics, and scientific advisor at SensorWise Inc.

Dr. Cheng received a B.Sc. with Distinction in Engineering Physics from Cornell University in 1973, and a Sc.D. in Geophysics from M.I.T. in 1978. He was the project leader of M.I.T.'s Borehole Acoustics and Logging Consortium from 1982 to 1996. He then joined Western Atlas Logging Services as Manager of Acoustic Science. He served in various managerial capacities in Borehole Acoustics and Borehole Geophysics at Western Atlas (later Baker Atlas) and Baker Hughes Inteq before leaving the management ranks for personal reasons in 2000. He is the author/co-author of over 80 papers and three patents. He is also the co-author with Frederick Paillet on the book "ACOUSTIC WAVES IN BOREHOLES" in 1990.

ACKNOWLEDGEMENTS

Writing a book in borehole acoustics requires both theoretical background and practice experience in this field. Having conducted "pure" research in academia and "applied" research in industrial organizations over almost two decades, the authors are fortunate to have interacted and cooperated with many talented individuals from both academia and industry. Their technical achievements and rich experience, the vast amount of data they provided, and even their criticisms, result in the broad range of theory, methods, and applications described in this book.

The Earth Resources Laboratory (ERL) at the Massachusetts Institute of Technology has been the unique environment to conduct active research in borehole acoustics, thanks to Nafi Toksöz, founder of ERL, whose foresight to form the Borehole Acoustic and Logging Consortium has produced a far-reaching impact in this field and fruitful results for years to come. ERL has attracted many bright graduate students and postdoctoral associates, as well as several brilliant scientists. Over nearly two decades, they studied almost all major aspects of borehole acoustics and their results have significantly advanced our understanding in this field. Most notable of them, in no particular order, are Mark Willis, Ken Tubman, Dan Burns, Frederic Mathieu, Benoit Paternoster, Marc Larrere, Fatih Guler, Dominique Dubucq, Roy Wilkens, Denis Schmitt, Karl Ellefsen, Lisa Block, Jeff Meredith, Ningya Cheng, Rick Gibson, Sergio Kostek, Bata Mandal, Chengbin Peng, Wenjie Dong, Steve Pride, Xiaomin Zhao, Zhenya Zhu, Bertram Nolte, Guo Tao, Rama Rao, and Xiaojun Huang. ERL has also attracted many internationally known scientists as short term visitors to work on borehole acoustic problems. Included in those are Fred Paillet of the USGS, Michel Bouchon of the Univ. of Grenoble, Jinzhong Zhang of the Xian Petroleum Institue, Albert Hsui of the Univ. of Illinois, Yunsheng Zhu of the Jianghan Petroleum Institute, and Kazuhiko Tezuka of Japex. Their results, including theoretical and numerical modeling of borehole acoustic wave propagation, poroelastic properties of formation rocks, acoustic well log processing and interpretation, and laboratory experimental measurements, etc., have greatly benefited this book. We would like to express special thanks to the many companies who have supported the Borehole Acoustics and Logging Consortium at ERL. Over the years over thirty companies, many of which have been merged or no longer exist, have supported the Consortium.

The authors have also benefited from technical exchange/interaction with several university professors and scientists. Among them are Prof. J.E. White of Colorado School of Mines, Prof. Xiaofei Chen of Beijing University, Prof. Kexie Wang of Jilin University, Andrew Norris of Rutgers University, Wayne Pennington of Michigan Technical University, Bjorn Ursin of Norwegian Technology University, and Max Peeters of the Colorado School of Mines.

We also benefited from many individuals and colleagues who sponsored, encouraged, and cooperated in borehole acoustic research. Among them are Joe Zemanek, Fred Paillet, John Minear, Mike Williams, Gopa De, Leon Thomson, Brian Hornby, Keith Kahatara, Dave DeMartini, Paul Hatchell, Matt Hauser, Rob Vines, Alain Brie, John Walsh, David Johnson, Steve Chang, Kai Hsu, Bikash Sinha, Tom Plona, Sergio Kostek, Ningya Cheng, Tanya Tamachenko, Bob Joyce, Carsten Pretzschner, Tianrun Zhang, Edy Purwoko, Xinhua Sun, Bob Lester, Stephan Gelinsky, Yuli Quan, Tim Geerits, Kurt Strack, Chris Payton, Jane Thomas, Mathew Schmidt, Tsili Wang, Howard Glassman, Dan Georgi, Doug Patterson, Georgios Varsamis, and Joakim Blanch.

Finally, we thank Baker Atlas for permitting the use of many data examples for the publication of this book. Arthur Cheng would like to thank SensorWise, Inc., and Halliburton Energy Services for their support in finishing this manuscript. Liz Henderson and Karen Bush provided technical and editorial assistance for the book. We also thank Dr. Sven Treitel and Prof. Klaus Helbig, and the Elsevier Science Publishing staff. Without their encouragement, assistance, and support, the publication of this book would have been impossible.

PREFACE

Borehole acoustic logging has been a highly specialized technology in the oil and gas industry. It has become an indispensable tool for petroleum reservoir exploration, reserve estimation, well completion, and hydrocarbon production. Early measurements utilized analogue technology for detecting the acoustic arrivals. These measurements were aimed at determining the acoustic transit time (or slowness) along the borehole, correlating it with formation porosity, and using it to locate and estimate petroleum reservoirs. In the late 1970s, the "digital era" brought forth "full waveform" acoustic logs – logs obtained by digitizing the complete acoustic waveform signal for an array of receivers on the tool. The needs to process and interpret full waveform acoustic array data stimulated many theoretical studies and the development of quantitative processing and analysis methods.

In recent years we have witnessed the rapid development of quantitative analysis and interpretation methods in acoustic logging applications. This development is the result of the demand in the acoustic logging marketplace, years of continuing research at various academia and industry organizations, and the evolution of computer technology. Analysis methods from elaborate theory and numerical modeling, coupled with advanced inversion theory and computing techniques, are implemented onto fast computers and workstations. This allows for effective and efficient processing of acoustic data even at the well site.

Many studies have been done on acoustic wave propagation in boreholes with different types of formations. The formations were modeled as elastic, porous, heterogeneous, and/or anisotropic. The complicated wave phenomena in boreholes include the excitation of wave energy, mode partition of the source energy, effects of casing and cement, effects of permeability and anisotropy, and the general character of borehole response for the full range of formation properties. A complicated theory for basic wave propagation has been developed, forming the basis for reliable, quantitative analyses of full waveform log data. The theory of acoustic wave propagation in a borehole is well documented in various technical papers and books (e.g., UNDERGROUND SOUND, White, 1983; ACOUSTIC WAVES IN BOREHOLES, Paillet and able, quantitative analyses of full waveform log data. The theory of acoustic wave propagation in a borehole is well documented in various technical papers and books (e.g., UNDERGROUND SOUND, White, 1983; ACOUSTIC WAVES IN BOREHOLES, Paillet and Cheng, 1990). The focus of this book is to bridge the gap between theory and field data, from the theoretical analysis of borehole acoustic wave phenomena to the measurement of formation properties from acoustic waveform logs.

Proper processing of acoustic data is crucial for estimating formation acoustic parameters. During acoustic logging in a borehole environment, acoustic wave signals are contaminated by various noises, including digitization error, reflection/scattering from rough borehole or bed boundaries, mode conversion/interference, etc. These noises make the processing of acoustic data a formidable task. Some specific requirements for data processing are accuracy, resolution, reliability (quality control), and efficiency. The key in acoustic array data processing is minimizing data noise and maximizing data information. Several quantitative methods have emerged to extract acoustic parameters (e.g., velocity, attenuation, etc.) from the digital array waveform data, yielding results of greater precision for characterizing the mechanical, acoustic, and petrophysical properties of petroleum reservoirs. These methods will be described in detail and their merits and shortcomings will be discussed.

An important result of full waveform analysis is the estimation of formation permeability from Stoneley wave data. The Stoneley wave is the most obvious mode that dominates the low-frequency portion of full waveform log data. This wave, being fluid-borne and sensitive to the borehole interface condition, has been considered as a candidate for permeability estimation for more than three decades. Unfortunately, early studies rendered inconclusive results for two reasons. First, the tools used then did not generate low-frequency Stoneley waves (their signal spectrum was above 4 kHz). Second, the Stoneley wave, besides being affected by permeability, is sensitive to other effects unrelated to permeability. These non-permeability effects include data noise, formation/borehole changes, mud cakes at the borehole, intrinsic attenuation, and anisotropy, etc. A challenge is thus the separation of the permeability effects from the non-permeability effects. With extensive theoretical studies and the availability of low-frequency Stoneley wave data from modern acoustic tools, a breakthrough was achieved during the past few years. The new approach models the non-permeability effect using elastic wave theory and relates the difference between the modeled and the measured data to formation permeability. The permeability profile derived from this approach shows remarkable agreement with those from other measurements, such as Nuclear Magnetic Resonance and core measurements.

The emergence of dipole acoustic logging technology created new interest in acoustic logging applications. The development of dipole technology was the result of the need to obtain shear wave velocity in soft sediments, especially in deep-water reservoirs around the world. However, many important applications followed almost as an afterthought. Dipole-shear wave logging, besides its ability to measure shear velocity in soft formations, is a directional measurement that is sensitive to the azimuthal variation of formation properties. The directionality of the dipole-shear wave was further utilized to develop cross-dipole technology, allowing for the determination of azimuthal shear wave anisotropy around the borehole. This development opened a new horizon in acoustic logging applications. By measuring the azimuthal anisotropy, we can now analyze the formation stress field, detect/characterize fractures, and even map hydraulic fractures behind casing. The development of cross-dipole applications will be described in this book.

Besides measuring azimuthal shear-wave anisotropy, acoustic logging can also determine vertical-versus-horizontal shear-wave anisotropy, i.e., Transverse Isotropy (TI). The TI anisotropy exists in many sedimentary rocks and plays an important role in the seismic exploration of petroleum reservoirs. Although shear waves (monopole or dipole) propagating along a vertical borehole can measure only vertical shear-wave velocity, the Stoneley wave is sensitive to the velocity of horizontally propagating shear waves and can thus be used to determine the latter velocity. This book will describe the theory and method for determining shear-wave TI anisotropy from acoustic logging. The obtained formation anisotropy profile provides valuable information for seismic exploration, as it indicates whether anisotropy needs to be considered when migrating seismic data to image subsurface reservoirs.

In the coming years, we expect to see the development in acoustic logging continue and the range of application broaden. It is worth mentioning that acoustic logging-while-drilling has already taken place. The logging-while-drilling environment calls for complicated theoretical modeling, elaborate data analysis, and valid data interpretation. Continuing support and investment for academic and industry research and development will be important to keep up the development of acoustic logging technology.

Table of Contents

CHAPTER 1: OVERVIEW OF ACOUSTIC LOGGING APPLICATIONS AND RECENT ADVANCES

1.1 Acoustic Well Logging Concept and Evolution of Acoustic Tools

Acoustic logging is one of the principal geophysical well logging disciplines. Because formation acoustic properties, such as velocity and attenuation, are closely related to rock type and formation fluid (e.g., hydrocarbons), measuring these properties can provide useful information for formation evaluation. In acoustic logging, a tool is lowered into the borehole and an acoustic or sound wave is generated by a source in the tool. The wave travels along the borehole and interacts with the formation, and is detected by a number of receivers located in the same tool (Figure 1.1). The measurement is made continuously as the tool is pulled up or lowered along the borehole by means of a wireline cable, which supplies power from, and transmits data to, a surface control system. Thus this type of logging is called wireline logging. The processing of the acoustic data acquired by the tool produces a log of the acoustic properties of the formation around the borehole.

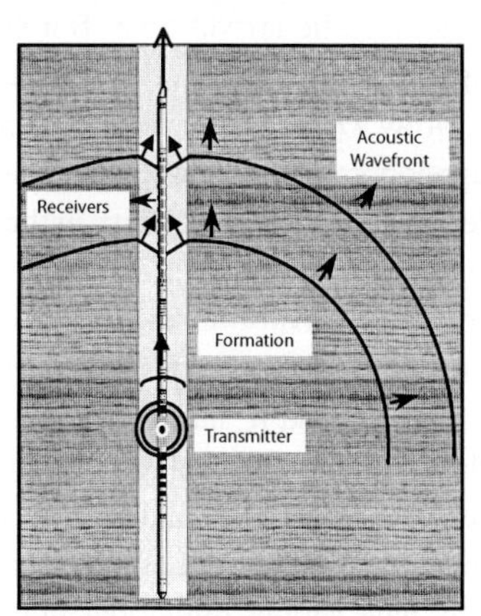

Figure 1.1 Logging of formation acoustic properties using an acoustic tool. As the tool is continuously raised (lowered) in the borehole, a log of formation acoustic parameters is obtained from processing the acoustic waves that are generated by a transmitter, propagate along the borehole, and received by a receiver array.

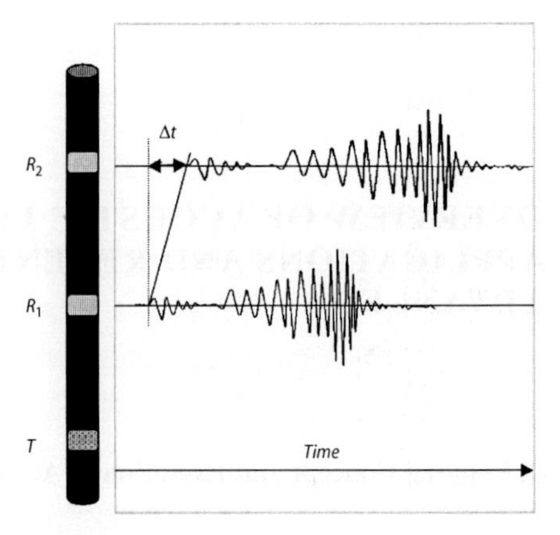

Figure 1.2. Early dual-receiver acoustic tool. The tool determines the formation compressional-wave slowness by measuring the wave's travel-time difference between the two receivers.

Early acoustic tools were usually dual-receiver tools (Figure 1.2). The tool measured the arrival-time difference of the sonic pulse generated by the transmitter source T and recorded by two receivers R_1 and R_2 located at different distances from the transmitter. The difference Δt in the travel time observed, divided by the separation of the receivers $\Delta R = R_2 - R_1$, gives the slowness (inverse of velocity) of the first (or primary) arrival of the acoustic wave traveling in the formation. This is commonly known as the P- or compressional-wave slowness of the formation.

Later measurements found that borehole variations (enlargement or cave-in) affected measurements of the acoustic arrival time. Borehole-compensated (BHC) acoustic tools were developed from two dual-receiver arrays, with the array $T'R_2R_1$ inverted with respect to the array TR_1R_2. The principle of borehole compensation is shown in Figure 1.3. The upper borehole is smaller than the lower borehole, with an abrupt change at the middle. Assume that the velocity in the formation is higher than in the borehole fluid. The dual-receiver system $T'R_2R_1$, with transmitter T' below the receivers R_2 and R_1, has a reduced arrival-time difference between R_2 and R_1 due to the additional time advance in the solid. The dual-receiver system TR_1R_2, with the transmitter T above the receivers R_1 and R_2 has an increased arrival-time difference between R_1 and R_2 due to the additional time delay in the fluid. Therefore, by averaging the measured time differences between R_2 and R_1 from source T, and between R_1 and R_2 from source T', the effect caused by the borehole change can be compensated. Dual-receiver tools, however, have serious limitations in the presence of significant noise in the data. For example, in a rugose borehole or a fractured formation, the first arrival part of the acoustic waveform is severely attenuated, making it difficult to pick or cross-correlate the wave signal at two receivers. As a result, many old sonic logs exhibit zones known as "cycle skipping", caused by erroneous picking of the first wave arrivals. To overcome the difficulty of two-receiver tools, array acoustic tools were developed.

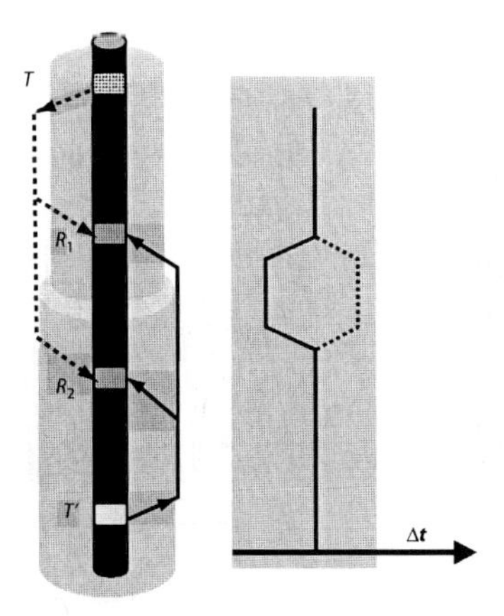

Figure 1.3. Early borehole-compensated (BHC) acoustic tool using two dual-receiver systems. One is inverted relative to the other. At the change of borehole diameter, one system measures a smaller, and the other, a larger travel-time difference between the two receivers. Averaging the two compensates the effect due to the borehole change.

The essence of using acoustic arrays is to utilize the redundancy of information in the array data so as to compensate for the possible loss or contamination of information at some receivers in the array. Because the extraction of the information from the redundant data must be processed using numerical methods, the acoustic waveform data must be recorded digitally. The development of digital array acoustic tools started in the 1970s. An acoustic array tool normally consists of four to twelve receivers, typically spaced 0.5 ft apart. The distance from source transmitter to first receiver is between 6 ft to 12 ft, depending on the depth penetration requirement. A longer spacing correponds to a deeper acoustic penetration in a formation with a velocity increasing away from the borehole. An acoustic isolator is placed between the transmitter source and the receiver array. Acoustic isolation is an important feature of the tool, which isolates, or attenuates, the acoustic wave signals traveling along the tool. The effectiveness of the isolator often determines the quality of the acoustic wave data acquired by an acoustic tool.

With digitization of the received waveforms, the complexity of the received acoustic signal began to emerge. Besides the first – or compressional wave – arrival, there are other wave types or wave modes being excited by the acoustic source. Figure 1.4 shows different acoustic modes arriving at the receiver in an elastic formation, with the formation compressional (P) and shear (S) wave velocities higher than the acoustic velocity in the borehole fluid. The compressional-wave arrival is followed by the shear wave arrival, and then by a large packet of guided wave modes. The guided waves consist of two different kinds of modes, the pseudo-Rayleigh mode, which is related to "trapped" acoustic energy and only exists when the formation shear-wave velocity is higher than the borehole fluid velocity, and the Stoneley mode, which is an interface wave guided by the borehole-formation boundary. A more detailed discussion regarding the properties of these guided wave modes is given in Chapter 2.

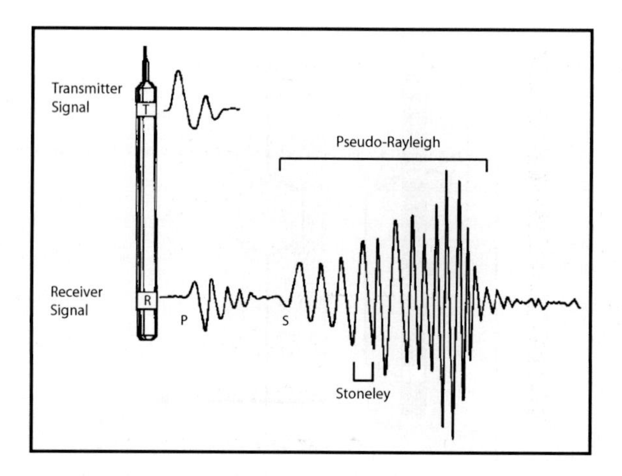

Figure 1.4. Acoustic wave modes at a receiver for a formation with a shear-wave velocity higher than the borehole fluid acoustic velocity.

A distinct advantage of the use of array receiver tools is that the different wave modes arrive at different times at different receivers. The moveout (or delay) across the receivers is seen in Figure 1.5. We make use of these different moveouts in the processing of the received array signals to analyze their propagation characteristics and to obtain formation properties. Various methods of array processing are given in Chapter 3. These methods can be categorized as frequency- and time-domain methods. The frequency-domain methods allow us to analyze the propagation and dispersion characteristics of a wave mode. The time-domain methods allow us to efficiently and effectively extract formation properties from the acoustic wave data.

Another advantage of the array receiver tools is that borehole compensation can be achieved by gathering data into a transmitter array and a receiver array and averaging the results from the two arrays (details of the data gathering schemes and their results are described in Chapter 3). This eliminates the need to use two separate transmitters for the borehole compensation, as were used in the earlier dual-receiver tools (Figure 1.3).

Conventional acoustic logging tools use sources that generate a pressure pulse in the borehole. This pressure pulse is usually omni-directional (at least with respect to the plane transverse to the borehole). Such a source is known as a monopole source. The compressional and shear-wave arrivals are refracted waves, traveling as head waves along the borehole boundary. Thus the shear-wave arrivals exist only in "fast" formations where the shear-wave velocity is higher than the velocity of the borehole fluid. When the formation shear-wave velocity is lower than the borehole fluid velocity, there is no critically refracted shear wave, therefore we cannot measure the shear-wave velocity in such a "slow" formation. However, many earth formations are slow formations, especially in places like the Gulf of Mexico, where high porosity and semi-consolidated sands are common, and where the shear-wave velocity can be significantly (about 2–6 times) smaller than the borehole fluid velocity.

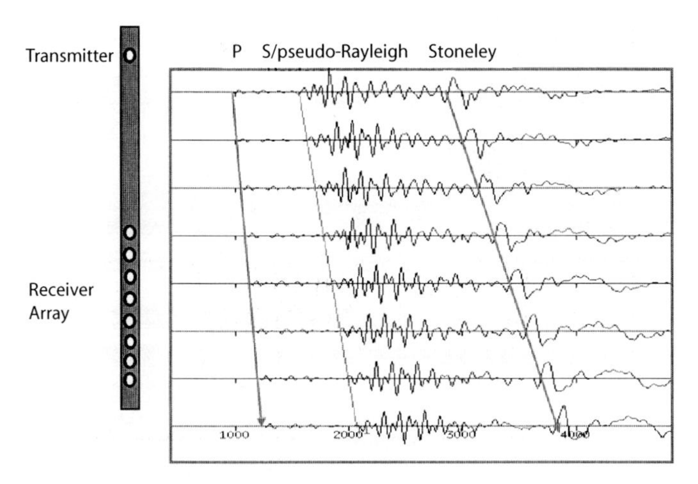

Figure 1.5. Acoustic waves moving out across a receiver array. Different waves have different moveouts because of their different propagation velocities.

To meet the requirement for determining shear-wave velocity in slow formations, dipole acoustic tools were developed. The principle of shear-wave logging was first proposed by J.E. White (1968) as the "Hula" log. This concept was later substantiated by Kitsunezaki (1980). J. Zemanek of Mobil pioneered practical shear-wave logging in the late 1970s. The result of the new shear-wave logging tool was first published by Zemanek et al. (1984). This shear-wave logging used a "bender" source to generate a uni-directional displacement in the fluid. This uni-directional wave motion generates a "flexural" wave mode in the formation, which travels along the borehole/formation interface and is detected by bender receivers located at some distance from the bender source. At high frequencies, this flexural mode is a surface-guided wave related to the Stoneley mode, and as such exists in both fast and slow formations (unlike refracted S waves). The flexural mode is dispersive, and has the formation shear-wave velocity as an upper bound for its propagation velocity at low frequencies. The uni-directional displacement source is commonly known as the "dipole" source. With this arrangement, it is possible to detect and measure shear-wave velocity in all types of formations.

To perform an efficient acoustic logging operation, the dipole acoustic transmitter and receiver system is combined with the monopole system to form a multipole array acoustic logging tool (Figure 1.6). For an acoustic logging tool, acoustic isolation is extremely essential for the quality of the dipole acoustic waveform data. When the dipole transmitter source is energized, it sets both the borehole and the tool into bending/flexural motion. Because of the low-frequency nature of the tool's flexural wave, the tool wave is very difficult to suppress or attenuate and may severely contaminate the formation flexural wave received by the receivers on the tool. A tool without effective tool-wave isolation can only measure formation shear-wave slowness in a limited range. When the formation shear arrival overlaps with the tool arrival, formation shear slowness determination either suffers from significant errors

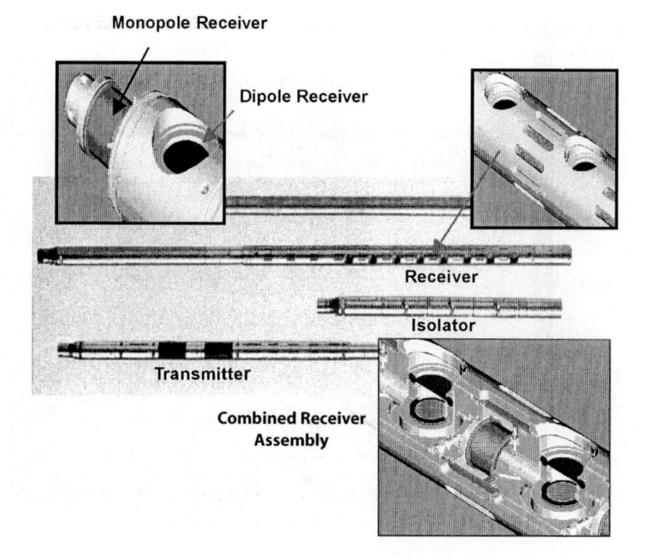

Figure 1.6. A multipole acoustic logging tool consisting of both monopole and dipole acoustic transmitter and receiver systems.

or becomes impossible. The logging industry has been struggling with the dipole acoustic isolator for over a decade. It has become evident that isolating the tool wave using soft materials (e.g., rubber) and decoupling each individual receiver from the rigid tool housing can achieve the best result (Cowles et al., 1994; see also Figure 1.7).

The dipole source has an additional advantage: because of the fact that it is uni-directional, it can be used to detect velocity variation as a function of azimuth around the borehole. This leads to the development of a cross-dipole acoustic tool. A cross-dipole acoustic tool consists of two orthogonal dipole systems, nominally facing the X and Y direction of a Cartesian coordinate system (Figure 1.7). During dipole excitation in the X – or Y – direction, the corresponding two receiver elements mounted on the opposite sides of the tool sense, respectively, a positive and a negative pressure wave. Thus, by numerically subtracting the outputs of the two elements, the signal amplitude is twice that of a single element alone. An additional benefit of this subtraction is the cancellation/reduction of common mode (e.g., monopole mode) that may be present in the data. The cross-dipole receiver system shown in Figure 1.7 can also operate as a monopole system. During monopole excitation, all four elements sense the same pressure signal. Thus, the monopole wave data are acquired simply by adding the four outputs from the four elements. This results in the crossed-multipole acoustic tool. With this tool, we can measure monopole, dipole, and cross-dipole array acoustic data in one logging pass. The advance in acquisition and processing of four-component cross-dipole acoustic data leads to the measurement of shear-wave azimuthal anisotropy and the interpretation of the anisotropy for formation evaluation, especially in fracture characterization and *in-situ* stress deter-

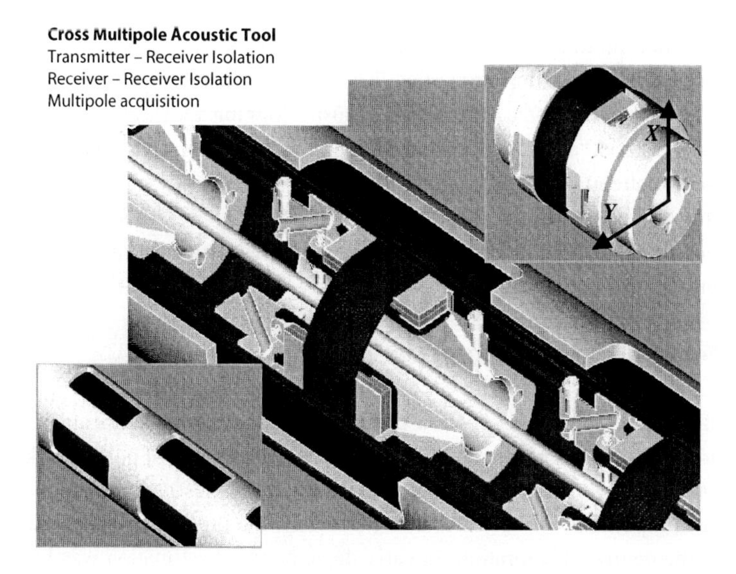

Figure 1.7. The receiver design of a crossed-multipole acoustic logging tool. The receiver consists of four elements whose outputs are combined according to the excitation mode (dipole or monopole). With this design, monopole, dipole, and cross-dipole acoustic data can all be acquired. Note that the receivers are also acoustically isolated from one another.

mination, as will be described in detail in Chapter 5.

In the following sections we give an overview of recent advances and applications of full-waveform and dipole acoustic logging. Details of the theory, measurement techniques, and case histories are presented in subsequent chapters.

1.2 Formation Elastic-Wave Property – Measurements and Applications

During acoustic logging using modern array tools, formation compressional (P) and shear (S) wave slowness (inverse of velocity) profiles are measured by the tool's monopole and/or dipole transmitter-receiver systems. The array acoustic waveform data are usually processed in the time domain using "coherency stacking" methods such as "semblance" or "*n*th root", or by a waveform inversion/matching method. The velocity profile can be determined at various resolutions using measurement apertures ranging from 3.5 ft, the typical acoustic array aperture, to 0.5 ft, the inter-receiver spacing. These processing methods and the associated resolution enhancement schemes will be described in Chapter 3. Processing dipole acoustic data for shear-wave velocity estimation may be affected by the dispersion effect of the flexural wave generated by the dipole tool. Chapter 3 describes a simple method for correcting the dispersion effect.

Formation elastic-wave velocity data have various applications in the petroleum industry. Velocity data are used to locate hydrocarbon reservoirs, estimate reserves, help map reservoir shapes, and determine the well-completion procedure best suited for hydrocarbon production. Besides formation P- and S-wave velocities, there are a number of other formation properties that we can measure or estimate from the waveform data of monopole and dipole acoustic logging tools. A commonly estimated parameter is "acoustic" porosity. Other important formation properties are P-and S-wave attenuation, permeability, shear-wave anisotropy and direction, fracture density and orientation, and *in-situ* stress, to list just a few. This book addresses the various methods for estimating these formation properties and the assumptions underlying the techniques used.

1.2.1 FORMATION POROSITY ESTIMATION

Most sedimentary formations contain porous – and sometimes permeable – rocks. Hydrocarbons in a reservoir are usually stored in the pore space of porous rocks. The estimation of porosity is thus an important task for reserve estimation. Elastic wave velocities of a porous rock are affected by porosity, the type of fluid saturating the rock, and the degree of saturation. In early days, porosity estimation was based on Wyllie's equation, which states that the total travel time in a formation is the sum of the travel time in the solid rock matrix and the travel in the fluid contained in the pores. Wyllie's equation can be written as:

$$S = (1-\phi)S_m + \phi S_f \ , \tag{1.1}$$

where ϕ is porosity, S is the observed formation (compressional-wave) slowness (or travel time divided by distance traveled, as measured, for example, by an acoustic logging tool), S_m is the slowness of the rock matrix, and S_f is the slowness of the pore fluid (whether it is oil, gas, or brine). Wyllie's equation can be treated as an empirical relation between porosity and slowness, in which the matrix slowness S_m is varied or calibrated with lithology (whether it is sandstone or limestone, etc.). There have been several modifications of Wyllie's equation, aimed at extending the application range of the slowness-porosity relation. For example, the $(1-\phi)$ term in Wyllie's equation can be modified to become $(1-\phi)^n$, where the exponent n is determined by the formation/lithology. Porosity ϕ can also be subdivided into fractions that contain movable fluids (oil, water, or gas) and fractions that contain irreducible materials (i.e., clay content). In any case, the measured formation slowness is an essential parameter for porosity estimation.

1.2.2 SEISMIC APPLICATIONS

In seismic exploration for hydrocarbon reservoirs, velocity profiles provide a constraint on the velocity model used to transform the seismic time section to the depth section. A convenient way to express the constraint is through the calculation of synthetic seismograms. Synthetic seismograms are calculated by convolving the reflectivity profile with a seismic wavelet of appropriate frequency comparable to that of the seismic data. A reflectivity profile is generated by first combining velocity and density information into acoustic impedance (density times velocity), and then calculating at each depth the normal-incidence reflection coefficient, which – for two layers in contact – is simply the difference of the two respective acoustic impedance values divided by their sum. By comparing the synthetically generated reflection trace with a stacked and moveout-corrected seismic trace, major reflection events can be identified, yielding the proper time-to-depth conversion.

Another important application is in seismic Amplitude-Versus-Offset (AVO) studies. The reflection coefficient (and thus the reflected amplitude) of the seismic wave depends on the angle of incidence as well as on the acoustic impedances. It also depends on the shear-wave velocities of the two layers in contact. By studying the AVO response of the seismic section, it is possible to estimate the P- and S-wave velocities and variations of the reflecting layer and thus infer the petrophysical properties of the formation. P- and S-wave velocity logs help constrain the AVO interpretation at the well location and help distinguish the variations caused by the petrophysical properties of the formation from those caused by P- and S-wave velocities.

1.2.3 FORMATION MECHANIC PROPERTIES

Elastic wave velocities can be used to estimate formation mechanic properties to help find the best procedure for well completion and production. The P- and S-wave velocity logs, combined with the density log, can be used to compute bulk modulus, shear modulus, and Young's modulus, as well as Poisson's ratio. These are the basic important parameters for characterizing the mechanical property of a formation. Although the moduli calculated from acoustic logs are "dynamic" moduli that may differ significantly from the "static" moduli that govern the long-term deformation behavior of the rocks (see, e.g., Cheng and Johnston, 1981), they can still be very useful in predicting the latter. Formation mechanical properties are crucial in estimating the stability of the borehole, the sanding potential, the fracture strength, and a number of other related parameters used in reservoir production and development.

1.2.4 ENHANCING SLOWNESS RESOLUTION

In formation evaluation, there is often a need to quantify the petrophysical and acoustic properties of laminated thin beds for better reserve estimation and reservoir characterization. Standard array acoustic processing obtains a slowness value over the length of the receiver array (typically 3.5 ft or 1.067 m), which may obscure the features that are smaller than the array aperture. To obtain higher vertical resolution of the slowness log profile, various techniques are sought for resolution enhancement. Enhancing resolution of slowness estimates from an array acoustic tool utilizes overlapping sub-arrays across the same depth interval with a thickness equal to

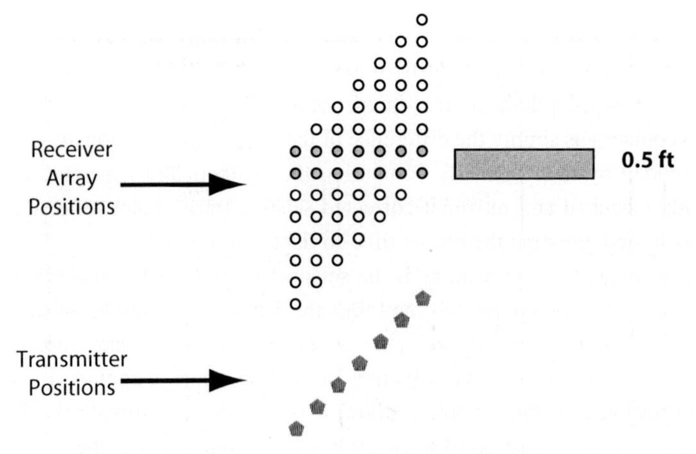

Figure 1.8. Gathering data of seven consecutive logging depths for an eight-receiver acoustic tool results in seven two-receiver subarrays that overlay the same depth interval of 0.5 ft thickness. This configuration has the highest vertical resolution for the acoustic slowness profile, if the acoustic data are appropriately analyzed.

the sub-array aperture. This is possible because in modern acoustic tools an array (typically eight receivers) of waveform data generated from the source transmitter is recorded repeatedly after the tool travels a depth interval equal to one inter-receiver spacing (typically 0.5 ft or 0.152 m). Consequently, the receiver arrays at successive source locations overlap. The thin-bed slowness analysis (Zhang et al., 2000; see also Chapter 3) uses redundant information in overlapping arrays to improve both the vertical resolution and the accuracy of the formation acoustic slowness estimation. The configuration in Figure 1.8 has the highest resolution with the processing aperture equal to the inter-receiver spacing (0.5 ft or 0.152 m) in the receiver array (a complete listing of all data-gather configurations for an eight-receiver array-acoustic tool and a validity condition for using them are given in Chapter 3). A pair-wise waveform inversion scheme is used to optimize the match between the waveform pairs across the common depth interval. This method substantially enhances the resolution of the slowness profile as compared to the conventional array processing. An example of enhanced vertical resolution in acoustic logs is given in Figure 1.9 and explained below.

Figure 1.9 compares P- and S-wave slowness profiles from the array (dashed) and thin-bed (solid) analyses (track 3). The corresponding dipole-shear waveforms (receiver 1) are shown in track 2, and the gamma ray and caliper log curves are shown in track 1. Compared to the array analysis results, the thin-bed slowness using the 0.5 ft processing aperture drastically enhances the vertical resolution of the P- and S-wave slowness profiles. To verify the enhanced resolution in the slowness profiles, a resistivity profile with 1 ft resolution (right curve) is plotted in track 3. The features on the three individual profiles have an excellent correspondence, giving a verification of the enhanced resolution. The enhanced vertical resolution allows us to resolve bedding and laminated features in formations that may be overlooked in

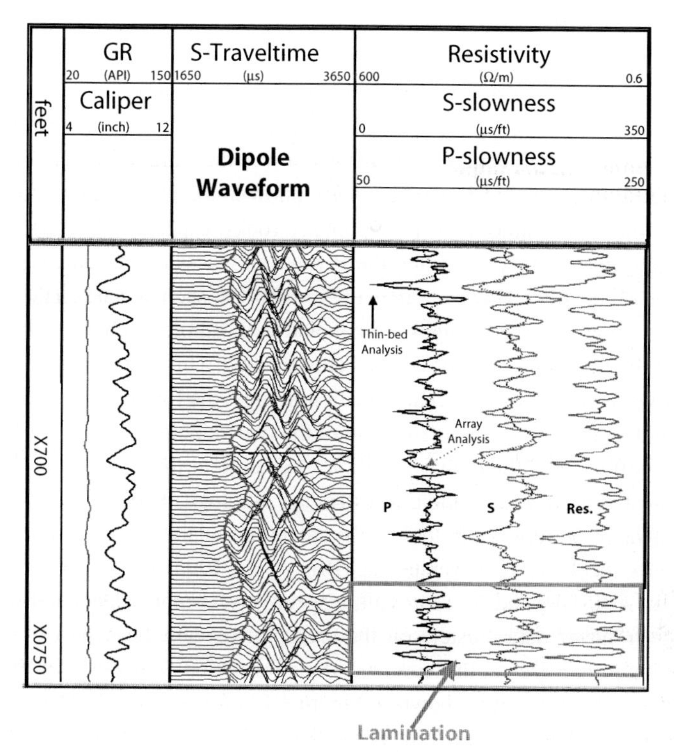

Figure 1.9. Comparison of P- and S-wave slowness profiles obtained using thin-bed (solid curves) and array (dashed curves) analysis methods (track 3) for a laminated formation. For verifying the enhanced resolution, a 1 ft resolution resistivity profile (right curve) is also plotted. Track 1 shows the caliper and gamma ray curves. Track 2 shows the dipole-shear waveforms (receiver 1).

conventional array analysis (e.g., the laminated features around X0750 ft in Figure 1.9). It can also enhance petrophysical analyses using P- and S-wave slowness and/or slowness (velocity) ratio, as will be discussed below.

1.2.5 Petrophysical Analysis Using P- and S-Wave Slowness Measurements

Formation hydrocarbon (oil and gas) saturation can significantly change the P-wave slowness and can thus be detected using the slowness measurement. The slowness change can be predicted by Gassmann's equation

$$K = K_d + \frac{(1 - K_d / K_s)^2}{\phi / K_f + (1-\phi)/K_s - K_d / K_s^2}, \tag{1.2}$$

where K is the rock bulk modulus, K_s is the modulus of the rock grain, K_d is the bulk modulus of the dry rock, ϕ is the porosity and K_f is the modulus of the pore fluid. The effect of pore-fluid saturation on P-wave velocity is calculated by

$$V_p = \sqrt{(K + \tfrac{4}{3}\mu)/\rho} \;, \tag{1.3}$$

where ρ is the rock density and μ is the shear modulus. The later is less affected by pore fluid. For porous rocks (e.g., sandstones) saturated with light hydrocarbons or gas, the value of K_f can be significantly lower than that of water. This results in a decrease (increase) in P-wave velocity (slowness). This hydrocarbon-induced change is more pronounced for high-porosity rocks, especially for gas. When the induced slowness change is strong, the P-wave slowness alone suffices to indicate the hydrocarbon effect. However, in many formations, such as sand-shale sequences, the hydrocarbon-induced changes are coupled with lithology changes. In this case, the use of the V_P/V_S ratio may provide a better method. This is so because, while both P- and S-wave velocities may change with lithology, the former is more sensitive to pore fluid than the latter. Taking the ratio of the two velocities can (at least partially) cancel the effects of lithological variation. A change in the V_P/V_S ratio is therefore more indicative of the pore fluid effect than the lithological effect.

An effective petrophysical tool for hydrocarbon analysis and saturation characterization is the V_P/V_S ratio versus Δt_P (Brie et al., 1995) or Δt_S (Williams, 1990) crossplots, where Δt_P and Δt_S represent, respectively, compressional and shear slowness. The trend of data clusters on the crossplot reveals the hydrocarbon effects and/or saturation state of the corresponding formation intervals. An example of V_P/V_S versus Δt_P crossplot is shown in Figure 1.10a for a sand-shale formation. The slowness data were obtained from array analysis (3.5 ft aperture). The shale, wet sand, and dry/gas sand trend curves in this figure (as well as in Figure 1.10b) are obtained using the method described by Brie et al. (1995). The crossplot shows that most data points cluster around the shale and wet sand trend curves, while quite a few data extend slantwise from the wet sand trend toward the dry/gas sand trend line, indicating the effects of light pore fluid (e.g., light hydrocarbon, gas, etc.) saturation. This can be understood from equations (1.1) and (1.2): as the pore fluid modulus is reduced, P-wave velocity decreases (slowness increases) while the S-wave velocity is nearly unaffected. This results in a decrease of the V_P/V_S ratio with increasing P-wave slowness on the crossplot (Figure 1.10). For a gas-bearing formation, the degree of decrease depends on the gas saturation. In this situation, the data points between the dry and wet trends represent sand intervals with various degrees of fluid saturation. This points out a way to quantify the pore fluid saturation (Brie et al., 1995). However, the exact relationship between the V_P/V_S ratio and saturation depends on the mixing law of a number of parameters, including properties of gas, oil, and brine, and is a subject still under study.

The use of high-resolution P- and S-wave slowness data can better define data trends or enhance the data correlation on the V_P/V_S ratio versus Δt_P or Δt_P crossplot, especially for formations with laminated sand-shale sequences. Figure 1.10b shows the same crossplot as in Figure 1.10a, plotted using high-resolution (0.5-ft processing aperture) P- and S-wave slowness profiles. Compared with Figure 1.10a, this plot shows a better-defined data trend extending all the way from the wet trend to

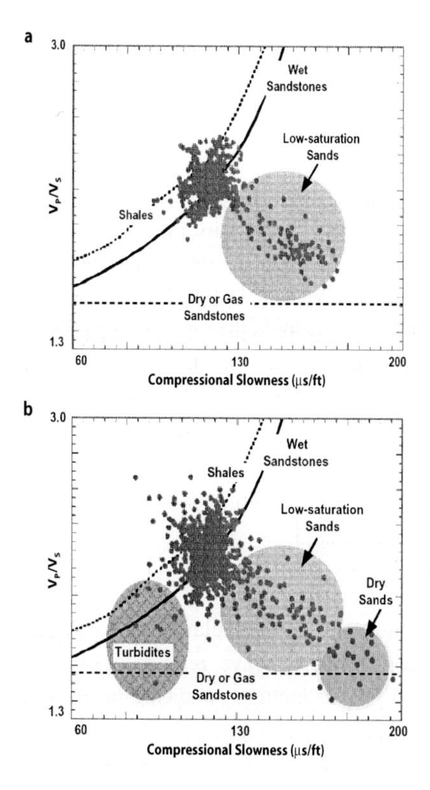

Figure 1.10. V_P/V_S versus Δt_P crossplot using array (a, 3.5 ft processing aperture) and thin-bed (b, 0.5 ft processing aperture) slowness profiles.

dry/gas line, covering a wider data range and indicating various saturation degrees of the corresponding sand intervals. This example demonstrates that the use of high-resolution acoustic slowness logs allows for a more accurate evaluation of formation hydrocarbon saturation properties.

1.2.6 STONELEY-WAVE REFLECTION ANALYSIS AND APPLICATIONS

In recent years the analysis of Stoneley waves has been a useful means of formation evaluation. The analysis of reflected Stoneley waves is of particular importance, especially for detecting fractures (Hornby et al., 1989) and thin gas beds (Tang and Patterson, 2001). The Stoneley (or tube) wave dominates the low-frequency portion of the monopole full-waveform acoustic log due to its relatively low velocity and large amplitude. Because this wave is an interface wave borne in the borehole fluid, the Stoneley wave is sensitive to formation and transport properties as elastic rigidity, permeability, and anisotropy (see Chapter 5). It is thus likely that any change of these properties results in the change of Stoneley-wave propagation characteristics. Particularly, the changes give rise to Stoneley-wave reflections that in turn can be used to characterize the changes.

At the interface between two formations (denoted by subscripts 0 and 1, respectively), the reflection of a low-frequency Stoneley wave is given by (White, 1983)

$$r = 2\left|\frac{k_1 - k_0}{k_1 + k_0}\right|, \tag{1.4}$$

where r is the reflection coefficient and $|\bullet|$ denotes taking the modulus value of a complex expression or quantity; k is the Stoneley wavenumber. If there is a significant difference between k_1 and k_0 in equation (1.4), then a significant reflection is generated from the interface.

For a thin formation layer, reflection occurs at the lower and upper layer interfaces. The total reflection caused by the layer (characterized by k_1) is given by (Tang and Cheng, 1993a):

$$r = \left|\frac{2i(k_1^2 - k_0^2)\sin(k_1 L)}{(k_1 + k_0)^2 e^{-ik_1 L} - (k_1 - k_0)^2 e^{ik_1 L}}\right|, \tag{1.5}$$

where L is the layer thickness. By calculating the wavenumber k_1 for the layer with different properties, Stoneley-wave reflection from a variety of formation features can be modeled. These include an elastic layer, a permeable porous layer, and even a fluid-filled fracture intersecting the borehole. Of particular interest is the scenario of a permeable porous layer. By using the poroelastic wave theory (described in Chapter 4) to calculate the Stoneley wavenumber for the layer, it can be shown that the Stoneley-wave characteristics (e.g., attenuation and dispersion) depend critically on the pore fluid (or saturant) properties, i.e., compressibility and viscosity (the viscosity effect can be expressed using fluid mobility, defined as the ratio permeability/viscosity). This suggests that a thin gas bed in a formation can produce a significant Stoneley-wave reflection because of its high fluid mobility and compressibility contrast to the surrounding formation.

Accompanying the theoretical analyses, data processing and interpretation techniques have also been developed for field Stoneley-waves. Field Stoneley-wave data can be processed using a wavefield separation technique (described in Chapter 4) in order to analyze the reflection events present in the data. This technique separates the Stoneley-wave data into transmitted, down-going reflected, and up-going reflected wave data. After the wave separation, various reflection events in the data can be clearly traced to their origin at the borehole. The merit of the wave-separation method is that it captures almost all reflection events, large and small, in the data. However, interpreting the cause of these reflection events is a formidable task, because many other factors, such as borehole changes (e.g., washouts/rugosity) and lithological boundaries, etc., also cause Stoneley-wave reflection. Numerical modeling (see Chapter 4) is used to reduce ambiguity in the interpretation. The modeling can realistically simulate Stoneley-wave propagation to account for borehole and formation changes along the wave path, provided that these changes, as input from

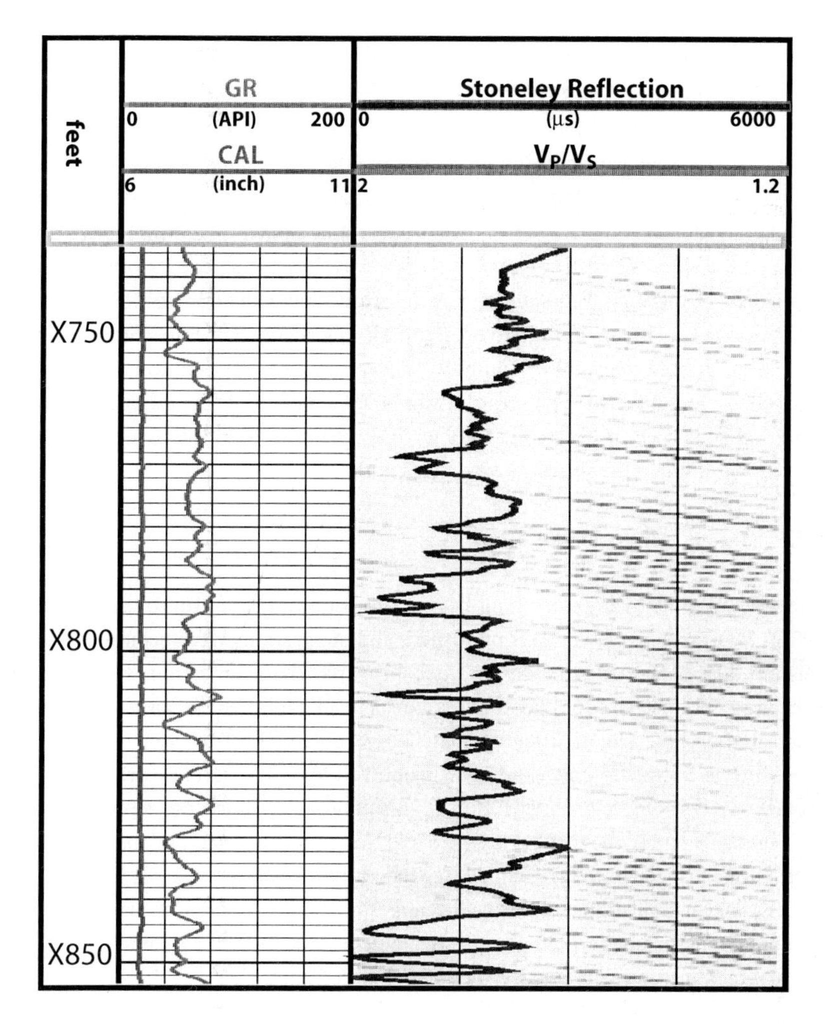

Figure 1.11. Gas detection in a formation with sand-shale sequences. Track 1 shows the caliper and gamma-ray curves. Track 2 shows the V_P/V_S ratio profile and the down-going Stoneley-wave reflection events. Note the correspondence between the low V_P/V_S ratio events and the Stoneley reflection events, which indicates the location of thin gas beds.

the log data (e.g., borehole caliper, shear and compressional slowness, and density logs), are accurate. In fracture characterization, the modeled reflection events can be compared with the measured events to distinguish reflections caused by borehole changes (rugosity) from those caused by borehole fractures (Tang, 1996b). In detecting thin gas beds in a formation, Stoneley-wave reflection analysis is combined with thin-bed slowness analysis to provide a joint interpretation method, because a thin gas bed can cause a substantial Stoneley-wave reflection and significant P-wave slowness changes.

Figure 1.11 is an example of the combination of Stoneley-wave reflection and thin-bed slowness analysis methods. The V_P/V_S ratio curve in track 2 shows many locally low and high values. Since gas saturation causes the V_P/V_S ratio to drop, the location with a low V_P/V_S ratio value may potentially be a gas bed. However, one cannot rule out that these low values are caused by lithological changes. Confirmation from Stoneley-wave analysis helps reduce this uncertainty. The measured Stoneley-wave reflection data in track 2 (plotted using a variable-density display) shows numerous events. More interesting, most Stoneley-wave reflection events originate from depths where the V_P/V_S ratio curve shows (locally) low values (~1.6–1.7). Since gas saturation causes the V_P/V_S ratio to drop and a thin gas layer generates Stoneley-wave reflection, a depth interval with a low V_P/V_S ratio *and* a reflection event corresponds most likely to a gas bed. Therefore, it can be concluded that gas production for the depth range shown in Figure 1.11 comes mostly from the locations indicated by a low V_P/V_S ratio and the Stoneley-wave reflection.

1.2.7 Seismic Wave Attenuation

Seismic wave attenuation has long been a subject of interest (Toksöz and Johnston, 1981). Many studies have been published, and suggestions have been made to use seismic attenuation as a means of rock property characterization. However, the measurement of seismic attenuation remains a formidable task because many factors (such as borehole rugosity and unmatched receivers) lead to poor accuracy and reliability in attenuation measurements from well logs. The recent work of Sun et al. (2000) addresses these problems. Sun et al.'s approach reduces the acoustic spectral data into two parts: one that varies only with logging depth and one that varies only with frequency. Since the frequency-dependent part is the same at all measurement depths, it can be estimated/removed using statistical averaging/mean methods to give a reliable estimate of the depth-dependent attenuation. This new attenuation estimation approach will be described in detail in Chapter 3.

An example of the application of attenuation measurements from acoustic logs is shown in Figure 1.12, in which the P- and S-wave attenuation logs appear to correlate with formation oil saturation quite well. Figure 1.12 shows a formation interval with two oil-bearing zones between 5730 m and 5740 m and between 5685 m and 5710 m (marked by their measured core permeabilities). An interesting phenomenon is that in these two oil zones the P-wave attenuation (solid curve) decreases and the S-wave attenuation (dashed curve) increases relative to their respective values outside the zones. Klimentos (1995) also reported a similar phenomenon. This example demonstrates the use of wave attenuation is a potentially useful tool for formation evaluation.

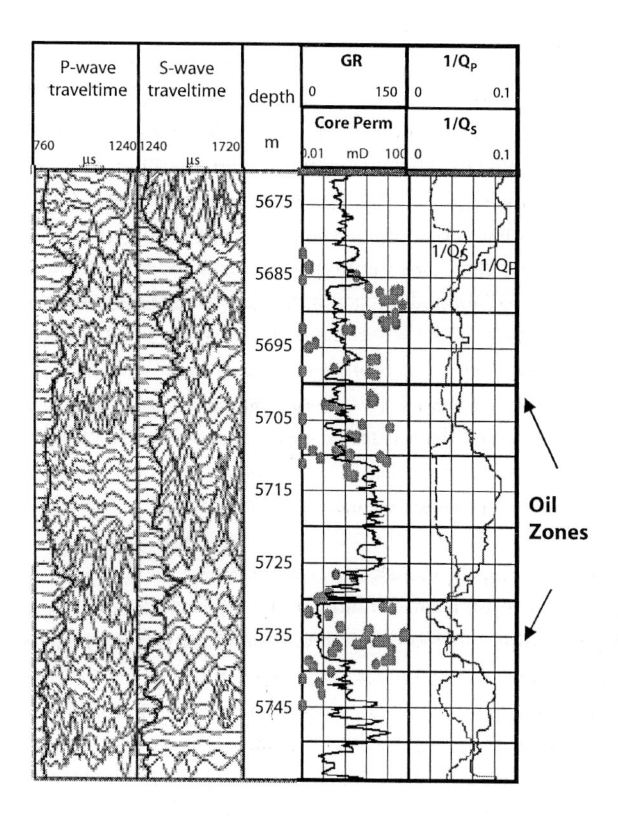

Figure 1.12. P- and S-wave attenuation profiles (track 3) across two oil-bearing zones. The zones are marked by core permeabilities in track 2. The acoustic waveform data are shown in track 1. Note the increase of P-wave attenuation and the decrease of S-wave attenuation across the oil zones.

1.3 Permeability Estimation

Besides seismic velocities and attenuation, other properties of the formation significantly affect the observed waveforms of acoustic monopole and dipole tools. The most significant property is formation permeability. The analysis of Stoneley waves in full waveform acoustic logs provides a means for determining formation permeability and permeability profiles. The derived permeabilities are related to fluid movements and pressure perturbations induced by the Stoneley-wave motion, which can be derived from theoretical principles. Early theoretical work (Biot, 1962; Rosenbaum, 1974) established the connection between permeability, Stoneley-wave attenuation, and velocity. Unfortunately, Rosenbaum's (1974) modeling results discouraged many workers because the modeling focused on frequencies common at that time in acoustic logging (~20 kHz) and assumed mud cakes at the borehole wall to be rigid and impermeable. Under these assumptions, the effects of permeability on acoustic wave data are exceedingly small and would not be detectable.

In 1984, Williams et al. published the first convincing field examples of permeability effects on Stoneley-wave data. The workers were successful because their acoustic tool was able to generate and respond to low-frequency (1–2 kHz) Stoneley waves. At such low frequencies, it was easier to measure the amplitude and velocity of the Stoneley waves. Also, the permeability effects on amplitude attenuation and velocity are greater than those at higher frequencies. Further, the results indicated that mud cake has a minimal impact on Stoneley-wave properties. In addition, there are suggestions that the Biot-Rosenbaum model, as commonly used, underestimated the effects of permeability on Stoneley-wave velocity and attenuation dispersion. A few years later, Winkler et al. (1989) performed the first laboratory measurements on Stoneley waves and demonstrated the connection between formation permeability and Stoneley-wave properties. These studies, especially the pioneering study of Williams et al. (1984), renewed interest in estimating permeability from Stoneley-wave measurements. Over the next ten years, several academic and industry research groups (for example, the Massachusetts Institute of Technology Borehole Acoustics and Logging Consortium and Schlumberger Doll Research), and D. Schmitt (Schmitt et al., 1988; Schmitt, 1989) did active research in this area. These studies significantly advanced our understanding of the interaction of formation permeability and Stoneley waves, and developed sophisticated numerical methods that allowed both rapid and accurate forward modeling.

Besides permeability, other factors contribute significantly to Stoneley-wave velocity and attenuation and must be accounted for if permeability is to be estimated from the Stoneley-wave data. Forward modeling allows for the separation of the dominant elastic and borehole-geometry effects from permeability effects. In the modeling, it is important to include borehole rugosity and layer-to-layer changes in elastic properties in order to accurately model the amplitude variation and dispersion (velocity change with frequency) of the Stoneley wave caused by these changes (see Chapter 4, as well as Tezuka et al., 1997, and Gelinsky and Tang, 1997). Besides forward modeling, Stoneley-wave data are also separated into transmitted, up-, and down-going reflected waves to remove interference effects of the reflected and transmitted waves. The wavefield separation and forward modeling make it possible to account for the dispersion and attenuation that are unrelated to the formation permeability.

The final step in estimating permeability from Stoneley-wave data is to account for the attenuation and velocity change associated with formation and borehole fluids. The transmitted waves from the modeling and from the measurement are compared. The attenuation and velocity of the measured wave relative to the modeled wave are calculated and are related to formation permeability. In practice, the relative attenuation and dispersion are, respectively, measured in terms of the center-frequency shift and travel time delay of the measured wave relative to the modeled wave.

Permeability-related Stoneley-wave attenuation and dispersion are approximately controlled by the following parameter combination:

$$\left.\begin{array}{l} Attenuation \\ Dispersion \end{array}\right\} \Leftrightarrow \frac{\kappa}{\eta\sqrt{K_{pf}}}, \tag{1.6}$$

where κ is permeability; η and K_{pf} are pore fluid viscosity and incompressibility, respectively. Thus, if the relative attenuation and dispersion (or the frequency shift and travel time delay) between the measured and modeled waves are caused by permeability, they should correspond and be correlated. This correlation provides a fundamental quality control for the derived permeability profile.

1.3.1 EXAMPLE OF PERMEABILITY ESTIMATION

Figure 1.13 shows the comparison of permeability profiles obtained from three different measurements: acoustic, nuclear magnetic resonance (NMR), and core studies. These measurements are made at different scales and are based on fundamentally different physical and measurement principles. The agreement of the results increases confidence in the validity of the permeability results. The measured and modeled Stoneley-wave data are shown, respectively, in tracks 4 and 5. The frequency shift and travel time delay of the measured data relative to the modeled data are shown in track 1. The correspondence between the delay (increasing from left to right) and shift (right to left) gives a clear indication of the permeability effects.

The frequency shift and travel time delay data are inverted to estimate permeability. The goodness of this inversion can be measured by overlaying the inversion-fitted time delay and frequency shift with the measured data, as shown in track 1. The estimated permeability is shown in track 2, along with a permeability log curve obtained using the NMR method. The core permeability values, averaged to correspond to the scale of the acoustic measurement, are also plotted (dots). The three permeability profiles agree quite well, demonstrating the validity of the results.

Details of the theory, processing, and applications of acoustic logging in a permeable formation are given in Chapter 4. Various effects such as gas saturation, fractures, mud cakes, etc., will be discussed and interpreted along with the data from other permeability measurements (e.g., NMR).

1.4 FORMATION ANISOTROPY MEASUREMENT AND APPLICATIONS

An important formation property that is measurable with modern cross-dipole acoustic logging tools is formation shear-wave anisotropy, which includes azimuthal anisotropy and vertical-versus-horizontal anisotropy. The ability to resolve formation shear-wave azimuthal anisotropy is one of the most significant recent advances in acoustic logging. Early studies were aimed at the anisotropy between vertical and horizontal directions modeled as Transverse Isotropy (TI). Of particular interest

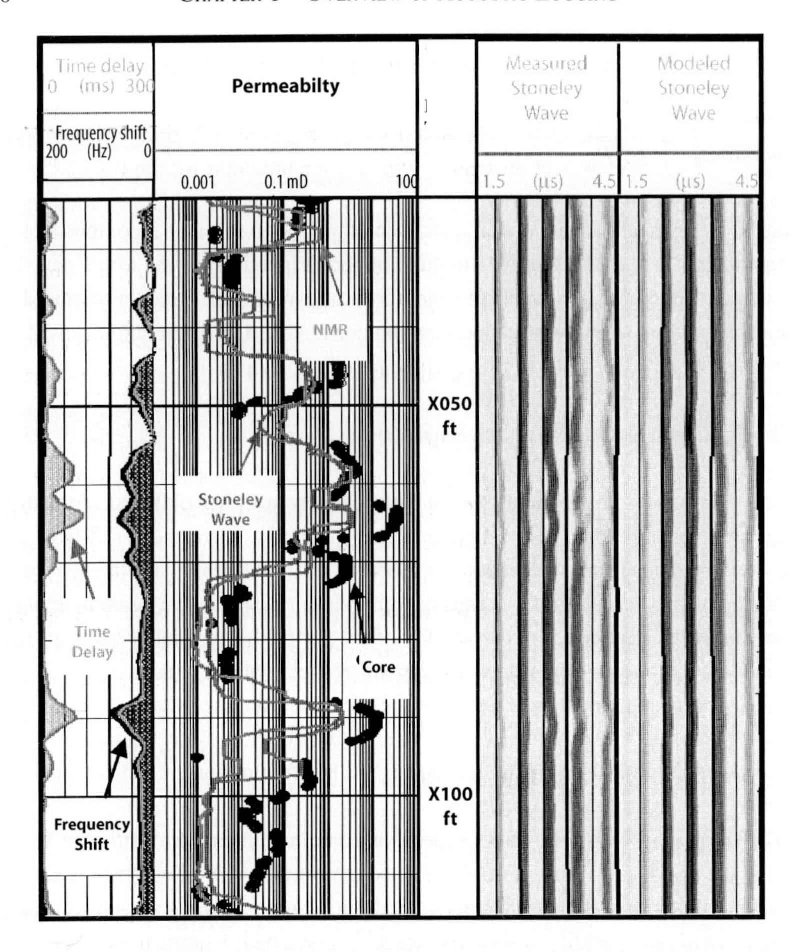

Figure 1.13. Comparison of permeability profiles from three different measurements: core, NMR, and Stoneley wave (track 2). The measured and modeled Stoneley waves are shown in tracks 3 and 4, respectively. The travel-time delay and frequency shift between the waves are given in track 1, with the correspondence between the two profiles indicating permeability effects.

to early investigators were transversely isotropic formations with a vertical axis of symmetry (VTI). This is because under such conditions the azimuthal symmetry of the borehole system is preserved and analytic solutions are possible. It is well known that some sedimentary rocks, such as shales, exhibit strong anisotropy that can be characterized as VTI. Acoustic wave propagation along a vertical well in a VTI formation was modeled first by White and Tongtaow (1981) and later by Ellefsen (1990). A detailed analysis is given in the last part of Chapter 2. The results showed that all acoustic waves in the borehole such as P and S waves and the dipole flexural wave, except the Stoneley wave, measure the vertical propagation velocity. Only the Stoneley wave is sensitive to the horizontal shear velocity and can be used to estimate shear-wave anisotropy of the VTI formation (an estimation analysis is described in Chapter 5). Schmitt (1989) developed the theory for permeable anisotropic formations.

The TI anisotropy becomes amenable to (dipole) acoustic logging measurement if its axis of symmetry is at an angle with, or perpendicular to, the borehole axis. In this situation, the borehole acoustic waves, especially the dipole-shear wave, become sensitive to the TI parameters. Specifically, a shear wave propagating along the borehole will measure different velocities depending on its polarization direction, resulting in shear-wave anisotropy around the borehole. This type of anisotropy is therefore called azimuthal anisotropy. A special case of azimuthal anisotropy is TI with horizontal axis of symmetry, called HTI. Based on the analysis of shear waves in the azimuthally anisotropic formation, cross-dipole tools have been developed in the past decade. These tools consist of two orthogonal dipole transmitter-receiver systems (see Figure 1.7 and details in Chapter 5) to allow for the measurement of shear-wave azimuthal anisotropy at a resolution impossible to achieve with seismic measurements. Many useful applications have followed the cross-dipole technology development. Specifically, the anisotropy has been used to measure the formation stress field, and to detect and characterize formation fractures. It can even be applied to post hydraulic fracture stimulation logging to map the orientation and extent of the fracture behind casing and cement. Details of the cross-dipole logging and processing are given in Chapter 5.

1.4.1 EXAMPLES OF CROSS-DIPOLE APPLICATIONS

1.4.1.1 Open-Hole Fracture Analysis

Characterization of fractures in earth formations is an important task in hydrocarbon exploration because fractures provide conduits for reservoir fluid flow. Natural or stimulated fractures parallel with and/or intersecting a borehole create azimuthal shear-wave anisotropy around the borehole (Figure 1.14a shows a diagramatic view of a fractured formation penetrated by a borehole). The degree of anisotropy gives a measure of fracture intensity, and the azimuth of the polarization of the fast shear wave gives the fracture strike (Esmersoy et al., 1995; Tang and Patterson, 2000). A simple physical explanation is that the shear stiffness along the fracture strike is primarily that of the rock matrix, while the stiffness in the fracture opening direction is weakened. Therefore, a shear wave polarized in the fracture strike direction sees a higher shear modulus and travels faster than a shear wave polarized in the opening direction of the fractures. For the dipole-logging configuration, this result has been confirmed by the numerical simulation of Xu and Para (1999) for the single fracture case, and by Tang and Patterson (2000) for the multiple fracture case. Furthermore, theoretical studies (e.g., Hudson, 1981; Schoenberg and Sayers, 1995) show that a crack- or fracture-induced anisotropy is proportional to crack density or fracture intensity. With the theory and modeling results, we can explain the cross-dipole analysis results in fractured formations.

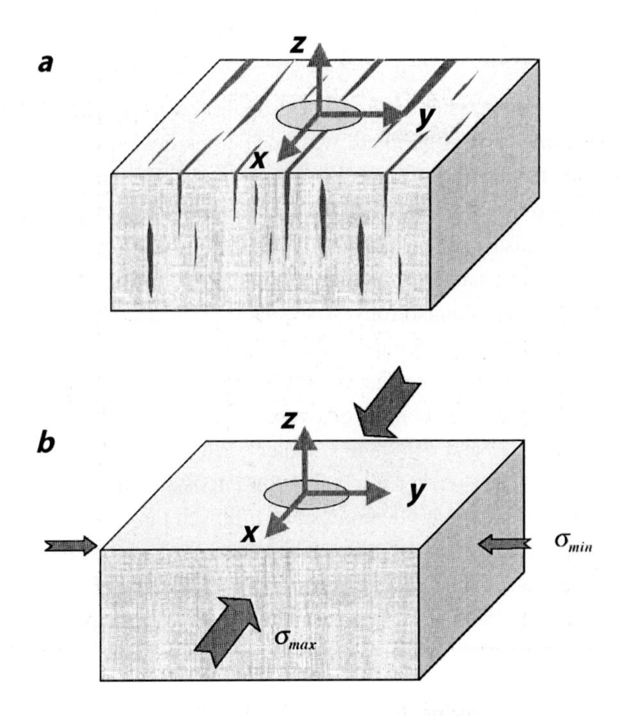

Figure 1.14. (a) A diagramatic view of a fractured formation penetrated by a borehole. Cross-dipole logging is used to determine the fracture intensity and orientation.
(b) A vertical borehole subjected to a horizontal stress system (σ_{max} and σ_{min}). Cross-dipole logging is used to find the orientation of the maximum stress σ_{max}.

A case study was conducted in a well in West Texas, USA (Joyce et al., 1998). The subject formation is a tight mixed carbonate from the Clear Fork limestone sequence in the central platform of the Permian Basin. The analysis is displayed in Figure 1.15. Track 2 shows the fast and slow dipole shear waves obtained from the data processing (see Chapter 5 for detailed analysis), along with the processing window used in the processing. Track 1 shows the fast and slow shear-wave slowness profiles, with their difference shaded, indicating anisotropy. In track 3 there are anisotropy profiles with two resolutions, each scaled to a maximum of 50 %. The higher resolution profile on the right represents data processed over the length of the receiver array (~3 ft) and the left-hand side shows the anisotropy over the acquisition length: transmitter to the middle of the receiver array (~12 ft). Over the lower dominant fracture interval, the anisotropy averages 15–20 %. It is interesting to observe that over the high-anisotropy interval (fracture location) the fast (solid curve) and slow (dashed curve) shear waves show a clear separation in time, demonstrating the well-known fracture-induced shear-wave splitting effect, as measured by the cross-dipole acoustic logging tool.

The magnitude of anisotropy is a direct measurement of the degree of fracturing, or fracture intensity. This is confirmed by a comparison of the anisotropy profile with

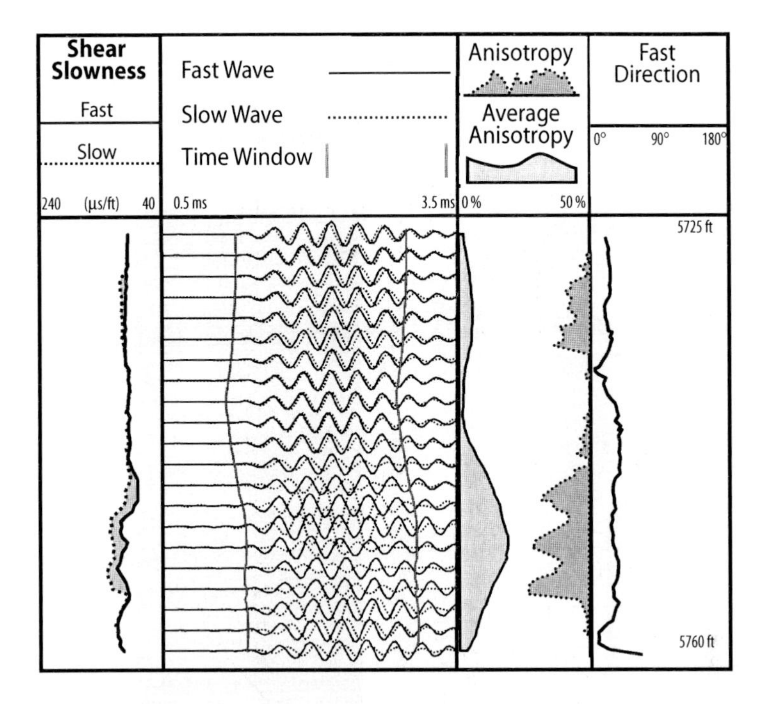

Figure 1.15. Cross-dipole analysis results across two fracture zones characterized by significant anisotropy values. Note the dramatic splitting of the fast and slow dipole-shear waves across the fracture zones (track 2). The anisotropy magnitude (track 3) characterizes the fracture intensity and the fast shear-wave polarization direction (track 4) corresponds to the fracture strike direction.

the acoustic image of the fractures shown in Figure 5.11. The image shows that the lower dominant fracture interval is more intensely fractured than the less dominant fracture interval above, where the fractures are just recognizable, with an average anisotropy of 6–10 %. The acoustic processing for a "fast shear angle" (the direction along the strike of a fracture) is shown in track 4. It can be seen that it is relatively stable and consistent over the fracture zones. Over the entire interval, the computed angle is in the vicinity of N30°E, in close agreement with the fracture strike direction obtained from the image analysis (Figure 5.11). This agreement confirms the theory that the fast shear wave polarization direction coincides with the fracture strike direction.

1.4.1.2 Evaluating Hydraulic Fractures Through Casing

Evaluating hydraulic fracture stimulation in cased boreholes presents an effective and economical application for cross-dipole technology. Hydraulic fracturing is often performed in the final stage of well completion. Pressurizing perforations through casing can effectively create hydraulic fractures (see the left panel of Figure 1.16). But evaluating the stimulation result presents a formidable task. The presence

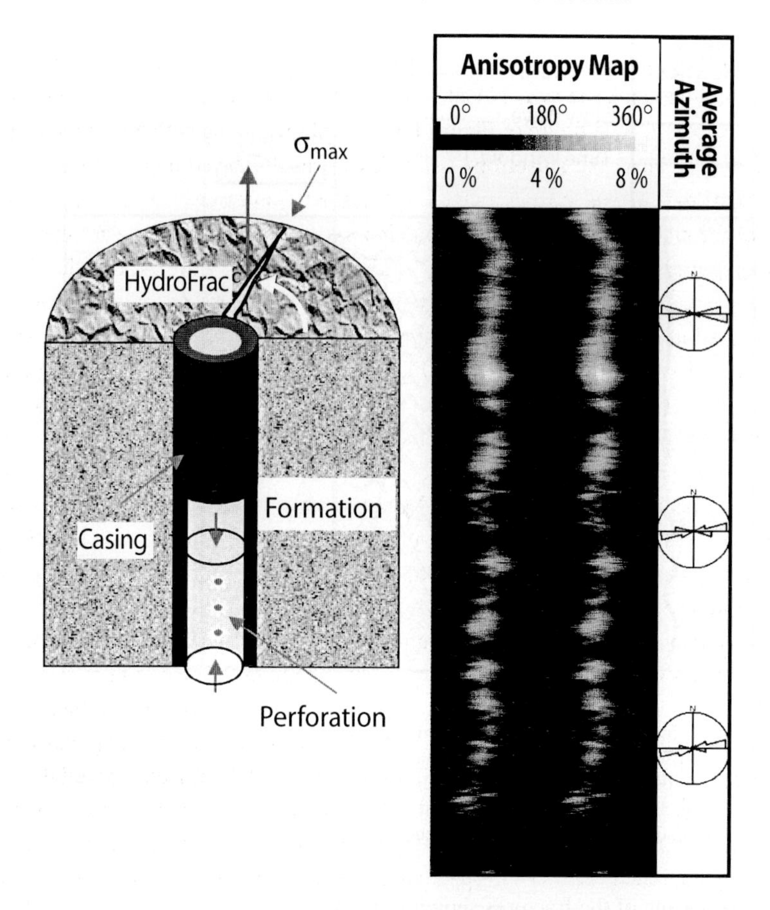

Figure 1.16. Left panel is a diagram illustrating the hydraulic fracture stimulation in a cased borehole. Right panel shows the cross-dipole-measured fracture trend behind casing after the stimulation.

of casing makes it difficult to evaluate and detect the vertical extent and azimuth of the stimulated fractures. In open-hole situations, the fracture-induced anisotropy can be effectively measured using cross-dipole acoustic logging. Application of this technology to cased holes has been hindered by two factors. The first is the unknown effect of casing and cement on the cross-dipole measurement, and the second is the lack of an effective device to measure the tool's orientation inside casing because magnetic compasses do not work inside a metal casing. Numerical modeling results (see Chapter 5) show that a cross-dipole tool can measure shear-wave anisotropy through casing and cement provided that they are well bonded with the formation. The tool azimuth measurement in cased holes is now made with a gyroscopic compass that maintains its orientation relative to the Earth's north while the compass frame moves or rotates with the tool.

An example of delineating hydraulic fractures behind casing is shown in Figure 1.16 (right panel). The cross-dipole results are presented using the *image presenta-*

tion or *anisotropy map*. The anisotropy map combines the derived average anisotropy and its azimuth to make an image display in the azimuthal range of 0–360°. Rose diagrams (see track 2, right figure of Figure 1.16) are plotted along with the image map to accurately indicate the anisotropy azimuth over each labeled depth interval. As can be seen from the anisotropy image display, the fracture stimulation creates dramatic anisotropy effects behind the casing. No measurable anisotropy was detected in a logging measurement before the stimulation (Tang et al., 2001). A high level of anisotropy is measured throughout the depth interval of interest. An important result of the post-stimulation cross-dipole measurement is the determination of the fracture and *in-situ* stress orientation. The overall anisotropy azimuth, as seen from the rose diagrams in track 2 (right figure), is in the N80°E direction. Perforation/fracture tests (Behrmann and Elbel, 1991) show that fractures initiated from perforations, regardless of their initial direction and perforation orientation, line up with the maximum *in-situ* stress direction within one borehole diameter. Since a dipole wave can penetrate through casing and cement and reach several borehole diameters into the formation, the anisotropy azimuth (i.e., fracture azimuth) seen in track 2 should correspond to the maximum stress direction.

Another important result is the mapping of fracture extent along the borehole. Since the shear-wave anisotropy in a fractured formation gives a measure of fracture intensity, the anisotropy magnitude and continuation can be used to evaluate the stimulated fractures. In the entire displayed interval, the anisotropy map shows a high-valued, almost continuous, feature. This suggests that the fracture is well developed across the high-anisotropy interval. The fracture may have been suppressed toward the lower interval, for the anisotropy map shows much reduced magnitude and distorted azimuth. The anisotropy analysis results give an effective mapping of fracture trends along the borehole.

1.4.1.2 Formation Stress Orientation Determination

Horizontal stress orientation is an important aspect of formation evaluation and a governing factor in the optimization of the development and drainage of a reservoir. This is especially true if the formation is hydraulically stimulated or is naturally fractured. In this case the stress orientation governs the directional aspects of permeability and hence production. The standard methods for determining formation stress include the evaluation of borehole breakout from image logs, packer and micro-fracture testing, stress-relief, and anelastic stress relaxation from whole core, etc. The development of cross-dipole technology allows for assessing the horizontal stress orientation using acoustic logging.

The foundation for determining formation stress from cross-dipole logging is the stress-induced shear-wave anisotropy. This effect shows that a shear wave polarized in the maximum stress direction travels faster than a shear wave polarized perpendicular to this direction (Rai and Hanson, 1988). Several authors (Sinha and Kostek, 1995; Tang et al., 1999) have modeled the result of this effect for the borehole configuration (see Figure 1.14b for this configuration). They demonstrated that an unbalanced formation stress field causes shear-wave splitting in the cross-dipole measured shear wave data. The fast shear polarization coincides with the maximum stress axis. Consequently, we can determine the formation maximum stress orientation using cross-dipole acoustic logging. It is also tempting to determine the stress magnitude from the cross-dipole data, as the theory (Sinha and Kostek, 1995; Tang et al., 1999) shows that the stress-induced anisotropy is proportional to the difference between the maximum and minimum stresses. In practice, however, the proportionality coefficients, called "nonlinear elastic constants" (Sinha and Kostek, 1995) or "stress-velocity coupling coefficients" (Tang et al., 1999), strongly depend on rock type and *in-situ* conditions. These parameters are practically unknown or unavailable. For this reason, the current interpretation focuses mainly on the determination of stress orientation.

An important aspect for interpreting cross-dipole measurement results is to distinguish the stress-induced anisotropy from other causes of anisotropy. A fundamental characteristic of stress-induced anisotropy is its strong near-borehole variation, caused by stress concentrations in the near-borehole region (Sinha and Kostek, 1995; Tang et al., 1999). The stress concentrations give rise to two wave phenomena that can be used to indicate the stress-induced anisotropy. The first is the crossover of dipole-flexural wave dispersion curves measured along and perpendicular to the maximum stress direction. The second is the splitting of shear waveforms measured by a monopole acoustic tool.

The crossover phenomenon of the dispersion-curve is demonstrated in Figure 5.20. At low frequencies, the slowness dispersion curve (solid) for a dipole polarization along the maximum stress direction has lower values than the curve along the minimum stress direction (dashed). The two curves cross each other in the medium frequency range where the dipole wavelength is approximately equal to the borehole diameter. Plona et al. (2000) demonstrated the use of the crossover characteristic to diagnose stress-induced anisotropy measured from cross-dipole logging.

The splitting of (monopole-generated) refracted shear waves in fast formations also provides a stress indication. The shear waves generated by a monopole tool usually have a much higher frequency (~7–10 kHz) than those generated by a dipole tool (~2–3 kHz). The monopole-generated wave therefore senses the near-borehole velocity variations caused by stress concentrations. The substantial velocity difference divides the circumference of the borehole into fast and slow regions. Consequently, the shear wave along the fast region arrives earlier than the wave along the slow region, giving rise to the monopole shear-wave splitting phenomenon. In practice, the monopole shear-wave splitting phenomenon is commonly observed in fast sandstone formations around the world. This wave characteristic provides an effective and straightforward diagnosis of stress effects.

Detailed analyses of stress-induced anisotropy and its effect on monopole and dipole acoustic logging are described in Chapter 5. The application to formation stress measurement is illustrated with field data examples.

1.5 Determining Shear-Wave Transverse Isotropy From Stoneley Waves

Many rocks in earth formations exhibit transverse isotropy with a vertical axis of symmetry (VTI). With the advancement of seismic data processing and interpretation, VTI information plays an increasingly important role in the accurate imaging of subsurface hydrocarbon reservoirs (Marrauld et al., 2000). It is desirable to obtain the VTI information from acoustic well logging, as the resulting continuous VTI profile indicates whether anisotropy needs to be considered for a particular depth interval of interest. This is especially important when migrating seismic data to image sub-surface reservoirs. For determining the VTI parameter using acoustic logging, Stoneley waves are the only borehole wave mode that has a significant sensitivity to TI effects, especially when the formation is acoustically slow compared to borehole fluid. The last part of Chapter 5 describes a method for deriving the formation shear-wave VTI parameter from the Stoneley-wave data acquired by a logging tool.

Although the effects of VTI on borehole Stoneley waves are well understood from theoretical analyses (e.g., Tongtaow, 1982; Ellefsen, 1990), application of the theory to acoustic logging data has in the past been hindered by two factors. The first is the neglection of the effect of a compliant logging tool on the Stoneley-wave propagation. The second is the lack of an efficient inversion method to estimate the TI effects from the Stoneley-wave data. The corresponding analysis in Chapter 5 describes how to model the tool effect in the analysis. In the inversion processing of Stoneley-wave data, the Stoneley-wave slowness is shown (in Chapter 3) to be a weighted average of the Stoneley-wave dispersion curve over the wave's frequency band. This provides a fast and effective method for the inversion.

Application of the method to Stoneley-wave logging data yields a continuous profile for the shear-wave TI parameter, commonly known as the Thomsen parameter γ (Thomsen, 1986). The results show that many shale formations exhibit a substantial transverse isotropy, with γ on the order of 20–30 %. A specific example for the Lewis shale formation in Braggs, Wyoming is given here. The well was drilled through the Lewis shale formation. The goal of the acoustic processing was to characterize the VTI property of this shale formation. Acoustic dipole and monopole waveform logging data were acquired throughout the formation. Figure 1.17 shows detailed analysis results for the Lewis shale formation. On top of the Lewis shale, the formation is characterized by shaly sandstone and sand-shale sequences. Track 1 of the figure shows the gamma ray curve. The high gamma ray value (> 120 API, where API is the unit for counting rate for gamma ray log) marks the beginning of the massive Lewis shale formation. Tracks 2 and 3 show, respectively, the Stoneley and dipole-shear wave data. Processing the data yields the Stoneley-wave slowness in track 4 (solid curve labeled '*DTST*') and the shear slowness in track 6 (dashed curve labeled '*DTSV*', standing for the slowness of the vertically propagating shear wave). As a quality control for the slowness curves, the curves are integrated over the transmitter-to-receiver-1 distance to give the travel-time curve for the Stoneley wave (track 2) and dipole-shear wave (track 3), respectively. The travel-time curves track the

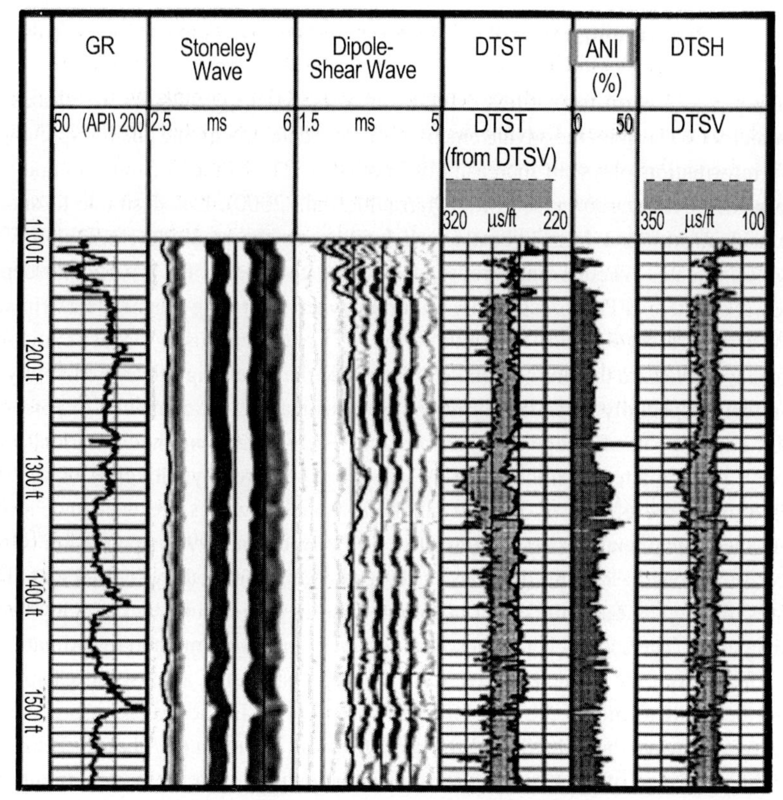

Figure 1.17. Stoneley-wave (TI) anisotropy analysis result for the Lewis shale formation (marked by high GR (> 120 API) values in track 1). The significant delay of the dipole wave (track 3) relative to the Stoneley wave (track 2) in the shale indicates the TI effect. Track 6 shows the estimated horizontal (solid) versus the vertical (dashed) shear slowness curves. Track 4 shows the measured (solid) Stoneley slowness versus the (hypothetical) isotropic slowness (dashed). Track 5 is the profile for the shear-wave (TI) anisotropy parameter γ.

respective waveforms quite well, indicating the validity of the slowness results.

An interesting feature is that the travel-time curve (or waveform) for the dipole wave shows more character/variation than that of the Stoneley wave. Relative to the Stoneley wave, the dipole wave is much delayed in the shale formation compared to the formation above it. This waveform/travel-time character/delay difference between the two types of waves provides a direct indication of the VTI effect. The dipole shear slowness (DTSV in track 6) and the Stoneley-wave slowness (DTST in track 4) are utilized to determine the horizontal shear slowness DTSH. The estimated horizontal shear slowness profile is shown in track 6 (solid curve labeled "DTSH"). This shear slowness profile shows a substantial difference (shaded area between DTSH and DTSV) against the vertical shear slowness DTSV. The two slowness curves are then used to calculate the shear-wave TI anisotropy parameter γ. The γ-profile is shown as the shaded curve in track 5, with a scale from 0–50 %. The anisotropy shows a massive, continuous feature below the top of the Lewis shale formation, but tends to vanish in the shaly-sand interval and sandstone streaks above the formation. The anisotropy effect can also be analyzed by comparing the measured Stoneley-wave slowness (having the TI effects) with a computed slowness without the TI effects. Assuming isotropy for the formation, the vertical shear slowness curve (DTSV in track 6) can be used to compute an isotropic Stoneley-wave slowness in track 4, the dashed curve labeled "DTST (from DTSV)". The difference between the measured and computed (isotropic) Stoneley-wave slowness curves (shaded area between the two curves) also indicates the presence of anisotropy.

Notice that substantial (TI) anisotropy is observed throughout this formation except in some thin streaks. The anisotropy is quite significant, generally on the order of 20–30 %. A prominent feature can be seen in a depth zone below 1300 ft, where the dipole wave (track 3) shows a significant delay, while the corresponding response on the Stoneley wave (track 2) is minimal. This feature corresponds to a significant increase in the difference between the two slowness curves in tracks 4 and 6, and the increase of anisotropy in track 5 (the anisotropy exceeds 30 % between 1300 ft and 1350 ft). The anisotropy estimation from the Stoneley-wave log data delineates the shear-wave (VTI) anisotropy magnitude and variation of the Lewis shale formation.

CHAPTER 2: ELASTIC WAVE PROPAGATION IN BOREHOLES

Theoretical modeling of acoustic propagation is an effective tool to study various wave phenomena in boreholes. Modeling helps us understand how different formation rock properties affect acoustic waveforms, and what type of wave modes can exist under certain source excitation and formation conditions. In a cased borehole, acoustic waves are significantly affected by the presence of casing. In this situation, modeling helps us understand the conditions under which we can still measure formation properties through casing and the type of waves used for the measurement.

Acoustic logging tools with multipole (e.g., monopole, dipole, quadrupole, etc.) data acquisition capabilities have been used in recent years. A theory for multipole elastic wave propagation in boreholes has also been developed (e.g., Cheng and Toksöz, 1981; Schmitt, 1988b). The theory lays the foundation for analyzing multipole wave phenomena present in the measured data. For example, we can use the theory to correctly model the effect of dispersion on the dipole acoustic-wave characteristics and to provide a method for correcting this effect. This chapter describes the analyses for modeling wave propagation in open and cased boreholes due to a multipole acoustic excitation. The analyses are also applied to model multipole acoustic propagation in the logging-while-drilling situation. The effects of attenuation and anisotropy (TI) will also be discussed. We focus on the general approaches in the theoretical analyses and present the main results of modeling that are relevant to common acoustic logging practices.

2.1 Borehole Source Formulation

Wave propagation in a borehole is conveniently analyzed using cylindrical coordinates (r, θ, z). The model consists of a fluid-filled borehole, extending to infinity in the z-direction. The borehole is surrounded by an elastic medium. To begin the wave propagation analysis, we place a point source at the location (r_0, θ_0, z_0) in the borehole fluid. The fluid has density ρ_f and acoustic velocity α_f. The wave motion excited in the fluid satisfies a wave equation, which, expressed in the frequency domain, can be written as

$$\nabla^2\Phi + k_f^2\Phi = 4\pi\delta(Z) \;, \tag{2.1}$$

where Φ is the wave displacement potential; δ denotes the Dirac delta function, whose argument Z denotes the source-to-receiver distance; $k_f = \omega/\alpha_f$; ω is the angular frequency, and

$$\nabla^2 = \frac{\partial^2}{\partial r^2} + \frac{1}{r}\frac{\partial}{\partial r} + \frac{1}{r^2}\frac{\partial^2}{\partial\theta^2} + \frac{\partial^2}{\partial z^2}$$

is the Laplace operator in cylindrical coordinates. The displacement potential in the borehole contains two parts: $\Phi = \Phi_d + \Phi_r$. The first part represents the direct wave emitted from the source, and the second represents the reflected wave from the borehole interface. The direct wave corresponds to the solution to equation (2.1) for an infinite medium, as given by

$$\Phi_d = \frac{\exp(ik_f Z)}{Z} = \frac{\exp\left(ik_f\left(r^2 + r_0^2 - 2rr_0\cos(\theta-\theta_0) + (z-z_0)^2\right)^{1/2}\right)}{\left(r^2 + r_0^2 - 2rr_0\cos(\theta-\theta_0) + (z-z_0)^2\right)^{1/2}} \;. \tag{2.2}$$

This solution is a spherical wave emitted from a point source. Because the reflected wavefield consists of cylindrical waves arising from the borehole interface, we want to express the spherical wave as the superposition of cylindrical waves, as given by the following expression

$$\Phi_d = \frac{1}{\pi}\int_{-\infty}^{+\infty} K_0\left(f\left(r^2 + r_0^2 - 2rr_0\cos(\theta-\theta_0)\right)^{1/2}\right)e^{ik(z-z_0)}dk \;, \tag{2.3}$$

where k is the axial wavenumber; $f = (k^2 - k_f^2)^{1/2}$ is the radial wavenumber, and K_n, ($n = 0, 1, ...$) is the modified Bessel function of the second kind and order n.

2.1.1 MULTIPOLE SOURCE IMPLEMENTATION

Different types of acoustic sources (monopole, dipole, quadrupole, etc.) excite different borehole-acoustic waves with different particle displacement patterns and dispersion characteristics. A multipole source is realized by decomposing equation (2.3) into azimuthal components using the Bessel addition theorem (Watson, 1944).

$$\Phi_d = \frac{1}{\pi}\int_{-\infty}^{+\infty}\sum_{n=0}^{+\infty}\varepsilon_n\cos\left(n(\theta-\theta_0)\right)\begin{Bmatrix} I_n(fr_0)K_n(fr), & r>r_0 \\ I_n(fr)K_n(fr_0), & r<r_0 \end{Bmatrix} e^{ik(z-z_0)}dk \,. \tag{2.4}$$

In the above expression, I_n ($n = 0, 1, ...$) is the modified Bessel function of the first kind and order n; ε_n is 1 for $n = 0$, and 2 for $n > 0$.

A monopole source corresponds to $n = 0$, which has no azimuthal (or θ) dependence. It is clear that n, called the azimuthal order number, controls the azimuthal variation of the wavefield. A multipole source of order n is constructed by using $2n$ point sources in a horizontal plane (e.g., $z_0 = 0$), each point source being described by equation (2.4). The point sources are distributed periodically along a circle of radius r_0 and alternate in sign. For example, the jth point has a sign of $(-1)^{j+1}$, a scaling factor of $1/(2n)$, and is located at an angle of $\theta_0 = \pi(j-1)/n + \phi$, $(j = 1, ..., n)$. Note that this angle is relative to some reference angle ϕ (Figure 2.1 gives an example for the cases $n = 0$, 1, and 2, corresponding to monopole, dipole, and quadrupole sources, respectively. The reference angle is $\phi = \pi/2$). By summing the contributions of all these point sources, the resulting potential, in terms of the integrand for the k-integration, is written as:

$$\tilde{\Phi}_d(k,\omega) = e^{ikz} \sum_{j=1}^{+\infty} \varepsilon_{(2j-1)n} I_{(2j-1)n}(fr_0)\, K_{(2j-1)n}(fr) \cos((2j-1)n(\theta-\phi)) \ . \tag{2.5}$$

Now, let r_0 approach zero to form a multipole point source. Then, take the leading order of the above expression in the limit of $r_0 = 0$. This yields the direct wavefield for the nth multipole,

$$\tilde{\Phi}_d(k,\omega) = \frac{\varepsilon_n}{n!}\left(\frac{fr_0}{2}\right)^n K_n(fr) \cos(n(\theta-\phi))\, e^{ikz} . \tag{2.6}$$

The reflected wavefield corresponding to the nth multipole has a similar functional form, as expressed by

$$\tilde{\Phi}_r(k,\omega) = \frac{1}{n!}\left(\frac{fr_0}{2}\right)^n A'_n I_n(fr) \cos(n(\theta-\phi))\, e^{ikz} \ . \tag{2.7}$$

where the Bessel function I_n is used to represent the incoming (instead of K_n as in equation (2.6), which describes outgoing – or radiating – waves). The coefficient A'_n is a function of both k and ω, which is to be determined by the boundary condition at the borehole interface.

From the displacement potentials in equations (2.6) and (2.7), the borehole-fluid radial displacement and stress (or negative pressure, $-p$) are given by

$$\begin{cases} u_f &= \dfrac{\partial \tilde{\Phi}_{d \text{ or } r}}{\partial r} \\ \sigma_{rrf} &= -\rho_f \omega^2 \tilde{\Phi}_{d \text{ or } r} \end{cases} \ . \tag{2.8}$$

The displacement potential at the right-hand side can be either that of the direct wave or that of the reflected wave, as denoted by the subscript d or r. The resulting radial displacement u_f and stress σ_{rrf} then correspond to the direct or the reflected wavefield, respectively.

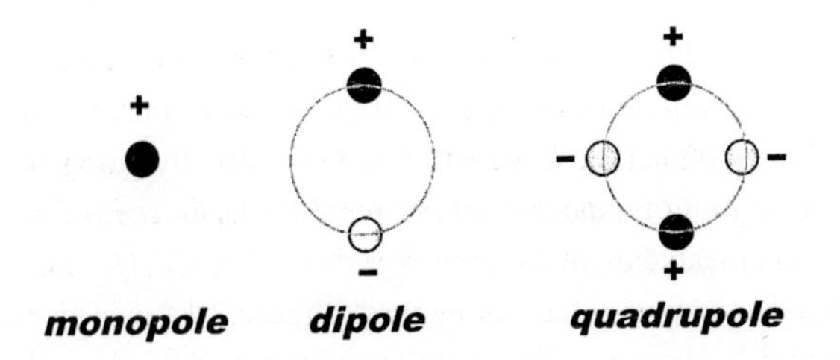

Figure 2.1. Point source combinations for monopole, dipole, and quadrupole sources.

2.2 Solution for the Elastic Formation

In an elastic formation with Lamé constants λ, μ and density ρ, the displacement vector **u** with radial, azimuthal, and vertical components (u, v, w) satisfies a vector wave equation, whose general solution, according to Helmholtz's theorem (Morse and Feshbach, 1953), can be given by three scalar potentials. The wave equation and the displacement solution are given as

$$(\lambda+\mu)\nabla(\nabla\cdot\mathbf{u})+\mu\nabla^2\mathbf{u}+\rho\omega^2\mathbf{u}=0,\quad \text{with: } \mathbf{u}=\nabla\Phi+\nabla\times(\chi\hat{\mathbf{z}})+\nabla\times\nabla\times(\Gamma\hat{\mathbf{z}}). \quad (2.9)$$

where

Φ is the compressional-wave potential;

$\hat{\mathbf{z}}$ is the unit vector in the z-direction;

Γ is the SV-type shear-wave potential;

χ is the SH-type shear-wave potential.

The polarization of an SV-type shear wave is in a vertical plane, while that of a SH-type shear wave is in a horizontal plane. Each of the potentials is the solution of a scalar wave equation

$$\begin{cases}\nabla^2\Phi+k_p^2\Phi=0\\ \nabla^2\chi+k_s^2\chi=0\\ \nabla^2\Gamma+k_s^2\Gamma=0\end{cases}, \quad (2.10)$$

where $k_\alpha=\omega/\alpha$ and $k_\beta=\omega/\beta$ are the compressional and shear wavenumber, and α and β are the compressional and shear velocity, respectively. Analogous to the solution for the fluid displacement potential of equation (2.7), the solution of equations (2.10) in the k-domain is

$$\left.\begin{matrix}\Phi\\ \chi\\ \Gamma\end{matrix}\right\}=e^{ikz}\frac{1}{n!}\left(\frac{fr_0}{2}\right)^n\begin{cases}(A_nI_n(pr)+B_nK_n(pr))\cos(n(\theta-\phi))\\ (C_nI_n(sr)+D_nK_n(sr))\sin(n(\theta-\phi))\\ (E_nI_n(sr)+F_nK_n(sr))\cos(n(\theta-\phi))\end{cases}, \quad (2.11)$$

where $p=(k^2-k_\alpha^2)^{1/2}$ and $s=(k^2-k_\beta^2)^{1/2}$ are the compressional and shear radial wavenumbers, respectively.

The solutions in equations (2.11) are for a formation with cylindrical layers that are concentric with the borehole. For such a layer, coefficients B, D, and F, which are associated with Bessel function K_n ($n = 0, 1, ...$) describe outgoing waves traveling from the inner to the outer boundary of the layer. Similarly, coefficients A, C, and E, which are associated with Bessel function I_n ($n = 0, 1, ...$) describe incoming waves traveling from the outer to the inner boundary. Both sets of coefficients are needed to model a layered structure surrounding the borehole (e.g., casing and cement). However, for a formation that extends from the borehole to infinity, the radiation condition (i.e., waves radiated from a finite source must vanish at infinity) implies that $A = C = E = 0$. Only B, D, and F are needed for modeling such a formation.

Using the potentials to express the displacement in equation (2.9), the displacement components are calculated by the following formulae:

$$\begin{cases} u = \dfrac{\partial \Phi}{\partial r} + \dfrac{1}{r}\dfrac{\partial \chi}{\partial \theta} + \dfrac{\partial^2 \Gamma}{\partial r \partial z} \\ v = \dfrac{1}{r}\dfrac{\partial \Phi}{\partial r} - \dfrac{\partial \chi}{\partial r} + \dfrac{1}{r}\dfrac{\partial^2 \Gamma}{\partial \theta \partial z} \\ w = \dfrac{\partial \Phi}{\partial z} + k_s^2 \Gamma + \dfrac{\partial^2 \Gamma}{\partial z^2} \end{cases} . \tag{2.12}$$

From the displacements, the elements of the strain tensor in the cylindrical coordinates can be calculated using

$$\begin{cases} e_{rr} = \dfrac{\partial u}{\partial r} \\ e_{\theta\theta} = \dfrac{u}{r} + \dfrac{1}{r}\dfrac{\partial v}{\partial \theta} \\ e_{zz} = \dfrac{\partial w}{\partial z} \\ e_{r\theta} = \dfrac{1}{2}\left(\dfrac{1}{r}\dfrac{\partial u}{\partial \theta} - \dfrac{v}{r} + \dfrac{\partial v}{\partial r}\right) \\ e_{\theta z} = \dfrac{1}{2}\left(\dfrac{1}{r}\dfrac{\partial w}{\partial \theta} + \dfrac{\partial v}{\partial z}\right) \\ e_{rz} = \dfrac{1}{2}\left(\dfrac{\partial w}{\partial r} + \dfrac{\partial u}{\partial z}\right) \end{cases} . \tag{2.13}$$

From the strain elements, the formation stress elements are obtained using Hooke's law:

$$\sigma_{ij} = \lambda \, e \, \delta_{ij} + 2\mu \, e_{ij} \tag{2.14}$$

where $\lambda = \rho(\alpha^2 - 2\beta^2)$ and $\mu = \rho\beta^2$; $e = (e_{rr} + e_{\theta\theta} + e_{zz})$ is the dilatation; $\delta_{ij} = 1$ if $i = j$, and 0 otherwise. By using the subscripts i and j to respectively denote r, θ, and z, we can calculate all six normal and shear stress elements in the cylindrical coordinates.

2.3 Employing the Boundary Condition at the Borehole

By employing the boundary condition at the borehole interface, we can relate the wave motion in the borehole to the wave motion in the formation. Mathematically, this allows us to determine the unknown coefficients in equations (2.7) and (2.11). We will consider the open-hole case (borehole embedded in an infinite formation) in this section.

The boundary condition at the borehole interface is the continuity of radial displacement and stress and the vanishing of shear stresses, as expressed below:

$$\begin{cases} u = u_f \\ \sigma_{rr} = \sigma_{rrf} \\ \sigma_{rz} = 0 \\ \sigma_{r\theta} = 0 \end{cases}, \quad (\text{at } r = R) , \qquad (2.15)$$

where R is the borehole radius. The other two displacement components, v and w, are not required in the boundary condition because a fluid-solid boundary condition requires only the continuity of the normal (radial) displacement. Notice that the borehole fluid displacement and stress contain both the direct and reflected contributions.

The above boundary condition leads to a matrix equation for the unknown coefficients:

$$\begin{pmatrix} M_{11} & M_{12} & M_{13} & M_{14} \\ M_{21} & M_{22} & M_{23} & M_{24} \\ M_{31} & M_{32} & M_{33} & M_{34} \\ M_{41} & M_{42} & M_{43} & M_{44} \end{pmatrix} \begin{pmatrix} A'_n \\ B_n \\ D_n \\ F_n \end{pmatrix} = \begin{pmatrix} u_f^d \\ \sigma_{rrf}^d \\ 0 \\ 0 \end{pmatrix}, \qquad (2.16)$$

where the two non-zero elements at the right hand side are the direct radial fluid displacement and stress generated by the source. They are calculated from equations (2.6) and (2.8) and evaluated at the borehole-formation interface, as given by

$$\begin{cases} u_f^d = \frac{\varepsilon_n}{n!}\left(\frac{fr_0}{2}\right)^n \left\{\frac{n}{R} K_n(fR) - fK_{n+1}(fR)\right\} e^{ikz} \cos(n(\theta - \phi)) \\ \sigma_{rrf}^d = -\frac{\varepsilon_n}{n!}\left(\frac{fr_0}{2}\right)^n \rho_f \omega^2 K_n(fR) e^{ikz} \cos(n(\theta - \phi)) \end{cases} . \qquad (2.17)$$

The elements of the matrix are given by the following expressions (note that there is a common term $(fr_0/2)^n \cos(n(\theta - \phi))\exp(ikz)/n!$ for all matrix elements; also, the matrix elements are grouped according to their associated displacement component or stress element).

$$
\begin{cases}
\text{For } u: \\
M_{11} = -\dfrac{n}{R} I_n(fR) - f I_{n+1}(fR) \\
M_{12} = -p Y_1(pR) \\
M_{13} = \dfrac{n}{R} K_n(sR) \\
M_{14} = -iks Y_1(sR)
\end{cases}
\qquad
\begin{cases}
\text{For } \sigma_{rr}: \\
M_{21} = \rho_f \omega^2 I_n(fR) \\
M_{22} = \rho(2k^2\beta^2 - \omega^2) K_n(pR) + \dfrac{2p\rho\beta^2}{R} Y_2(pR) \\
M_{23} = -\dfrac{2n\rho s\beta^2}{R} Y_3(sR) \\
M_{24} = 2ik\rho\beta^2 s^2 K_n(sR) + \dfrac{2iks\rho\beta^2}{R} Y_2(sR)
\end{cases},
$$

$$
\begin{cases}
\text{For } \sigma_{r\theta}: \\
M_{31} = 0 \\
M_{32} = \dfrac{2pn\rho\beta^2}{R} Y_3(pR) \\
M_{33} = -s^2 \rho\beta^2 Y_4(sR) \\
M_{34} = \dfrac{2iksn\rho\beta^2}{R} Y_3(sR)
\end{cases}
\qquad
\begin{cases}
\text{For } \sigma_{rz}: \\
M_{41} = 0 \\
M_{42} = -2ikp\rho\beta^2 Y_1(pR) \\
M_{43} = \dfrac{ikn\rho\beta^2}{R} K_n(sR) \\
M_{44} = (k^2 + s^2) s\rho\beta^2 Y_1(sR)
\end{cases}
\tag{2.18}
$$

where $Y_1 - Y_4$ denote the following combinations of Bessel functions:

$$
\begin{cases}
Y_1(x) = -\dfrac{n}{x} K_n(x) + K_{n+1}(x) \\
Y_2(x) = \dfrac{n(n-1)}{x} K_n(x) + K_{n+1}(x) \\
Y_3(x) = \dfrac{1-n}{x} K_n(x) + K_{n+1}(x) \\
Y_4(x) = \left[1 + \dfrac{2n(n-1)}{x^2}\right] K_n(x) + \dfrac{2}{x} K_{n+1}(x)
\end{cases}
$$

2.4 Full Waveform Synthetic Seismograms

Solving equation (2.16) for coefficient A'_n, we obtain the reflected wavefield in the borehole, which, together with the direct wavefield radiated from the source, gives the acoustic wavefield in the borehole. To simulate the logging of an acoustic tool centered at the borehole, we study the wavefield on the axis of the borehole. We also assume that pressure is measured if the source is monopole, displacement is measured if the source is dipole, and the spatial derivative of displacement is measured if the source is quadrupole, etc.

For a monopole source this results in:

$$
P(z,t) = \int_{-\infty}^{+\infty} S(\omega) D^{(0)}(\omega) e^{-i\omega t} d\omega + \int_{-\infty}^{+\infty} \int_{-\infty}^{+\infty} A'_0(k,\omega) S(\omega) e^{ikz} e^{-i\omega t} dk\, d\omega \ . \tag{2.19}
$$

For a multipole source of nth order, this results in:

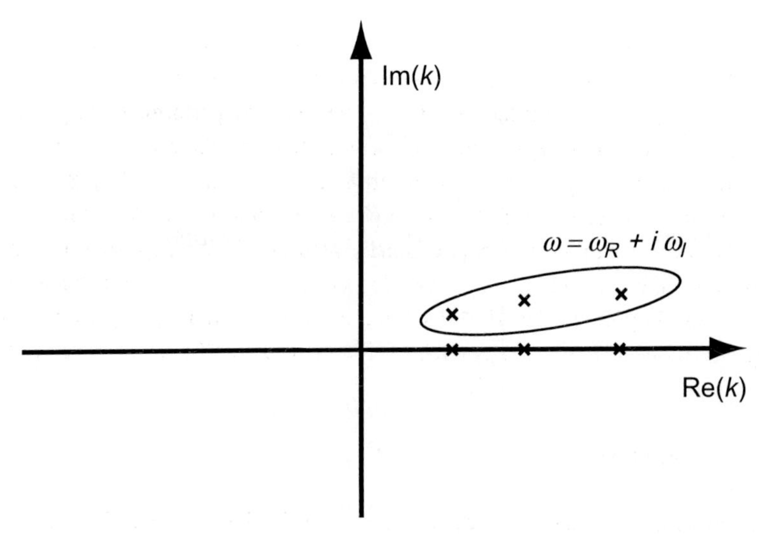

Figure 2.2. By adding a small imaginary part to the angular frequency ω, singularities that originally are located on the real wavenumber (k) axis are shifted into the complex k domain, as indicated by the encircling ellipse. This allows the direct integration along the real k-axis without special treatment of the singularities.

$$M^{(n)}(z,t)=\int_{-\infty}^{+\infty}S(\omega)D^{(n)}(\omega)e^{-i\omega t}d\omega+\frac{r_0^n}{2^{2n}n!}\int_{-\infty}^{+\infty}\int_{-\infty}^{+\infty}f^{2n}A_n'(k,\omega)S(\omega)\,e^{ikz}e^{-i\omega t}\,dk\,d\omega\,, \quad (2.20)$$

where $D^{(n)}(\omega)$ is the contribution arising directly from source radiation; $S(\omega)$ denotes the source spectrum, whose functional form can be arbitrarily chosen. For modeling convenience, a Ricker wavelet with center frequency ω_0 is often used. $S(\omega)$ and $D^{(n)}(\omega)$ are given by

$$S(\omega)=\left(\frac{\omega}{\omega_0}\right)^2 e^{-(\omega/\omega_0)^2};\;\; D^n(\omega)=\frac{\pi\varepsilon_n r_0^n}{2^{2n}n!}\sum_{m=0}^{n}C_m^n(-2ik_f)^m\frac{(2n-m)!}{z^{2n-m+1}}e^{ik_f z}\;,$$

where $C_m^{\;n}$ is the binomial factor. The double integration in the above equations is performed using a discrete wavenumber summation and a Fast Fourier Transform (FFT) method. For each given frequency, the wavenumber axis is discretized into small intervals and the contributions (integrand values) from the intervals are summed. In the wavenumber integration, there are singularities of the integrand at the real wavenumber axis. The locations of the singularities are $k = k_\alpha$, $k = k_\beta$ and $k = k_f$, respectively. Although these singularities are integrable in the Cauchy principal value sense, they pose a problem for the numerical integration, because a fine discretization is needed near the singularities. An effective scheme based on a Fourier transform property is used to avoid the problem. In the wavenumber integration for each frequency, a small imaginary part is added to the frequency, as

$$\omega = \omega_R + \omega_I\,,\;\; \omega_I > 0\,.$$

With this complex frequency, the singularities are shifted away from the integration path (the real k axis) into the complex k plane, as illustrated in Figure 2.2. The wavenumber integration along the real k axis can now be performed straightforwardly without special treatment of the singularities. However, the wave propagation computed this way is damped because of the artificial attenuation caused by ω_I. The undamped solution is recovered following the frequency integration. After the complex wave spectrum over all frequencies is calculated, the FFT is applied to obtain the synthetic seismogram (i.e., acoustic waveform) for the given receiver distance(s). The effect caused by the artificial attenuation is removed by multiplying the waveforms by an exponential factor $\exp(\omega_I t)$, as follows from a well-known property of Fourier transform (Bracewell, 1965).

2.4.1 EXAMPLES OF SYNTHETIC SEISMOGRAMS

Examples of synthetic micro-seismograms are given in Figures 2.3 and 2.4 for a fast and a slow formation, respectively. In the language of the acoustic logging community, a formation is fast (or slow) if its S-wave velocity is greater (or smaller) than the acoustic velocity in the borehole fluid. The P- and S-wave velocities and density for both the fast and slow formations are listed in Table 2.1. The borehole has a diameter of 0.2 m, and the fluid acoustic parameters are given in Table 2.1. For the fast-formation example (Figure 2.3a, with a source-excitation center frequency of 8 kHz), the monopole wavetrain contains P, S, and Stoneley waves. A high-frequency dispersive wave packet following the S wave is the pseudo-Rayleigh wave, which will be described in more detail in the next section. The dipole wavetrain shown in Figure 2.3b is calculated with a source excitation frequency of 3 kHz. The onset of the wave propagates with the S-wave velocity of the formation. The later portion of the wave shows dispersive behavior, which will be discussed in more detail in the next section and in Chapter 3.

The slow-formation results in Figures 2.4a and b are calculated with the same source excitations as their counterparts in Figures 2.3a and b, respectively. The waveforms in Figure 2.4a show only the P and Stoneley waves, because no refracted shear-wave energy can be measured when formation S-wave velocity falls below the velocity of the borehole fluid. The P wavetrain in this case is an attenuative and dispersive wave packet. In fact, this wave is called the "leaky-P wave" because it radiates energy into the formation and consequently attenuates along the borehole. The dipole-flexural wave in Figure 2.4b travels at the S-wave velocity of the formation; the wave is much less dispersive than its fast-formation counterpart (Figure 2.3b). At higher frequencies, a dipole source can also generate a P wave in the slow formation situation (see Figure 2.4c, where the source frequency is 6 kHz). The various wave phenomena will be better understood by analyzing their dispersion characteristics in the frequency domain, as we describe in the following section.

Table 2.1. Formation, casing, cement, and borehole fluid elastic/acoustic properties used in the calculation for Figures 2.3 to 2.10

	P-velocity	S-velocity	density
Fast Formation	4000 m/s	2300 m/s	2500 kg/m^3
Slow Formation	3000 m/s	1200 m/s	2500 kg/m^3
Casing	6098 m/s	3354 m/s	7500 kg/m^3
Cement	2823 m/s	1729 m/s	1920 kg/m^3
Fluid	1500 m/s	–	1000 kg/m^3

2.5 Analysis of Wave Modes in a Borehole

The theoretical borehole acoustic wave spectrum $X(\omega, z)$ can be generally written as (see equations (2.19) and (2.20)):

$$X(\omega,z) = S(\omega)\int_{-\infty}^{+\infty} A(k,\omega)\, e^{ikz} dk \ , \tag{2.21}$$

where $A(k, \omega)$ now denotes the frequency-wavenumber response of the borehole and formation due to a multipole acoustic source. The straightforward wavenumber integration, as described previously, is along the real wavenumber axis. An alternative approach involves contour integration in the complex wavenumber plane. The contributions to $X(\omega, z)$ now come from the poles which are enclosed by the contour and the branch lines (see Figure 2.5). For this alternative approach the above integral may be written:

$$X(z,\omega) = 2\pi i \sum_{l} S(\omega)\, \text{Res}\left(A(k,\omega)e^{ikz}\right)_{k_l} + \sum_{j=p,s,f} \int_{bl_j} S(\omega)A(k,\omega)e^{ikz}dk \ , \tag{2.22}$$

where "Res" denotes taking the residue of $A(k, \omega)\exp(ikz)$ at the wavenumber value k_l at the lth pole The notation $\int_{blj}$ denotes the contribution that comes from the jth branch-line integrals. There are three branch points in the complex k-plane: $k = k_\alpha$, k_β, and K_f, respectively, as denoted by the summation indices in the above equation. The branch points arise from the radial wave numbers, $p = (k^2 - k_\alpha{}^2)^{1/2}$, $s = (k^2 - k_\beta{}^2)^{1/2}$, and $f = (k^2 - k_f^2)^{1/2}$ (see equations (2.16) through (2.18)). It is interesting to note that the branch-line integration of the reflected wavefield associated with k_f precisely cancels the direct wavefield from the source for any azimuthal order n. The branch-line contribution from k_α gives the refracted compressional wave, while that from k_β gives the refracted shear wave (Tsang and Radar, 1979).

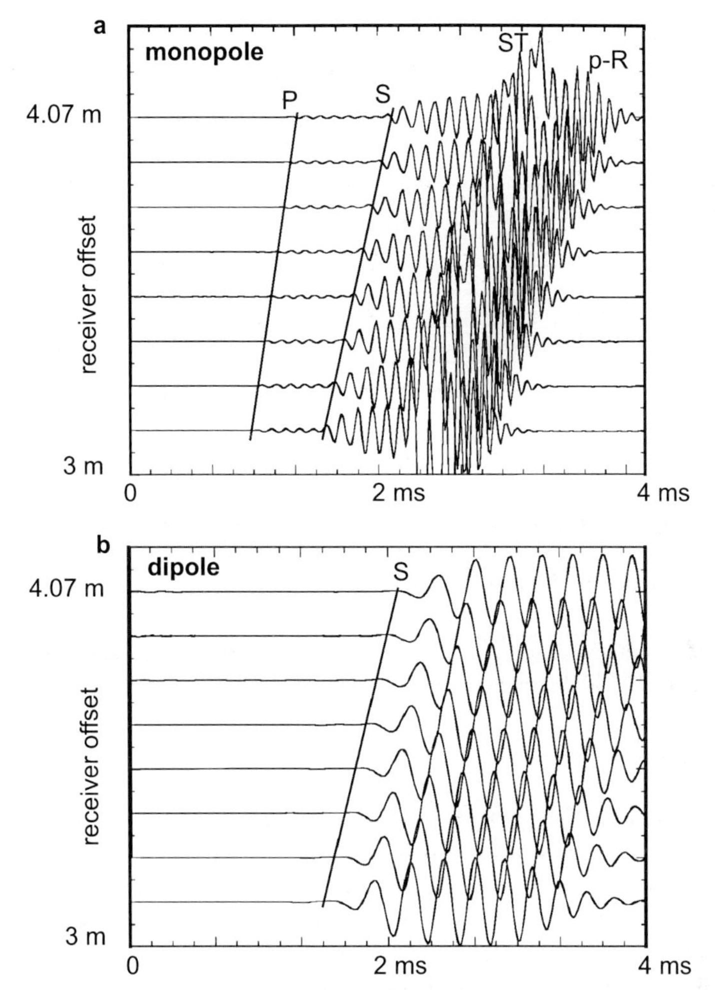

Figure 2.3. Synthetic monopole (a) and dipole (b) array waveforms for a fast formation surrounding a borehole. Note the presence of P, S, Stoneley (ST), and pseudo-Rayleigh (p-R) waves in (a).

To analyze the poles of $A(k, \omega)$, we solve it from equation (2.16) ($A(k, \omega)$ now denotes A'_n) as

$$A(k,\omega) = \frac{u_f^d \det \mathbf{M}^{11} - \sigma_{rrf}^d \det \mathbf{M}^{21}}{\det \mathbf{M}} , \tag{2.23}$$

where "det" denotes taking the determinant of a matrix. Matrix $\mathbf{M}^{ij}$ is the residue matrix of **M**, obtained by removing the ith row and jth column from **M**. The poles are found by setting

$$D(k,\omega) = \det \mathbf{M}(k,\omega) = 0 \ . \tag{2.24}$$

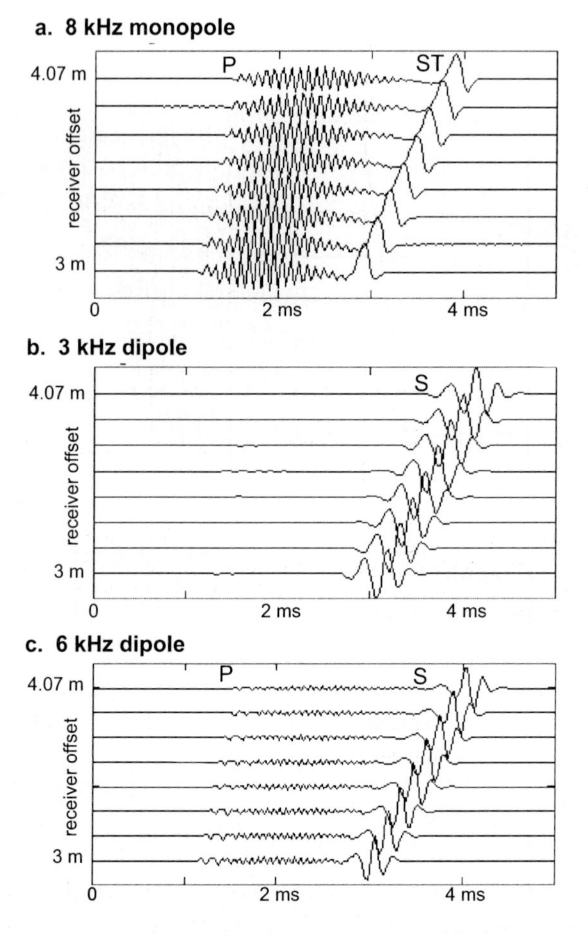

Figure 2.4. Synthetic monopole (a) and dipole (b, 3 kHz source and c, 6 kHz source) array waveforms for a slow formation surrounding a borehole.

Equation (2.24) is known as the period- or dispersion equation. For a fixed ω, this equation is a nonlinear function for k and should be solved numerically. A numerical algorithm using the Newton-Raphson method consists of the following steps:

1. Choose a starting frequency ω_S with a known initial-guess velocity v_I, such that $k_I = \omega_S/v_I$ (for example, to compute the dispersion curve for the dipole-flexural wave, we can start near the cutoff frequency of the mode and assign formation shear velocity β as the initial guess).
2. With the known root at ω, we extrapolate it to $\omega + \Delta\omega$ using

$$k(\omega+\Delta\omega) = k(\omega) + \Delta\omega\frac{dk}{d\omega} = k(\omega) - \Delta\omega\left(\frac{\partial D}{\partial \omega}\right)\bigg/\left(\frac{\partial D}{\partial k}\right),$$

where the partial derivatives of the dispersion equation D with respect to ω and k can be calculated numerically, e.g., using finite differences.

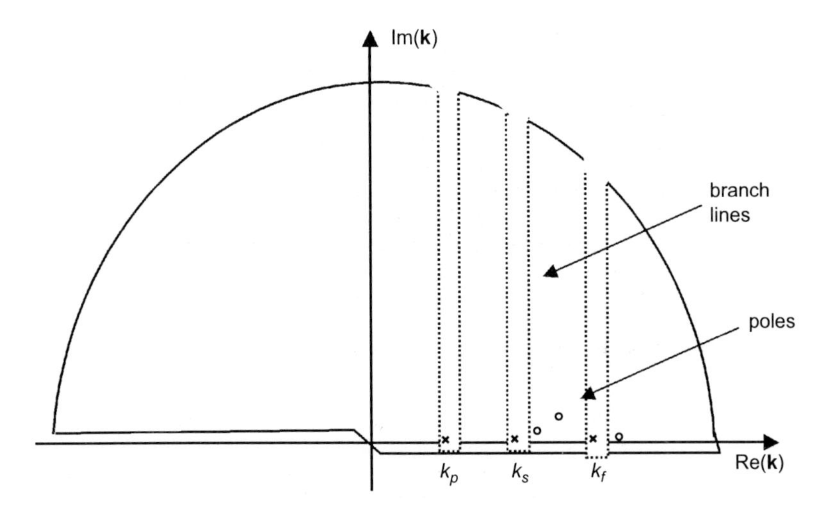

Figure 2.5. Replacing the real k-integration path with a contour in the upper k-plane and the lines cut from the branch points (crosses). Since the contribution from the contour vanishes, the real k-axis integration is equivalent to the contributions from the branch lines and the poles (open circles).

3. The extrapolated result is then used as the initial guess in a root-finding method, such as the Newton-Raphson method (see, e.g., Press et al., 1989), to refine the result in an iterative manner.
4. Repeat steps 2 and 3 at the increment $\Delta\omega$ for all frequencies until roots of k are found for the given frequency range.

We next discuss the modes associated with poles. The period equation (2.24) can be used to investigate the properties of the various modes by tracking the poles in the k plane for each frequency. The phase and group velocities of the wave modes are given by

$$\begin{cases} v_{phase}(\omega) = \dfrac{\omega}{k(\omega)}, \\ v_{group}(\omega) = \dfrac{d\omega}{dk} = -\left(\dfrac{\partial D}{\partial k}\right) \Big/ \left(\dfrac{\partial D}{\partial \omega}\right). \end{cases} \tag{2.25}$$

A conventional way to calculate the group velocity is by differentiating the phase velocity with respect to k, as $v_{group} = v_{phase} + k\, dv_{phase}/dk$. With known dispersion equation D, we can use equation (2.25) to directly calculate the group velocity from D.

The wave associated with a pole is called a wave mode. The wave mode's spectral amplitude, as given by the contribution from the pole, is called the wave mode excitation function $E(\omega)$, or borehole response function, for the wave mode. This function is obtained by evaluating the residue of the wave's spectral amplitude function $A(k, \omega)$, as in equation (2.23). By setting $z = 0$ for a wave mode (identified by subscript l) in equation (2.22) and using equation (2.23), the excitation function for the wave mode is given by

$$E(\omega) = \left(\frac{u_f^d \det \mathbf{M}^{11} - \sigma_{rrf}^d \det \mathbf{M}^{21}}{\partial D / \partial k} \right)_{k=k_l} , \tag{2.26}$$

where wavenumber k is evaluated at wavenumber k_l of the lth pole, as found from equation (2.24).

2.5.1 WAVE MODE DISPERSION CHARACTERISTICS

We now show examples of the phase and group velocity curves as a function of frequency for various wave modes. Two cases are considered. The first is a fast formation, where the formation shear velocity is greater than the borehole fluid velocity (same parameters as for Figure 2.3). The second is a slow formation, where the shear velocity is lower than the fluid velocity (same parameters as for Figure 2.4). The curves are given in Figures 2.6a, b, and c. The dispersion curves in these figures show that the properties of the borehole wave modes depend strongly upon the S-wave velocity of the formation. These wave modes are guided by the borehole only when their phase velocity is lower than the S-wave velocity. Above this threshold, the wave modes radiate energy into the formation and become leaky modes. The characteristics of the various wave mode dispersion curves are described below.

2.5.1.1 Monopole

For an axi-symmetric source excitation (azimuthal order number $n = 0$), the period equation (2.24) yields characteristic velocities and cutoff frequencies. For waves with cutoff frequencies (e.g., pseudo-Rayleigh and leaky P), there is an infinite number of such waves along the frequency axis. However, for the frequency range commonly used in acoustic logging, only one or two such modes are encountered.

Pseudo-Rayleigh wave: In a fluid-filled borehole in a formation with a shear velocity that is greater than the fluid acoustic velocity, these modes exist and have the combined effects of reflected waves in the fluid and critical refraction along the borehole wall. In some literature these modes are simply called "shear normal modes" to distinguish them from the Rayleigh waves on a half-space or plane layers. The term "normal mode" is used to describe wave propagation along a low-velocity channel or wave guide (the borehole). When the formation shear velocity falls below the acoustic velocity of the borehole fluid, the pseudo-Rayleigh wave does not exist because the borehole can no longer trap the shear energy.

As Figure 2.6a shows, the pseudo-Rayleigh wave is strongly dispersive. The phase velocity drops from the formation shear velocity at the cutoff frequency to approach the fluid velocity at high frequencies. The group velocity, which is the velocity for the energy of a finite wave packet composed of many Fourier (frequency)

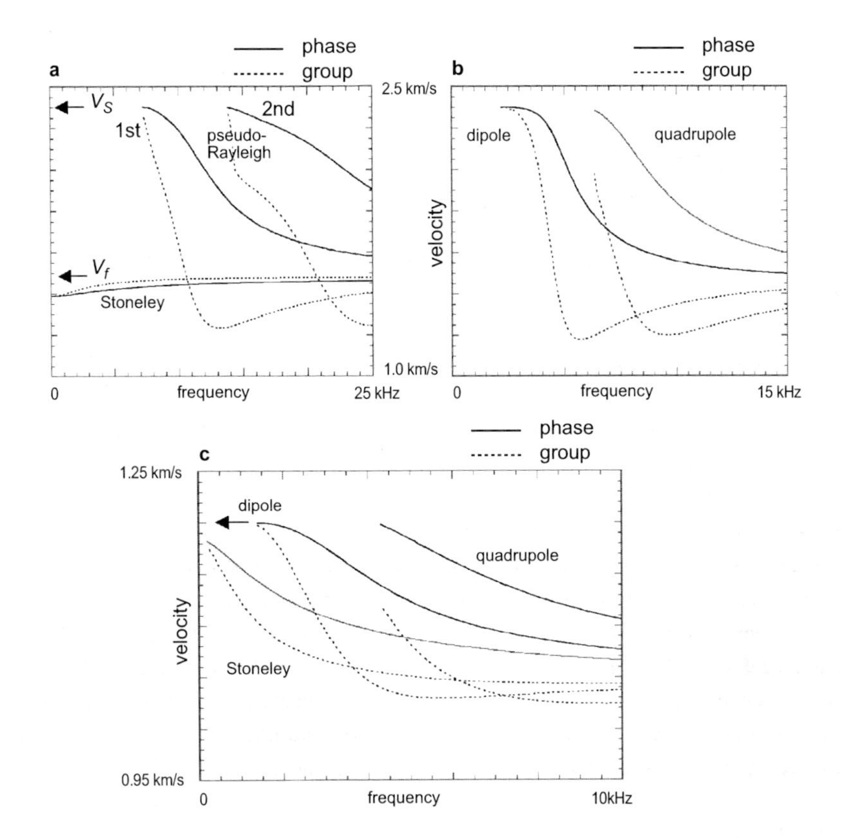

Figure 2.6. Wave mode phase (solid curves) and group (dashed curves) dispersion curves in fast (a and b) and slow (c) formations. (a) Stoneley and first and second pseudo-Rayleigh modes, (b) dipole and quadrupole wave modes, and (c) Stoneley, dipole, and quadrupole wave modes.

components, shows an even more drastic dispersive behavior. The group velocity decreases rapidly from the shear velocity at the cutoff frequency to less than the fluid velocity, and then increases slowly back towards the fluid velocity as frequency increases. There is a group velocity minimum in the medium frequency range. The presence of the group velocity minimum produces a phenomenon called the "Airy phase". The wave energy associated with the Airy phase lags significantly behind the onset of the wave mode, as shown in Figure 2.3a.

Stoneley wave: Historically, the Stoneley wave denotes the waves propagating along a planar interface between two elastic solids. A special case of the Stoneley wave is a wave traveling along the interface between a fluid and a solid, called the *Scholte wave*. In the borehole case, the Stoneley wave refers to the wave along the borehole interface. At low frequencies, the Stoneley wave is also known as the tube wave because the "hoop" stress effects of the cylindrical wall become prominent at large wavelengths. As shown in Figures 2.6a (fast formation) and 2.6c (slow formation), the Stoneley wave exists at all frequencies. The Stoneley-wave phase and group velocities increase or decrease, depending on formation type (fast or slow), from the low-frequency limit to a high-frequency limit value called the Scholte-wave velocity, which is the Stoneley-wave velocity at a planar solid-fluid interface. This property is understandable because at short wavelengths, the cylindrical surface appears to be planar to the wave.

Leaky P: When the formation shear velocity falls below the acoustic velocity of the borehole fluid, trapped modes are dominated by critical refraction of compressional waves at the borehole wall. These wave modes are called "leaky P" or PL (partially leaky) waves. The waves are leaky because they lose energy by conversion to shear, which radiates into the formation and carries wave energy away from the compressional waves. The wave characteristics were shown in Figure 2.4a.

2.5.1.2 Dipole

The wave mode with azimuthal order number $n = 1$ is the flexural wave. The flexural wave mode generated from a dipole source exhibits a behavior similar to that of the pseudo-Rayleigh wave, especially in fast formations (Figure 2.6b). Like the pseudo-Rayleigh waves, there is an infinite number of flexural modes along the frequency axis. For both fast (Figure 2.6b) and slow (Figure 2.6c) formations, the phase velocity of the flexural wave drops from the formation shear velocity at the cutoff frequency to approach the high-frequency limit of the Stoneley-wave velocity, i.e., the Scholte-wave velocity. The group velocity shows a well-defined minimum associated with the Airy phase, indicating that the flexural wave can be very dispersive if the wave excitation is in the frequency range of the Airy phase. The cutoff frequency of the lowest-order mode is in the low-frequency range of the acoustic logging measurement. For this reason, acoustic dipole tools are designed to measure the low-frequency flexural waves. At low frequencies, the flexural wave velocity is very close to the formation shear-wave velocity and the wave's dispersion effect is minimal. As the synthetic waveform examples (Figures 2.3b and 2.4b, for both fast and slow formations) show, the onset of low-frequency flexural waves travel at the formation S-wave velocity with negligible dispersion.

2.5.1.3 Quadrupole

The wave mode with azimuthal order number $n = 2$ is called the "quadrupole wave". The quadrupole wave is also called screw wave because its particle motion exhibits an alternating oblate/prolate pattern. The dispersion characteristic of the quadrupole wave is similar to that of the flexural wave, except that the former wave has a higher cutoff frequency (Figures 2.6b and c). Because the wave's phase velocity near the cutoff frequency is the shear-wave velocity of the formation, the quadrupole wave can also be utilized to perform shear-wave logging in fast and slow formations (Winbow et al., 1991). Recent studies (e.g., Tang et al., 2002a) have found that quadrupole waves are particularly advantageous for the shear-wave velocity measurement in the logging-while-drilling (LWD) environment. The LWD quadrupole wave, when excited at low frequencies, is free of the drilling-tool-wave interference and travels at the formation shear velocity. The modeling of the LWD acoustic system, however, requires the analysis of the multi-layered structure containing concentric elastic and fluid layers, as will be described in the following section.

2.6 Modeling Multi-Layered Formations

The effects of casing and cement can be modeled by considering them as concentric, cylindrical layers between the borehole and the formation. An acoustic logging tool can also be modeled as a cylindrical structure in the borehole (e.g., Tang et al., 2002b). The Thomson-Haskell propagator matrix method is used to connect the wavefield through the layers (Schmitt, 1988b). The displacement-stress vector for the cylindrical system is defined as:

$$\mathbf{S} = (u, v, w, \sigma_{rr}, \sigma_{r\theta}, \sigma_{rz})^T \ , \qquad (2.27)$$

where the superscript T denotes transpose. Using equations (2.11) through (2.14), we can write the vector as

$$\mathbf{S} = \mathbf{T}\mathbf{X} \ , \qquad (2.28)$$

where $\mathbf{X} = (A_n, B_n, C_n, D_n, E_n, F_n)^T$. Note that both incoming and outgoing waves are now involved to model the wavefield in a layer. The matrix $\mathbf{T}$ is a 6×6 complex matrix. Compared with equation (2.16), the matrix dimension is increased by two because now the displacement components v and w are involved in the boundary condition at the boundaries of the solid layer. The expressions for many elements of the $\mathbf{T}$ matrix are the same or similar to those of the $\mathbf{M}$ matrix in equations (2.18) and can be abbreviated or simplified. The elements of $\mathbf{T}$ are given below:

$$u: \begin{cases} T_{11} = M_{12}(K \to I) \\ T_{12} = M_{12} \\ T_{13} = M_{13}(K \to I) \\ T_{14} = M_{13} \\ T_{15} = M_{14}(K \to I) \\ T_{16} = M_{14} \end{cases} \quad v: \begin{cases} T_{21} = -I_n(pR)n/R \\ T_{22} = -K_n(pR)\,n/R \\ T_{23} = -I_n(sR)n/R - sI_{n+1}(sR) \\ T_{24} = -K_n(sR)n/R + sK_{n+1}(sR) \\ T_{25} = -I_n(sR)ikn/R \\ T_{25} = -K_n(sR)ikn/R \end{cases} \quad w: \begin{cases} T_{31} = ikI_n(pR) \\ T_{32} = ikK_n(pR) \\ T_{33} = 0 \\ T_{34} = 0 \\ T_{35} = -s^2 I_n(sR) \\ T_{36} = -s^2 K_n(sR) \end{cases} \qquad (2.29)$$

$$\sigma_{rr}: \begin{cases} T_{41} = M_{22}(K \to I) \\ T_{42} = M_{22} \\ T_{43} = M_{23}(K \to I) \\ T_{44} = M_{23} \\ T_{45} = M_{24}(K \to I) \\ T_{46} = M_{24} \end{cases} \quad \sigma_{r\theta}: \begin{cases} T_{51} = M_{32}(K \to I) \\ T_{52} = M_{32} \\ T_{53} = M_{33}(K \to I) \\ T_{54} = M_{33} \\ T_{55} = M_{34}(K \to I) \\ T_{56} = M_{34} \end{cases} \quad \sigma_{rz}: \begin{cases} T_{61} = M_{42}(K \to I) \\ T_{62} = M_{42} \\ T_{63} = M_{43}(K \to I) \\ T_{64} = M_{43} \\ T_{65} = M_{44}(K \to I) \\ T_{66} = M_{44} \end{cases}$$

In the above expressions, the notation $M_{ij}(K \rightarrow I)$ means taking the expression for M_{ij} as given in equations (2.18), e.g., M_{12}, and substituting the associated Bessel function K_n with I_n and K_{n+1} with $-I_{n+1}$ (for example, making these substitutions in T_{24} results in T_{23}). This $(K \rightarrow I)$ substitution applies to the Bessel function combinations (Y_1 through Y_4).

2.6.1 WELL-BONDED BOUNDARY CONDITION

The wavefield at both sides of a layer interface is connected by boundary conditions. For a welded solid-solid contact, the boundary condition is that the components of the displacement-stress vector must be continuous at the interface, as

$$\mathbf{S}(j,r) = \mathbf{S}(j+1,r) \ , \tag{2.30}$$

where the index j denotes the jth layer ($j = 1, ..., N-1$) (suppose there are $N-1$ layers between borehole and formation, with the formation being indexed by N). Using equations (2.27) and (2.28), we can relate the **S** vector at the borehole, to the **S** vector at the outmost layer interface, as

$$\mathbf{S}(1,R_{bh}) = \left(\prod_{j=1}^{N-1} \mathbf{T}(j,R_{in})\mathbf{T}^{-1}(j,R_{out}) \right) \mathbf{S}(N,R_{fm}) = \left\{ \left(\prod_{j=1}^{N-1} \mathbf{g}_j \right) \mathbf{T}(N,R_{fm}) \right\} \mathbf{X}(N) = \mathbf{G}\,\mathbf{X}(N) \ , \tag{2.31}$$

where the matrix $\mathbf{g}_j$, which connects wave propagation through the jth layer, is called the *propagator matrix*; the inner and outer radius of the layer is denoted by R_{in} and R_{out}, respectively. The product of the **g** matrices for all layers and $\mathbf{T}(N, R_{fm})$ of the formation is denoted by **G**, which is a 6×3 complex matrix because $\mathbf{X}(N)$ contains only three coefficients. This matrix governs the wave propagation from the borehole radius R_{bh} across the layers to the starting radius R_{fm} of the formation. Notice that the radiation condition reduces the coefficient vector of the formation to $\mathbf{X}(N) = (B_n^{(N)}, D_n^{(N)}, F_n^{(N)})^T$ (note that n denotes the order of the multipole and N is the total number of layers). The elastic parameters in $\mathbf{T}(N, R_{fm})$ are those of the formation.

It should be pointed out that a numerical problem occurs when calculating the matrix **T** for very high frequencies and/or layer boundaries that are far from borehole. Under these circumstances, the argument z of the modified Bessel function $I_n(z)$ becomes large; this causes an overflow problem in the function (notice that z may denote either pR or sR, as in equation (2.29)). To solve this problem, Chen et al. (1994) proposed the use of renormalized Bessel functions (notice that Chen et al. used Hankel functions, instead of modified Bessel functions in their calculations). Employing Chen et al.'s approach, we use the renormalized functions $\bar{K}_n(z) = e^z K_n(z)$ and $\bar{I}_n(z) = e^{-z} I_n(z)$ to respectively replace their counterparts $K_n(z)$ and $I_n(z)$ in **T**, resulting in a modified matrix $\bar{\mathbf{T}}$. The renormalization factors e^z and e^{-z} can also be factored out to form a diagonal matrix that multiplies $\bar{\mathbf{T}}$. With this treatment, the propagator matrix $\mathbf{g}_j$ of the jth layer, as appears in equation (2.31), becomes

$$\mathbf{g}_j = \overline{\mathbf{T}}(j, R_{in}) \begin{pmatrix} e^{-p\Delta R} & & & & & \\ & e^{p\Delta R} & & & & \\ & & e^{-s\Delta R} & & & \\ & & & e^{s\Delta R} & & \\ & & & & e^{-s\Delta R} & \\ & & & & & e^{s\Delta R} \end{pmatrix} \overline{\mathbf{T}}^{-1}(j, R_{out}) ,$$

where $\Delta R = R_{out} - R_{in}$ is the layer thickness. The above approach stabilizes the calculation of $\mathbf{T}$ for high frequencies and/or distant layers. To prevent the possible overflow of $e^{p\Delta R}$ or $e^{s\Delta R}$ at high frequencies and/or large layer thickness, the exponential function $e^{p\Delta R}$ is factored out from the above diagonal matrix. Each element of the matrix, after division by $e^{p\Delta R}$, has a modulus less than unity, thus eliminating the overflow problem. Combining the factor $e^{p\Delta R}$ for all layers using equation (2.31) results in a product of N–1 exponentials, as $\exp(p\Delta R)_1 \ldots \exp(p\Delta R)_{N-1}$. This product and the renormalization factors for the matrix $\mathbf{T}(N, R_{fm})$ of the formation are incorporated into the to-be-determined coefficient vector $\mathbf{X}(N) = (B_n^{(N)}, D_n^{(N)}, F_n^{(N)})^T$. These manipulations, therefore, make the propagator matrix operation of equation (2.31) numerically stable and accurate.

The boundary condition at the borehole wall involves only u, σ_{rr}, $\sigma_{r\theta}$ and σ_{rz} (see equations (2.15)), the last two being zero for the borehole fluid. The azimuthal and vertical displacement components v and w (matrix elements G_{2j} and G_{3j} ($j = 1, \ldots, 6$)) are not involved. With the borehole boundary condition and equation (2.29), we finally obtain a 4×4 matrix system, as

$$\begin{pmatrix} M_{11} & G_{12} & G_{14} & G_{16} \\ M_{21} & G_{42} & G_{44} & G_{46} \\ M_{31} & G_{52} & G_{54} & G_{56} \\ M_{41} & G_{62} & G_{64} & G_{66} \end{pmatrix} \begin{pmatrix} A'_n \\ B_n^{(N)} \\ D_n^{(N)} \\ F_n^{(N)} \end{pmatrix} = \begin{pmatrix} u_f^d \\ \sigma_{rrf}^d \\ 0 \\ 0 \end{pmatrix}, \qquad (2.32)$$

where the elements M_{ij} and the right hand side quantities are the same as in equation (2.16). The elements G_{ij} are those of the product matrix $\mathbf{G}$ in equation (2.31). The above equation reduces to equation (2.16) if no layer exists between borehole and formation. The calculation of synthetic micro-seismograms and dispersion curves follows the same steps as described in sections 2.3 and 2.4.

2.6.2 Unbonded Boundary Condition

The effects of a poorly cemented cased hole are commonly modeled using a thin fluid annulus layer between casing and cement or between cement and formation. The case with the fluid annulus between casing and cement (but good bonding between cement and formation) is known as the "free-pipe situation".

The propagator matrix calculation with a fluid layer (or fluid layers) is somewhat complicated. Let us consider a fluid layer with subscript L within N – 1 elastic layers.

The fluid-layer field quantities involved in the boundary condition are radial displacement and stress (shear stress is zero in fluid). Thus the potential Φ in equations (2.11) with incoming and outgoing wave coefficients A and B suffices to specify the solution. Using equation (2.8), the fluid displacement and stress are derived as

$$\left.\begin{matrix} u \\ \sigma_{rr} \end{matrix}\right\} = e^{ikz} \frac{1}{n!}\left(\frac{fr_0}{2}\right)^n \cos(n(\theta-\phi)) \begin{cases} A_n^L[nI_n(fr)/r + fI_{n+1}(fr)] + B_n^L[nK_n(fr)/r - fK_{n+1}(fr)] \\ -\rho_f\omega^2[A_n^L I_n(fr) + B_n^L K_n(fr)] \end{cases} . \quad (2.33)$$

The material parameters in the above expressions are those of the fluid layer. At the inner and outer boundaries of the fluid layer, the boundary condition is the continuity of radial displacement u and stress σ_{rr}, and the vanishing of the other two shear stresses $\sigma_{r\theta}$ and σ_{rz}. Again, displacement components v and w are not involved. This results in two sets of equations

$$T_{L-1}^* \begin{pmatrix} A_n^{(L-1)} \\ B_n^{(L-1)} \\ C_n^{(L-1)} \\ D_n^{(L-1)} \\ E_n^{(L-1)} \\ F_n^{(L-1)} \end{pmatrix} = \begin{pmatrix} u \\ \sigma_{rr} \\ 0 \\ 0 \end{pmatrix}, \text{ (inner)}; \quad \begin{pmatrix} u \\ \sigma_{rr} \\ 0 \\ 0 \end{pmatrix} = T_{L+1}^* \begin{pmatrix} A_n^{(L+1)} \\ B_n^{(L+1)} \\ C_n^{(L+1)} \\ D_n^{(L+1)} \\ E_n^{(L+1)} \\ F_n^{(L+1)} \end{pmatrix}, \text{ (outer)} . \quad (2.34)$$

The radial fluid displacement and stress at the inner and outer boundaries of the fluid layer are calculated using equations (2.33). The $\mathbf{T}^*$ matrices of layers L–1 and L+1 are a 4×6 matrix reduced from the matrix $\mathbf{T}$ of equations (2.28). The reduction consists in removing from $\mathbf{T}$ the elements T_{2j} and T_{3j} (j = 1, ..., 6) that relate to v and w, respectively.

Using the propagator matrix method, the $\mathbf{S}$ vector of the L–1th layer can be propagated to the borehole to relate the wave coefficient A'_n for the borehole fluid. Similarly, the $\mathbf{S}$ vector of the L+1th layer can be propagated to the formation to relate the three outgoing wave coefficients $B_n^{(N)}$, $D_n^{(N)}$, and $F_n^{(N)}$. The final system of equations to solve is:

$$\mathbf{HO} = \begin{pmatrix} u_f^d \\ \sigma_{rrf}^d \\ 0 \\ 0 \end{pmatrix}, \quad (2.35)$$

where $\mathbf{O} = (A'_n, A_n^{(L-1)}, B_n^{(L-1)}, C_n^{(L-1)}, D_n^{(L-1)}, E_n^{(L-1)}, F_n^{(L-1)}, A_n^{(L)}, B_n^{(L)}, B_n^{(N)}, D_n^{(N)}, F_n^{(N)})^T$ is a 12-element coefficient array. Notice that the coefficients for the L+1th layer are not involved because, after using the propagator matrices, they can be expressed in terms of the coefficients $B_n^{(N)}$, $D_n^{(N)}$, and $F_n^{(N)}$. $\mathbf{H}$ is a 4×12 complex matrix. The right-hand side is the same as equation (2.32). The matrix elements of $\mathbf{H}$ can be derived following the above procedure and are not listed here for reasons of brevity.

2.6.3 DISPERSION ANALYSES AND SYNTHETIC WAVE CALCULATION FOR A MULTI-LAYERED SYSTEM

For a fluid-filled borehole with a point acoustic source, the calculation of synthetic full waveform seismograms and dispersion analyses for a multi-layered structure, either for the well-bonded (equation (2.32)) or unbonded case (equation (2.35)), follow the procedures similar to those outlined previously, except that extra calculation of the matrix **G** or **H** is required. However, if the dimension of the source on an acoustic tool (e.g., the LWD acoustic tool) cannot be ignored and the effects of the tool need to be analyzed, the use of a point source, as described by equation (2.4), is not applicable. In this case, we use an acoustic ring source and apply it to the unbonded boundary condition (equation (2.34)) to calculate the synthetic seismograms corresponding to the source on the acoustic tool.

For an acoustic tool modeled as a cylindrical structure of outer radius a, a multi-pole acoustic ring source is represented by a distribution of the point sources along a circle of radius a. The source intensity varies azimuthally, as given by $\cos(n(\theta_0 - \phi))$, where θ_0 is the azimuth of a point source on the circle with respect to some reference angle ϕ. Multiplying $\cos(n(\theta_0 - \phi))$ with the right-hand-side of equation (2.4) and integrating the resulting expression along the circle ($r_0 = a$; $0 \le \theta_0 \le 2\pi$) gives the fluid acoustic displacement potential for the ring source. Then, using the potential in equation (2.8) gives the (radial) displacement at the source location, which, expressed in k-domain, is given by

$$u_f^d = \varepsilon_n\big(nK_n(fa)/a - fK_{n+1}(fa)\big)I_n(fa)\cos(n(\theta - \phi)), \tag{2.36}$$

where ε_n is 1 for $n = 0$, and 2 for $n > 0$. Clearly, the azimuthal order number n specifies the source type, with $n = 0, 1, 2$ corresponding to monopole, dipole, and quadrupole sources, respectively. Finally, we use the inner boundary condition in equation (2.34) to model the source excitation (the outer boundary condition of the equation now represents the boundary condition at the borehole interface). Assuming that the tool, modeled as a cylindrical layered structure of outer radius a, is represented by the left-hand-side of the (inner) boundary condition equation, we add the source displacement u_f^d to the borehole fluid displacement at the right-hand-side of the equation. This assigns the source to the rim of the tool. The resulting system of equations to solve is similar to equation (2.35), except that the source term at the right-hand-side is replaced by $(u_f^d, 0, 0, 0)^T$. Moreover, we now use the coefficients $A_n^{(L)}$ and $B_n^{(L)}$ instead of A'_n, to calculate the acoustic wavefield in the borehole fluid annulus (the coefficient A'_n specifies the acoustic wavefield in the innermost fluid column, if any, of the tool). For example, the pressure waveform at the rim of the tool can be calculated using

$$P(z,t) = \int_{-\infty}^{+\infty} \rho_f \omega^2 S(\omega) e^{-i\omega t} d\omega \int_{-\infty}^{+\infty} \left(A_n^{(L)} I_n(fa) + B_n^{(L)} K_n(fa) \right) e^{ikz} dk \ . \tag{2.37}$$

It is worthwhile to mention that – although the dispersion analysis of a multi-layered system does not depend on the source (whether it is a ring source at the rim of the tool or point source in the innermost fluid column) – the wave amplitude excitation and resulting waveforms strongly depend on the source and its location in the system. Furthermore, the presence of the acoustic tool introduces acoustic wave modes that travel along the tool. The tool-wave phenomenon is particularly pronounced in the acoustic logging-while-drilling environment where the tool occupies a large portion of the borehole. For example, Tang et al. (2002b) applied the above described ring-source formulation to analyze the problem of multipole acoustic logging-while-drilling. The results demonstrate that the LWD acoustic wavefield has characteristics and complexity that are quite different from its wireline counterpart (see also section 2.8).

2.7 Multi-Layered Formation and Cased-Hole Acoustic Logging Synthetic Seismograms

Having described the theory and methods for modeling wave propagation in a borehole with multiple concentric layers, we can now demonstrate the acoustic wave characteristics. Specific examples include formation alteration, well-bonded cased hole, poorly-bonded cased hole, and dipole logging with poor casing bonding. The modeling examples assume a point source centered in the borehole fluid.

2.7.1 Modeling of Formation Alteration

As an application for the multi-layer modeling, we can study the effect of formation alteration on borehole acoustic dipole waveforms. An altered formation model is shown in Figure 2.7a, where the altered zone is modeled as a cylindrical layer around the borehole. The shear velocity of the zone can be either greater or smaller than that of the virgin formation. The shear velocity of the virgin formation is assumed to be 2500 m/s. The thickness of the altered zone is 0.08 m and the borehole diameter is 0.2 m. Two cases are considered. In the first case, the shear velocity of the zone is decreased from 2500 m/s to 2300 m/s (an 8 % decrease). In the second case the shear velocity of the zone is increased from 2500 m/s to 2700 m/s (an 8 % increase). The modeling results are shown in Figure 2.7b for the dispersion characteristics and in Figure 2.7c for the array acoustic waveform features. In the array waveform modeling, the source excitation frequency is about 4 kHz. We notice that the waveform corresponding to the decreased velocity model is more dispersive and lags behind that of the increased velocity model. However, the early portions of the two sets of waveforms travel across the array at the shear velocity of the virgin formation.

The waveform characteristics shown in Figure 2.7c can be explained using the

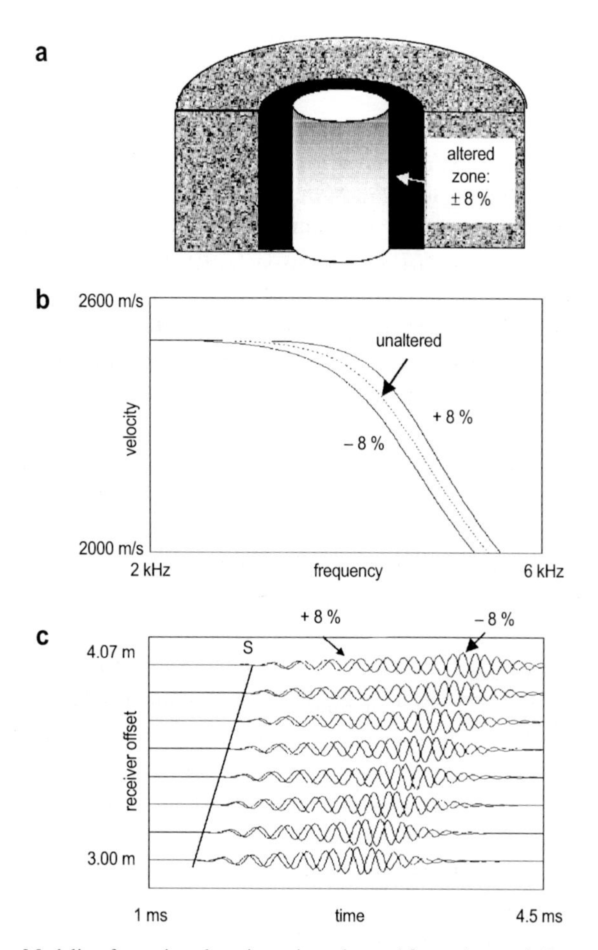

Figure 2.7. Modeling formation alteration using a layered formation model.
(a) Altered formation model. Shear velocity in the altered zone is increased/decreased by 8%.
(b) Dipole-flexural dispersion curves with/without alteration.
(c) Synthetic dipole array waveforms for the increased/decreased velocity case. The time line S is the arrival time of the virgin formation shear wave.

dispersion curves in the frequency domain. Figure 2.7b shows the phase velocity curves versus frequency for the altered zone model with increased and decreased velocity scenarios, as indicated in the figure. The dispersion curve for an unaltered (virgin) formation is also shown (dashed curve). Compared with the unaltered formation, the velocity curve of an altered formation shows more dispersion if there is a decrease in formation shear velocity, and less dispersion if there is an increase. However, toward low frequencies, the dispersion curves approach the virgin formation shear velocity value, showing that a low-frequency flexural wave has a deeper penetration, and is less affected by the alteration. This modeling example shows that the dispersion characteristics of the flexural wave can be used to diagnose formation alteration. Tang (1996a) described a procedure for this diagnosis.

2.7.2 EFFECTS OF CASING ON MONOPOLE LOGGING: WELL-BONDED CASE

The presence of casing and cement, as well as a fluid layer between casing and cement, can have significant effects on acoustic logging waveforms. We first study the case where the casing is well-bonded with the formation by a cement layer. For modeling this situation, we use the well-bonded boundary condition (equation (2.30)). The casing thickness is 1 cm and the cement thickness is 2 cm. The P- and S-wave velocities and the densities of casing and cement are given in Table 2.1. Figure 2.8a shows synthetic monopole seismograms in a fast formation with casing. The open-hole counterpart of this figure was shown in Figure 2.3a. The characteristics of the two sets of seismograms are quite different. In the cased-hole example, the Stoneley-wave arrival is much stronger than in the open hole. In contrast, the P- and S- (and pseudo-Rayleigh-) wave packets are smaller in amplitude than in the open-hole case. The cause of these phenomena is the presence of casing and cement, which effectively decrease the borehole radius. For a borehole of reduced radius, the effective excitation of the P- and shear- (pseudo-Raleigh-) waves is shifted to a higher frequency range than for a borehole with a larger radius. Notice also that the wave arrivals in the cased-hole scenario, compared to the open-hole scenario, are slightly advanced due to decreased travel time through the higher-velocity cement and steel that replace a portion of the borehole fluid. Note especially that the formation P- and S-wave arrivals can still be measured despite the presence of casing and cement.

In a slow formation, casing changes the character of acoustic waveforms even more significantly. Figure 2.8b shows synthetic monopole waveforms in a slow formation with casing. The open-hole counterpart of this figure was shown in Figure 2.4a. The waveform is dominated by a Stoneley-wave arrival. Moreover, the Stoneley-wave arrival is similar to that in a fast formation. This shows that the Stoneley wave is largely controlled by the – highly rigid – steel casing, with very little sensitivity to the formation. In contrast, the waveform for the open borehole shows a prominent P-wave arrival and a prominent Stoneley wave of smaller amplitude. The Stoneley-wave velocity is much lower in the open hole than in the cased borehole and is controlled by the formation shear velocity. An important aspect concerning acoustic logging through casing is that the formation P-wave arrival is still discernable, even though its amplitude is much reduced (in Figure 2.8b the waveform amplitude is enlarged in order to see the small-amplitude P-wave arrival).

2.7.3 EFFECTS OF CASING ON MONOPOLE LOGGING: POORLY-BONDED CASE

When casing and formation are poorly bonded, the acoustic waveform characteristics are almost entirely controlled by casing and fluid. This situation can be modeled by placing a fluid layer between casing and cement (this is also known as the "free pipe situation", as modeled by Tubman et al. (1986) for the monopole case). For modeling this situation, we use the unbonded boundary condition (equations (2.34)). A fluid layer of 0.5 cm thickness is introduced between casing and cement (the fluid acoustic

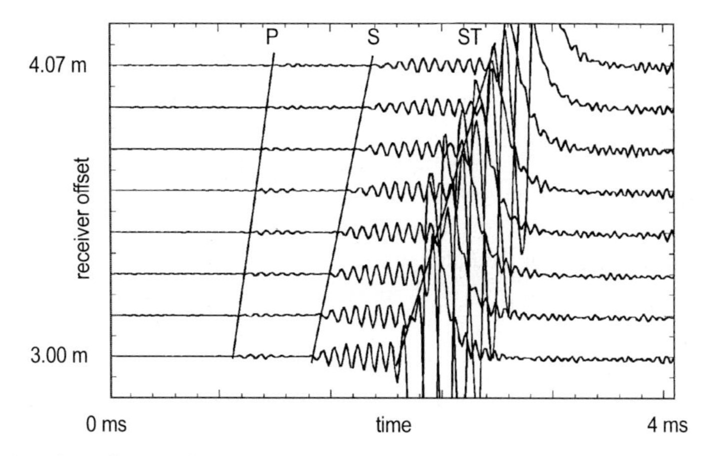

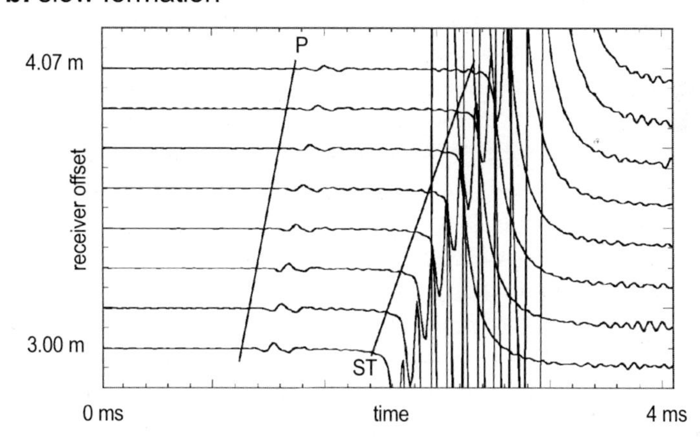

Figure 2.8. Synthetic monopole array waveforms in a well-bonded cased borehole.
(a) Fast formation case.
(b) Slow formation case. In either case, formation P- and/or S-waves can still be detected. Note that the wave amplitude is enlarged in order to see the low-amplitude formation P-wave arrival.

parameters are given in Table 2.1). Figure 2.9 shows the synthetic array waveforms for the free pipe situation. The formation is a fast one, having the same parameters for calculating the waveforms as in Figure 2.8a. There is a high-amplitude ringing P wavetrain, with a moveout velocity almost identical to the plate velocity of the steel. There is also an emergent wavetrain related to the formation shear/pseudo-Rayleigh waves. This wavetrain is not very coherent due to the interaction with other wave modes caused by the fluid layer. The Stoneley-wave arrival is still prominent and is largely influenced by the steel casing. In the free-pipe situation, it is difficult to pick the formation velocities, even with the help of sophisticated array processing techniques.

2.7.4 MODELING CASED-HOLE DIPOLE LOGGING

The above examples show that monopole logging through casing to obtain formation velocities is not a problem when the casing is well-bonded with the formation. However, it is problematic in the poorly-bonded case. This problem is largely related to the nature of the wave type used. For example, for a compressional wave propagating along the borehole, the polarization of the wave is primarily in the borehole axial direction. As a result, the acoustic disturbance associated with the wave is much smaller in the radial direction than in the axial direction. In the free pipe situation, the axial disturbance causes ringing casing waves that mask the small formation arrival associated with the radial disturbance. The above analysis suggests that a transversely polarized wave, such as a dipole-shear wave (see Figure 2.10a), should penetrate more effectively through a (poorly bonded) casing than an axially polarized wave. With this concept, acoustic logging through casing using dipole tools is evaluated with numerical modeling examples.

We use the free-pipe case to demonstrate the feasibility, as well as a potential problem, of cased-hole dipole-acoustic logging. Figure 2.10 shows synthetic dipole-acoustic waves for the free-pipe situation. The formation for this modeling is identical to that of Figure 2.9. The synthetic dipole waveforms are computed for two source excitation frequencies: 6 kHz (Figure 2.10b) and 3 kHz (Figure 2.10c). The higher-frequency case clearly shows a prominent P-wave arrival and a flexural-wave arrival, traveling at the formation P- and S-wave velocities, respectively. This proves that a transversely polarized wave can "see" through casing/cement better than an axially polarized wave, even in the free-pipe situation (in practice, this advantage of dipole logging has recently been utilized to determine compressional-wave slowness in cased holes with poor cement bonding (Pampuri et al., 2003)). However, the seismogram also shows an emergent event trailing the formation waves. This wave is the casing flexural wave. When the casing is not bonded with the formation, it flexes when it is excited by a dipole source. At a lower excitation frequency (3 kHz, Figure 2.10c), the P-wave arrival almost disappears, and the waveform is dominated by the formation flexural wave and the casing flexural wave with strong amplitude. The presence of a casing flexural wave may pose a problem for cased-hole dipole logging, particularly when the formation and casing flexural waves have similar frequencies and velocities.

The presence of tube waves (i.e., low-frequency Stoneley waves) in cased-hole logging may also cause a problem. The tube waves cannot be excited by a dipole source if the source and receivers are perfectly centered in the borehole, as is the case in synthetic modeling. In an actual logging situation, however, it is difficult to maintain perfect tool centering. A cased borehole is a good wave guide for the tube waves. Even a slightly off-center dipole tool tends to generate tube waves that interfere with the formation arrival. The situation becomes serious when the formation flexural wave and the tube wave have similar frequency content and their velocities are close.

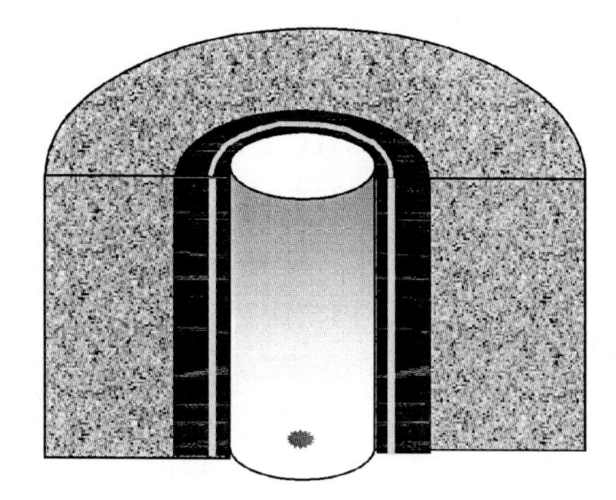

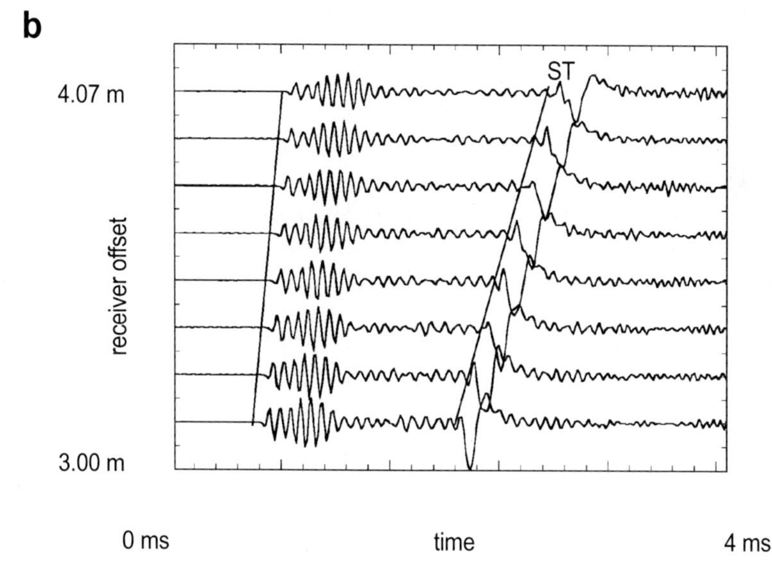

Figure 2.9. (a) The free-pipe scenario modeled as a thin fluid layer behind casing. A monopole source is placed inside the borehole.
(b) Synthetic monopole array waveforms for the model in (a). The wave arrivals are dominated by casing waves and the Stoneley wave that is largely controlled by the casing.

The situation with the fluid annulus between cement and formation (but good bonding between casing and cement) can be modeled using the same approach as in equation (2.34). Prof. Kexie Wang and his group at Jilin University, China, have conducted the modeling (personal communication). The results show that the multipole wave phenomena are more complex than those of the free-pipe case. Although the results discourage the attempt to derive formation properties in the presence of poor cement-formation bonding, the modeled wave phenomena can be used to diagnose the condition of cement-formation bonding.

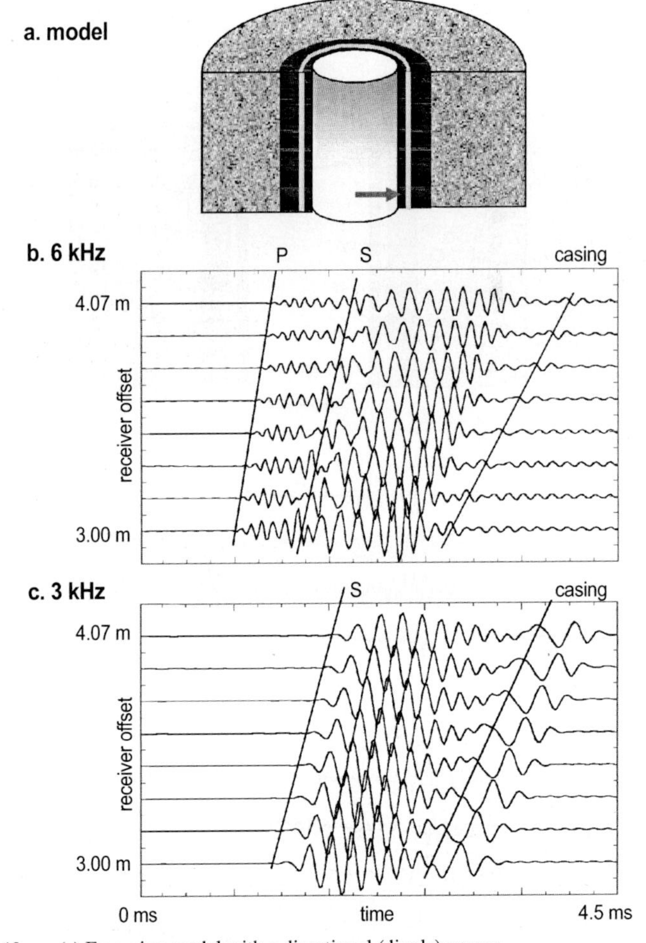

Figure 2.10. (a) Free-pipe model with a directional (dipole) source.
(b) Synthetic dipole array waveforms with an excitation source center frequency of 6 kHz.
Note the excitation of formation P wave and an emergent casing flexural wave. (c) Dipole array waves with an excitation frequency of 3 kHz. Note the strong casing flexural wave besides the formation flexural wave.

2.8 Modeling Logging-While-Drilling Multipole Wave Propagation

In this section, we use the acoustic wave theory for a multi-layered system to model multipole acoustic wave propagation for the logging-while-drilling (LWD) configuration. In the LWD situation, the drill collar occupies a large portion of the borehole (see Table 2.2 and Figure 2.11a) and substantially influences the acoustic wave propagation characteristics. Consequently, measurement principles that are proven in wireline logging may suffer drawbacks or even become invalid. For example, the measurement principle for shear-wave logging using dipole sources needs to be investigated for the LWD condition. Using acoustic wave modeling and

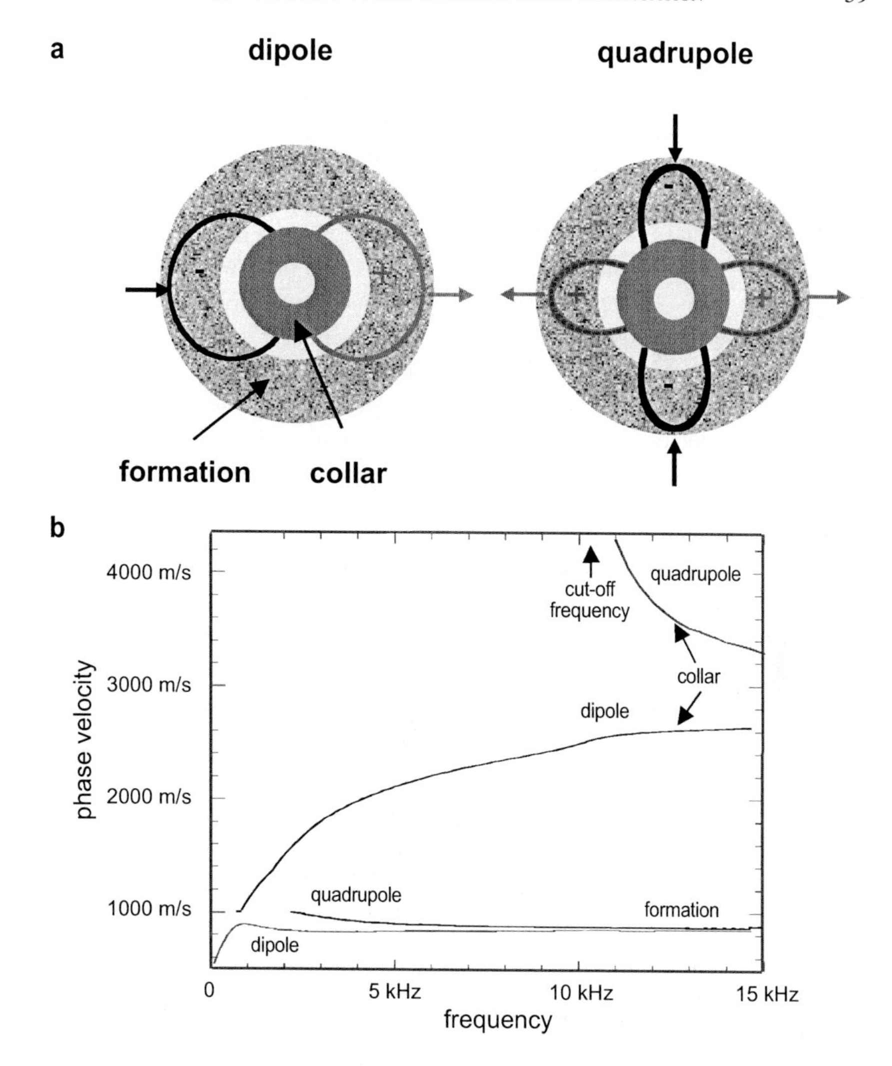

Figure 2.11. Analysis of dipole and quadrupole waves in the logging-while-drilling configuration. (a) The azimuthal wave-amplitude patterns of the dipole and quadrupole sources. (b) Dipole and quadrupole wave dispersion curves.

analysis, we can evaluate the LWD shear-wave velocity measurement with multipole acoustic sources, particularly, the dipole and quadrupole sources. Understanding the LWD acoustic-wave phenomena can help design an LWD acoustic tool to measure the formation acoustic properties.

Figure 2.11a is a schematic view of an LWD multipole acoustic source built into a drill collar. This figure shows the characteristic patterns of azimuthal variation for the dipole and quadrupole sources, respectively. We now discuss the characteristics of LWD acoustic waves for the slow formation in Table 2.2. The wave velocity dispersion calculation uses the determinant of the **H** matrix of equation (2.35), which involves the model parameters given in Table 2.2. The calculation of synthetic acoustic waves uses the previously described multipole acoustic ring source formulation in equation (2.36), which is applied to the right-hand-side of the (inner) boundary condition in equation (2.34).

The velocity dispersion characteristics for the dipole and quadrupole wave modes are shown in Figure 2.11b, for the frequency range of 0–15 kHz. The upper two curves are the collar dipole and quadrupole modes, and the lower two curves are the formation dipole (or flexural) and quadrupole modes. The formation and collar dipole-wave modes coexist almost for the entire frequency range, except at very low frequencies where the collar flexural mode appears to terminate at the formation shear velocity. Below the frequency where the collar mode terminates, the formation flexural-mode velocity appears to have the collar flexural-mode behavior that would exist in the absence of the formation, the velocity decreasing to zero at zero frequency. Unlike the situation in the absence of the collar (see Figure 2.6c), the formation flexural-wave velocity differs significantly from the shear velocity. This means that the dipole wave, when used in LWD, does not directly yield the formation shear velocity. In comparison, the quadrupole source shows a better characteristic. Although the collar-quadrupole waves can also be excited, the collar waves exist only above a certain frequency called the cutoff frequency. Below the cutoff frequency, there is only one mode, the formation quadrupole wave mode. This wave travels at the formation's shear-wave velocity at low frequencies, similar to the situation in the absence of the collar (Figure 2.6c). This characteristic, when used in LWD, allows the measurement of formation shear-wave velocity using quadrupole waves.

It is also interesting to note that the velocities of formation flexural- and quadrupole waves approach each other at high frequencies. In fact, in the high-frequency limit, the two wave velocities approach the Scholte-wave velocity for a planar solid-fluid interface. This common characteristic results from the fact that the high-frequency dipole and quadrupole waves, similar to the Stoneley wave, are interface waves guided along a fluid channel, regardless of their azimuthal variation patterns.

The waveform characteristics of the dipole (Figure 2.12a) and quadrupole (Figure 2.12b) sources are now discussed. The acoustic source and an array of receivers are placed on the rim of the collar. The source excitation center frequency is 2 kHz, the amplitude spectrum of the source wavelet diminishing to zero at about 5 kHz. The acoustic pressure waveforms are calculated for the receiver locations ranging from 2 m to 3.07 m with a 0.152 m spacing. For the dipole source, the first wave to arrive is the collar flexural wave. This wave has a high velocity and is strongly dispersive, as can be seen from the wave's rapid phase variation across the array. The strong dispersion corresponds to the rapid increase of collar flexural-wave velocity in the 2 kHz frequency range (see Figure 2.11b). The last event across the array is the formation flexural wave. The formation wave, however, lags significantly behind the expected arrival time (dashed line) of the formation shear wave. The formation flexural wave does not have a clear onset. It appears to trail behind a low-frequency, low-amplitude wave component. In fact, this wave component corresponds to the

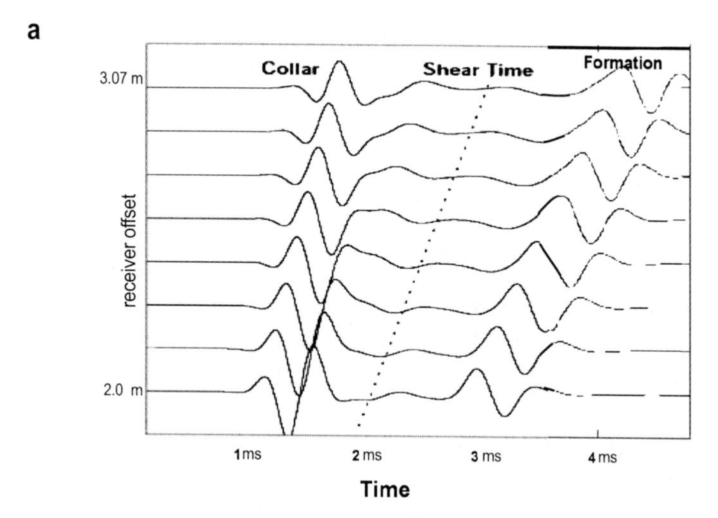

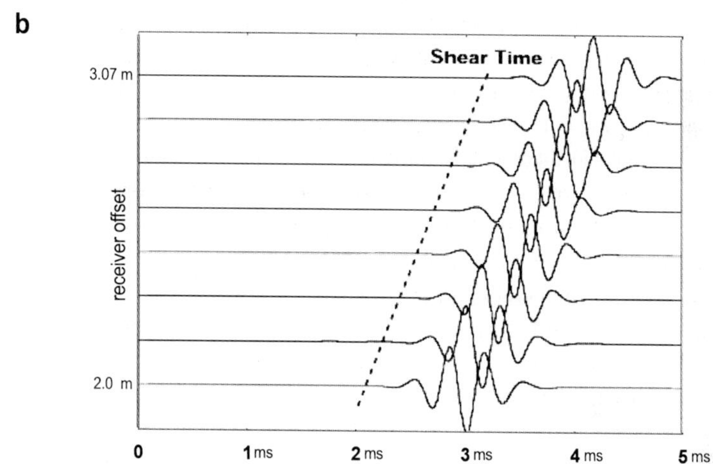

Figure 2.12 Modeling dipole and quadrupole wave propagation for the logging-while-drilling configuration.
(a) Array waveforms for a 2 kHz dipole source.
(b) Array waveforms for a 2 kHz quadrupole source.

(formation and collar) wave energy near the cross-over of the collar and formation curves (Figure 2.11b). The later formation flexural wave, which is only slightly dispersive, corresponds to the flat portion of the formation flexural-wave dispersion curve at higher frequencies.

The quadrupole waveform shows characteristics suitable for LWD shear velocity measurement. As shown in Figure 2.12b, there is only one wave, the formation quadrupole wave, traveling across the array at the formation shear velocity. This wave simulation result is consistent with the frequency-domain dispersion analysis result of Figure 2.11b. Because the frequency range of the wave simulation (< 5 kHz) is

Table 2.2 *LWD-acoustic model used in theoretical modeling. The model parameters are P- and S-velocities, densities and outer radii for inner and outer fluid annuli, drill collar and formation*

	P-velocity	S-velocity	density	outer radius
inner fluid	1 470 m/s	–	1 000 kg/m^3	0.027 m
drill collar	5 860 m/s	3 130 m/s	7 850 kg/m^3	0.090 m
outer fluid	1 470 m/s	–	1 000 kg/m^3	0.117 m
slow formation	2 300 m/s	1 000 m/s	2 000 kg/m^3	∞

well below the cutoff frequency of the collar quadrupole wave (~10 kHz), the collar wave is not excited. Further, at low frequencies (~2–3 kHz) the quadrupole velocity is equal to the formation shear velocity.

These modeling examples demonstrate that using an acoustic dipole in LWD has two drawbacks: (1) there is a significant collar-wave contamination, and (2), the formation flexural wave velocity differs significantly from the formation shear wave velocity. The above analysis also demonstrates that the quadrupole wave in LWD is better suited for the shear velocity measurement because (1) the collar quadrupole wave is absent when operating in the low-frequency range, and (2), the quadrupole wave in a slow formation travels at formation shear velocity at low frequencies.

2.9 Modeling the Effect of Attenuation

During wave propagation through the borehole fluid and the earth formation, a portion of the wave energy is lost due to the anelasticity, or intrinsic damping, of the media. In other words, the wave amplitude is attenuated and the wave velocity changes with frequency (i.e., there is dispersion associated with intrinsic absorption). The dispersion and attenuation can be modeled using a complex velocity method (Aki and Richards, 1980). By defining a complex wavenumber $k(\omega)$, the phase velocity and attenuation (defined as inverse of quality factor Q) are calculated using

$$\begin{cases} v(\omega) = \dfrac{\omega}{\mathrm{Re}\{k(\omega)\}} \\ Q^{-1} = 2\dfrac{\mathrm{Im}\{k(\omega)\}}{\mathrm{Re}\{k(\omega)\}} \end{cases}, \tag{2.38}$$

where Re and Im denote taking the real or imaginary part, respectively, of the complex wavenumber. A commonly used model for intrinsic attenuation and dispersion is the

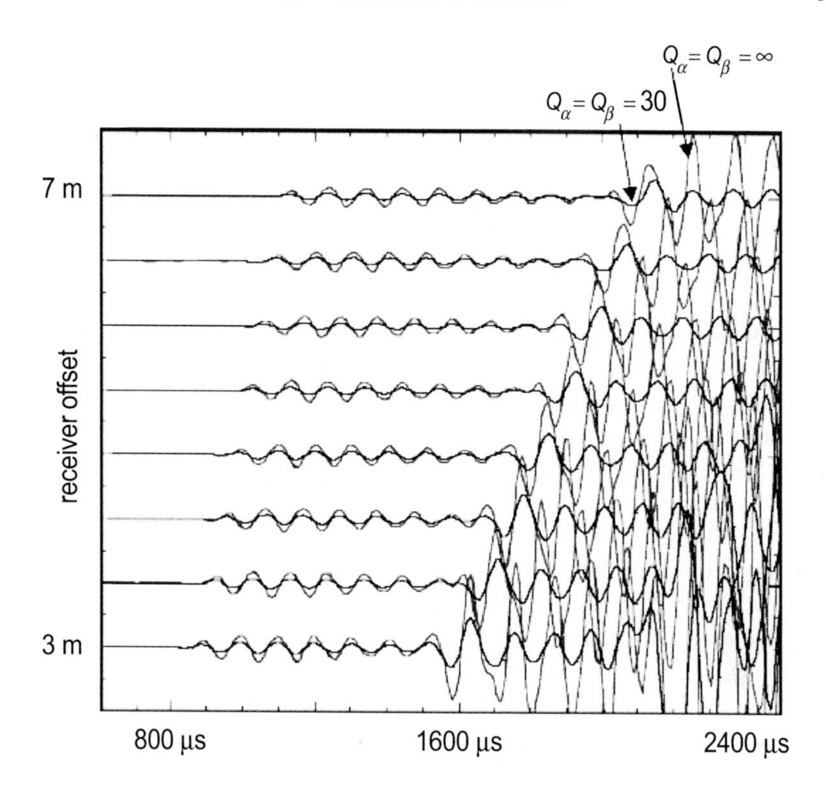

Figure 2.13. Synthetic micro-seismograms calculated with and without attenuation. In the former case, the P- and S-wave quality factors are both set to 30. Notice the significant amplitude reduction caused by the attenuation.

constant-Q model (Aki and Richards, 1980), which assumes that the quality factor Q is independent of frequency. This model is given by

$$c(\omega) = \frac{c_0}{\left\{1 + \frac{1}{\pi}\ln\left(\frac{\omega}{\omega_0}\right)\right\}\left\{1 + \frac{i}{2Q}\right\}}, \tag{2.39}$$

where c_0 is the reference velocity at a reference angular frequency ω_0. The phase velocity $c(\omega)$ and attenuation Q^{-1} can be used to model the complex velocity and attenuation corresponding to the formation compressional wave (α, Q_α), the shear wave (β, Q_β), and the borehole fluid acoustic wave (α_f, Q_f), respectively. With the complex phase velocity in equation (2.39), the wave propagation exponential $\exp(ikz)$ becomes:

$$\exp\left(ik(\omega)z\right) = \exp\left(i\frac{\omega z}{c(\omega)}\right) = \exp\left(-\frac{\omega z}{2Qc_0} + i\frac{\omega z}{c_0}\left(1 + \frac{1}{\pi Q}\ln\left(\frac{\omega}{\omega_0}\right)\right)\right), \tag{2.40}$$

where we have neglected $O\{Q^{-2}\}$ terms. The first term in the bracket is the wave amplitude decay with distance z, and the second term is the phase change associated with a dispersive phase velocity. The velocity varies with frequency in a logarithmic manner. Thus the velocity dispersion relation of a constant-Q model is also called the logarithmic dispersion law. The amplitude attenuation model in equation (2.40) will be used in Chapter 3 to estimate attenuation from acoustic waveform log data.

Attenuation can have a substantial influence on wave amplitude. Figure 2.13 shows the synthetic seismogram calculated with ($Q_\alpha = Q_\beta = 30$) and without attenuation. The borehole and formation parameters are identical to those used for Figure 2.3a. The waveform amplitudes are scaled by a common factor to compare the effects with and without attenuation. The P- and S-wave amplitude is significantly reduced by the attenuation. However, the waveform in the absence of the attenuation still exhibits some amplitude decay with increasing receiver offset. This decay is caused by geometric spreading. As we will show in Chapter 3, this wave geometric spreading effect needs to be accounted for in order to estimate the true attenuation caused by intrinsic damping.

2.10 Acoustic Logging in a Transversely Isotropic Formation

In this section, we consider the formation as a transversely isotropic (TI) solid and discuss its effect on acoustic logging using multipole (e.g., monopole and dipole) acoustic wave sources. This anisotropy has a symmetry axis such that along any direction transverse to this axis, one sees the same material property (velocity/slowness). Between the direction of the symmetry axis and the direction perpendicular to it, the material properties change. For seismic exploration in general, and for acoustic logging in particular, transverse isotropy is a common and most important type of anisotropy, because many sedimentary rocks, such as shales, exhibit TI characteristics. Besides, the TI formation, when its axis of symmetry coincides with the borehole axis, is the only anisotropy case amenable to the exact analytical treatment (other cases of anisotropy, or even the simple case of TI with the symmetry axis tilted from the borehole, require either sophisticated numerical analyses, e.g., finite-difference, or approximate/asymptotic techniques, e.g., perturbation methods). The result presented in this section will be used to provide a theoretical basis for estimating formation TI property from acoustic logging data, as will be discussed in Chapter 5.

2.10.1 Elastic Waves in a TI Solid

In the mathematical treatment of a TI solid, the elastic properties of the solid are expressed by five elastic constants:

$(c_{11}, c_{13}, c_{33}, c_{44}, c_{66})$.

If the TI symmetry axis is along the z-axis, then we can write the stress-strain relationship for the cylindrical coordinates in the following form:

$$\begin{pmatrix} \sigma_{rr} \\ \sigma_{\theta\theta} \\ \sigma_{zz} \\ \sigma_{\theta z} \\ \sigma_{rz} \\ \sigma_{r\theta} \end{pmatrix} = \begin{pmatrix} c_{11} & c_{11}-2c_{66} & c_{13} & 0 & 0 & 0 \\ c_{11}-2c_{66} & c_{11} & c_{13} & 0 & 0 & 0 \\ c_{13} & c_{13} & c_{33} & 0 & 0 & 0 \\ 0 & 0 & 0 & c_{44} & 0 & 0 \\ 0 & 0 & 0 & 0 & c_{44} & 0 \\ 0 & 0 & 0 & 0 & 0 & c_{66} \end{pmatrix} \begin{pmatrix} e_{rr} \\ e_{\theta\theta} \\ e_{zz} \\ e_{\theta z} \\ e_{rz} \\ e_{r\theta} \end{pmatrix}, \tag{2.41}$$

where σ_{ij} and e_{ij} ($i, j = r, \theta, z$) are stresses and strains in the TI solid, respectively.

We now describe how to solve the wave propagation problem in the frequency domain. Expressing the displacement **u** using the three displacement potentials Φ, χ, and Γ, as in equation (2.9), and calculating the radial, azimuthal, and vertical displacement components u, v, and w using equation (2.12), we substitute u, v, and w into equation (2.13) to calculate the strain elements. From the strain elements, we can express the stress elements using the stress-strain relationship in equation (2.41). Then, we substitute the stress and displacement into the general elastic wave equation

$$\nabla \cdot \sigma + \rho\omega^2 \mathbf{u} = 0 \ , \tag{2.42}$$

where σ denotes the 3×3 stress tensor, as given by

$$\begin{pmatrix} \sigma_{rr} & \sigma_{r\theta} & \sigma_{rz} \\ \sigma_{r\theta} & \sigma_{\theta\theta} & \sigma_{\theta z} \\ \sigma_{rz} & \sigma_{\theta z} & \sigma_{zz} \end{pmatrix}.$$

The above procedure results in three individual wave equations for Φ, χ, and Γ, as in the following:

$$\begin{cases} c_{11}\nabla^2\Phi + (c_{13}+2c_{44}-c_{11})\dfrac{\partial^2\Phi}{\partial z^2} + \rho\omega^2\Phi + \\ \qquad \dfrac{\partial}{\partial z}\left((c_{11}-c_{13}-c_{44})\nabla^2\Gamma + (c_{13}+2c_{44}-c_{11})\dfrac{\partial^2\Gamma}{\partial z^2} + \rho\omega^2\Gamma\right) = 0 \\ \dfrac{\partial}{\partial z}\left((c_{13}+2c_{44})\nabla^2\Phi + (c_{33}-c_{13}-2c_{44})\dfrac{\partial^2\Phi}{\partial z^2} + \rho\omega^2\Phi\right) + \\ \qquad (\dfrac{\partial^2}{\partial z^2} - \nabla^2)\left(c_{44}\nabla^2\Gamma + (c_{33}-c_{13}-2c_{44})\dfrac{\partial^2\Gamma}{\partial z^2} + \rho\omega^2\Gamma\right) = 0 \\ c_{66}\nabla^2\chi + (c_{44}-c_{66})\dfrac{\partial^2\chi}{\partial z^2} + \rho\omega^2\chi = 0 \end{cases} . \tag{2.43}$$

The first two equations are two coupled partial differential equations. The third equation describes a pure SH-type shear-wave equation that is decoupled from the first two equations.

For the multipole borehole acoustic wave source (order n) excitation given in equation (2.6), the solution for Φ, χ, and Γ in an unbounded TI formation is written in the following form:

$$\left.\begin{matrix}\Phi\\ \chi\\ \Gamma\end{matrix}\right\} = e^{ikz}\frac{1}{n!}\left(\frac{fr_0}{2}\right)^n \begin{cases} B_n K_n(qr)\cos(n(\theta-\phi)) \\ D_n K_n(qr)\sin(n(\theta-\phi)) \\ F_n K_n(qr)\cos(n(\theta-\phi)) \end{cases} . \tag{2.44}$$

Substituting the above expressions into equations (2.43) results in a system of three equations for the wave amplitude coefficients B_n, D_n, and F_n. A non-trivial solution of the system leads to the following period equation to determine the eigenvalues of the radial wavenumber q:

$$(q^2-k^2)(c_{66}q^2-c_{44}k^2+\rho\omega^2)(Uq^4+V\omega^2q^2+W\omega^4)=0 \ , \tag{2.45}$$

where

$$\begin{cases} U = c_{11}c_{44} \\ V = \rho(c_{11}+c_{44})-(c_{11}c_{33}-c_{13}^2-2c_{13}c_{44})(k^2/\omega^2) \\ W = c_{33}c_{44}(\rho/c_{44}-k^2/\omega^2)(\rho/c_{33}-k^2/\omega^2) \end{cases} .$$

Equation (2.45) is factorized into three equations. The first equation, as given by the expression in the left parenthesis, is satisfied when $q = \pm k$, which does not correspond to a wave motion and is thus ignored. The second equation, given by the expression in the middle parenthesis, has a solution

$$q = q_{sh} = \sqrt{\frac{c_{44}k^2-\rho\omega^2}{c_{66}}} \ , \tag{2.46}$$

which corresponds to a pure SH-type shear wave, with a solution given by

$$\chi = e^{ikz}\frac{1}{n!}\left(\frac{fr_0}{2}\right)^n D_n K_n(q_{sh}r)\sin[n(\theta-\phi)] \ . \tag{2.47}$$

The last equation, given by the expression in the last parenthesis in equation (2.43), has two positive roots (the other two negative roots are neglected because they cause the solution in equation (2.44) to diverge at large radial distances). The two roots are given by

$$\begin{cases} q = q_p = \omega\sqrt{\dfrac{-V+\sqrt{V^2-4UW}}{2U}} \\ q = q_{qv} = \omega\sqrt{\dfrac{-V-\sqrt{V^2-4UW}}{2U}} \end{cases}, \quad (2.48)$$

The two coupled equations in (2.43) and the corresponding eigenvalues in equations (2.48) indicate that the P- and SV-type of wave motions in a TI solid, unlike their counterparts in an isotropic solid, cannot be resolved into independent compressional and shear waves. Instead, the waves are designated as "quasi-P" and "quasi-S" waves. One has to combine these two wave motions to solve equations (2.43). Mathematically, this means that the general solution for the first two equations in (2.43) contains both eigen-wave modes, as

$$\left.\begin{matrix}\Phi \\ \Gamma\end{matrix}\right\} = e^{ikz}\frac{1}{n!}\left(\frac{fr_0}{2}\right)^n \cos(n(\theta-\phi))\begin{cases}(B_n K_n(q_p r) + b' F_n K_n(q_{sv} r)) \\ (a' B_n K_n(q_p r) + F_n K_n(q_{sv} r))\end{cases}, \quad (2.49)$$

where

$$\begin{cases} a' = -\dfrac{1}{ik}\dfrac{(c_{13}+2c_{44})k^2 - c_{11}q_p^2 - \rho\omega^2}{c_{44}k^2 - (c_{11}-c_{13}-c_{44})q_p^2 - \rho\omega^2} \\ b' = -ik\dfrac{c_{44}k^2 - (c_{11}-c_{13}-c_{44})q_{sv}^2 - \rho\omega^2}{(c_{13}+2c_{44})k^2 - c_{11}q_{sv}^2 - \rho\omega^2} \end{cases}.$$

The above two coefficients result from finding an eigensolution $(B_n, F_n)^T$ for the eigenvalues q_P and q_{SV}. Equation (2.49) shows that the anisotropy induces a coupling of shear vibration to the compressional wave, and a coupling of compressional vibration to the shear wave. The magnitude of the former coupling is controlled by b', while that of the latter coupling is controlled by a' (when the solid is isotropic, $a' = b' = 0$).

Using the solutions in equations (2.47) and (2.49) and the boundary condition in equation (2.15) results in the following matrix equation for determining the unknown coefficients

$$\begin{pmatrix} Q_{11} & Q_{12} & Q_{13} & Q_{14} \\ Q_{21} & Q_{22} & Q_{23} & Q_{24} \\ Q_{31} & Q_{32} & Q_{33} & Q_{34} \\ Q_{41} & Q_{42} & Q_{43} & Q_{44} \end{pmatrix}\begin{pmatrix} A'_n \\ B_n \\ D_n \\ F_n \end{pmatrix} = \begin{pmatrix} u_f^d \\ \sigma_{rrf}^d \\ 0 \\ 0 \end{pmatrix}, \quad (2.50)$$

where A'_n, as in the isotropic case (equation (2.7)), is the amplitude coefficient of the reflected wavefield in the borehole fluid. The borehole source quantities, as in the right-hand-side of the above equation, are given in equation (2.17). Analogous to equation (2.18), the elements of the **Q** matrix are given by

$$\begin{cases} \text{For } u: \\ Q_{11} = -\dfrac{n}{R} I_n(fR) - f I_{n+1}(fR) \\ Q_{12} = -(1 + ika') q_p Y_1(q_p R) \\ Q_{13} = \dfrac{n}{R} K_n(q_{sh} R) \\ Q_{14} = -(ik + b') q_{sv} Y_1(q_{sv} R) \end{cases},$$

$$\begin{cases} \text{For } \sigma_{rr}: \\ Q_{21} = \rho_f \omega^2 I_n(fR) \\ Q_{22} = (c_{11} q_p^2 - c_{13} k^2 + (c_{11} - c_{13}) ika' q_p^2) K_n(q_p R) + \dfrac{2c_{66} q_p}{R}(1 + ika') Y_2(q_p R) \\ Q_{23} = -\dfrac{2c_{66} n q_{sh}}{R} Y_3(q_{sh} R) \\ Q_{24} = ((c_{11} q_{sv}^2 - c_{13} k^2) b' + (c_{11} - c_{13}) ikq_{sv}^2) K_n(q_{sv} R) + \dfrac{2c_{66} q_{sv}}{R}(ik + b') Y_2(q_{sv} R) \end{cases},$$

(2.51)

$$\begin{cases} \text{For } \sigma_{r\theta}: \\ Q_{31} = 0 \\ Q_{32} = \dfrac{2c_{66} n q_p}{R}(1 + ika') Y_3(q_p R) \\ Q_{33} = -c_{66} q_{sh}^2 Y_4(q_{sh} R) \\ Q_{34} = \dfrac{2c_{66} n q_{sv}}{R}(ik + b') Y_3(q_{sv} R) \end{cases},$$

$$\begin{cases} \text{For } \sigma_{rz}: \\ Q_{41} = 0 \\ Q_{42} = -c_{44} q_p (2ik - a'(k^2 + q_p^2)) Y_1(q_p R) \\ Q_{43} = \dfrac{iknc_{44}}{R} K_n(q_{sh} R) \\ Q_{44} = c_{44} q_{sv} ((k^2 + q_{sv}^2) - 2ikb') Y_1(q_{sv} R) \end{cases},$$

where the Bessel function combinations $Y_1 - Y_4$ are given in equation (2.18). By solving equation (2.50), we find the coefficient A'_n for the reflected wavefield in the borehole. By taking the determinant of the $\mathbf{Q}$ matrix in equation (2.50), we obtain the dispersion equation for borehole wave modes in a TI formation, as:

$$D(k,\omega) = \det \mathbf{Q}(k,\omega) = 0 \, .$$

The calculations of synthetic seismograms and dispersion curves follow the same steps as for the isotropic situations (equations (2.16–24)).

2.10.2 WAVEFORM CHARACTERISTICS

To demonstrate the effects of anisotropy (TI) on acoustic logging waveforms, we calculate the monopole and dipole synthetic micro-seismograms for a TI formation (the TI parameters are given in Table 2.3). The formation is acoustically slower than the borehole fluid, because both velocities of the vertically propagating S wave ($V_{Sv}{}^2 = c_{44}/\rho$) and the horizontally propagating S wave ($V_{Sh}{}^2 = c_{66}/\rho$) are lower than the borehole fluid's acoustic velocity V_f (which is equivalent to α_f used previously). The center frequencies of the monopole and dipole wave source are 8 kHz and 3 kHz, respectively. Figure 2.14a illustrates a TI formation model as fine sedimentary layers surrounding the borehole, because many layered/laminated sediments exhibit TI characteristics. The monopole waveforms in Figure 2.14b contain leaky P and Stoneley waves similar to the isotropic situation (Figure 2.4a). The leaky P waves exhibit the same characteristics as the isotropic example in Figure 2.4a. The first arrival of the wave packet is at the vertically propagating P-wave velocity ($V_{Pv}{}^2 = c_{33}/\rho$). If the wave propagation were at the horizontally propagating P-wave velocity ($V_{Ph}{}^2 = c_{11}/\rho$), the arrival time and moveout would be the dashed line shown in Figure 2.14b. Trailing the leaky P wave is the Stoneley wave (labeled 'ST'). Compared to the isotropic example in Figure 2.4a, the Stoneley wave in this TI formation shows a quite dispersive characteristic: the low-frequency portion of the Stoneley wave travels faster than the high-frequency portion of the wave. In fact, the dispersion characteristic contains the information about the TI formation, which we discuss in more detail in the following dispersion and sensitivity analysis.

The synthetic waveforms for the dipole source are shown in Figure 2.14c. The propagation of the dipole-flexural waves exhibits a characteristic similar to the P-wave situation in Figure 2.14b. The solid line in Figure 2.14c indicates that the onset of the flexural wave is at the vertically propagating S-wave velocity ($V_{Sv}{}^2 = c_{44}/\rho$), instead of the horizontally propagating S-wave velocity, ($V_{Sh}{}^2 = c_{66}/\rho$), as indicated by the dashed line in the figure.

The theoretical result shown in Figure 2.14 shows that monopole/dipole acoustic logging in the TI formation can only determine the velocity of vertically propagating P- and S waves. Only the Stoneley wave shows sensitivity to the TI formation. We demonstrate the dispersion characteristics of the guided waves (e.g., flexural- and Stoneley-wave modes) and their sensitivity to the TI formation in the following analysis.

Table 2.3. Formation TI constants, borehole fluid modulus K_f, and hole radius used in Figures 2.14 and 2.15

c_{11}	c_{33}	c_{13}	c_{44}	c_{66}	K_f	R
23.87 GPa	15.33 GPa	9.79 GPa	2.77 GPa	4.27 GPa	2.25 GPa	0.1 m

2.10.3 DISPERSION CURVE AND SENSITIVITY ANALYSIS

Figure 2.15a shows the dipole-flexural- and Stoneley-wave dispersion curves for the TI formation (Table 2.3). The phase and group velocities of the waves are plotted using solid and dashed curves, respectively. The frequency range is from 0–8 kHz. The curves with markers (open circles) belong to the flexural wave, those without markers belong to the Stoneley wave. The phase velocity of the flexural wave is quite flat, especially toward low frequencies (below 4 kHz). The Airy phase in the group velocity curve is not well defined, for the curve does not show a well-defined minimum. As a result, the wave's dispersion effect is minimal, as can be seen from the synthetic waveform data in Figure 2.14c. Most importantly, both phase and group velocities of the flexural wave approach V_{Sv} towards the cutoff frequency, as indicated by an arrow in Figure 2.15a (the value of V_{Sh} is also indicated by an arrow in this figure). This shows again that shear-wave velocity logging using low-frequency flexural waves can only determine the c_{44} or V_{Sv} parameter of the TI formation.

Stoneley-wave dispersion exhibits interesting characteristics. Below about 3 kHz, the phase and group velocities increase significantly with decreasing frequency. This explains why the low-frequency portion of the Stoneley wave arrives earlier than the high-frequency portion of the wave, as shown in the synthetic waveform data in Figure 2.14b. More interesting, the group velocity curve appears to have two branches that are separated by a sharp cusp at about 3.4 kHz. In fact, this cusp occurs at the frequency point where the Stoneley-wave phase velocity is equal to V_{Sv}, the vertical S-wave velocity. Below this frequency, the phase velocity curve values are higher than V_{Sv} and the group velocity shows a minimum (Airy phase), resulting in a dispersive, low-frequency Stoneley-wave energy that travels faster than high-frequency energy (Figure 2.14b). Moreover, in the low-frequency range below the cusp, the wave becomes partially leaky because its phase velocity is higher than V_{Sv}. This happens because a portion of the wave's energy sensitive to V_{Sv} (see the sensitivity analysis in Figure 2.15c and associated discussions) tends to radiate away from the borehole into the formation. However, this energy-leaking or attenuation phenomenon is generally insignificant (see the synthetic waveform in Figure 2.14b), because the Stoneley-wave sensitivity to V_{Sv} is small in the low-frequency range. In fact, in the low frequency limit, the Stoneley-wave velocity is (White and Tongtaow, 1981)

$$V_{ST}\big|_{\omega=0} = V_f \,/\, \sqrt{1+K_f \,/\, c_{66}} \;, \quad (K_f = \rho_f V_f^2) \;, \tag{2.53}$$

where K_f is the borehole fluid modulus. This result shows that, at low frequencies, the sensitivity of the wave to the formation TI parameters is controlled mostly by the velocity of the horizontally rather than vertically propagating shear waves. This is substantiated by the following sensitivity analysis.

In the theoretical modeling of guided-wave propagation in a borehole, many parameters can influence the wave propagation characteristics. Sensitivity analysis is

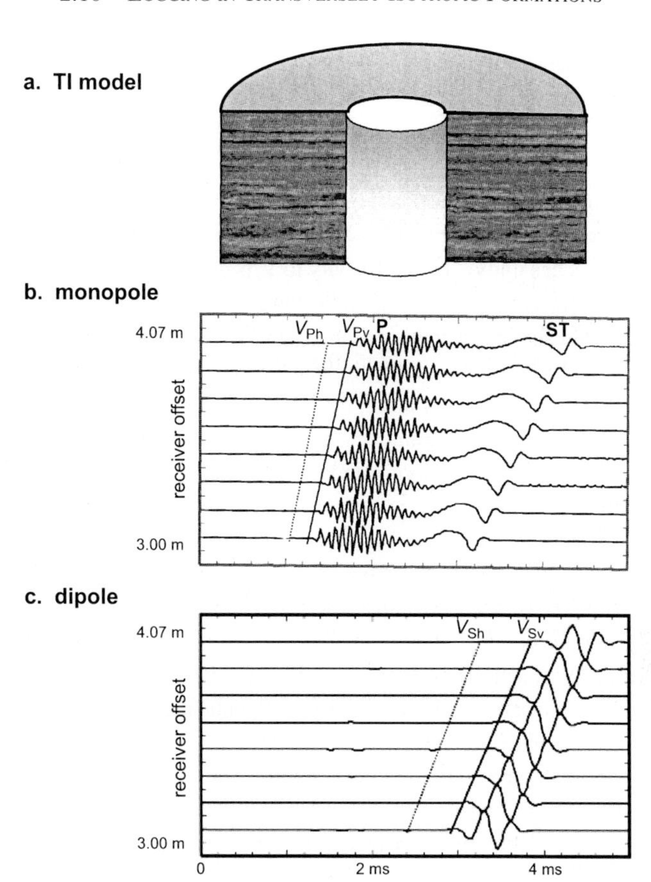

Figure 2.14. Synthetic micro-seismograms calculated for a TI formation model (a) with an 8 kHz monopole (b) and a 3 kHz dipole (c) source, respectively. Note that the propagation of the compressional (b) and flexural (c) waves is at the velocity of the vertically propagating compressional (V_{Pv}) and shear (V_{Sv}) waves, respectively. The moveout for the horizontally propagating P (V_{Ph}) and S (V_{Sh}) waves is also shown (dashed lines).

commonly used to analyze the importance of each parameter in affecting the wave propagation. Sensitivity analysis is particularly helpful in the TI-related wave propagation problem, because the problem involves quite a few model parameters. The sensitivity is simply defined as the normalized partial derivative of the wave's phase velocity with respect to a model parameter p as given by

$$\text{Sensitivity} = \left(\frac{p}{V_{\text{phase}}(\omega)} \right) \frac{\partial V_{\text{phase}}(\omega)}{\partial p} \quad . \tag{2.54}$$

The sensitivity is also called the partition coefficient (Cheng et al., 1982) because it describes the partitioning of the total wave energy into smaller fractions belonging to individual model parameters. Because of the conservation of energy, the sum of the partition coefficients must be equal to one.

Applying the sensitivity analysis to the borehole and TI formation parameters (Table 2.3), we obtain the various sensitivity curves for the flexural and Stoneley waves in Figures 2.14b and c, respectively. Specifically, sensitivity is calculated for three model parameters: V_{Sh}, V_{Sv}, and V_f. As Figure 2.15b shows, the sensitivity of the flexural wave to the TI formation is largely controlled by V_{Sv} in the 0–8 kHz range, and is completely controlled by this parameter near the cutoff frequency (where sensitivity approaches unity). The sensitivity to V_f is quite small in this example. The sensitivity to the horizontally propagating S-wave velocity V_{Sh} is zero near the cutoff frequency and increases appreciably with frequency. The sum of the three sensitivity curves is close, but not equal to, unity, indicating that other model parameters, such as c_{11}, c_{13}, c_{33}, etc., can minimally affect the propagation of flexural waves. Although sensitivities to V_{Sh} and other parameters may change as the formation TI property varies, the sensitivity is mostly dominated by V_{Sv}. This fact shows that it is difficult, if not impossible, to derive V_{Sh} using flexural waves from dipole acoustic logging.

The Stoneley-wave result, as in Figure 2.15c, shows the possibility of using this wave to determine the formation S-wave TI property (the sensitivity result is shown in 0–5 kHz to emphasize the low frequency range). This figure shows that sensitivity to the TI formation is dominated by V_{Sh} at low frequencies, as can be anticipated from equation (2.53). With increasing frequency, V_{Sv} becomes an important parameter to affect the wave propagation. The sensitivity to V_f indicates that it is a dominant parameter. This is well understood since the Stoneley wave is a fluid-borne wave and is naturally controlled by the borehole fluid property. The sum of the three sensitivities is close to unity, suggesting that Stoneley-wave sensitivity to other TI parameters, e.g., c_{11}, c_{13}, c_{33}, is small in this slow formation situation.

2.10.4 REMARK ON TI-FORMATION MODELING AND RESULTS

The above modeling results allow us to discuss wave propagation in a TI solid and its relation to acoustic logging in a borehole along the TI-symmetry axis. Although in general there is coupling of compressional and shear wave motions in the TI solid, the two wave motions decouple in the limit of wave propagation along or transverse to the axis of symmetry. As a result, acoustic logging in a borehole along the TI-symmetry axis measures the propagation of 'pure' compressional and/or shear waves. The measurement of these waves can only determine their vertical propagation velocity, regardless of the type of sources used: monopole, dipole, or quadrupole. The vertical propagation velocity is given by $V_{Pv}^2 = c_{33}/\rho$ for the P wave, and by $V_{Sv}^2 = c_{44}/\rho$ for the S wave (including flexural and screw waves). This is because most borehole waves (monopole, dipole, etc.) involve wave motion or vibration either parallel (P wave) or transverse (S wave) to the borehole. The only exception is the Stoneley wave. At low frequencies, the Stoneley wave, or tube wave, involves radial displacements that distort the circumference of the borehole. This circumferential distortion is controlled by the shear modulus in the horizontal plane, corresponding to c_{66} in the TI situation. This provides the theoretical basis for inverting the shear-

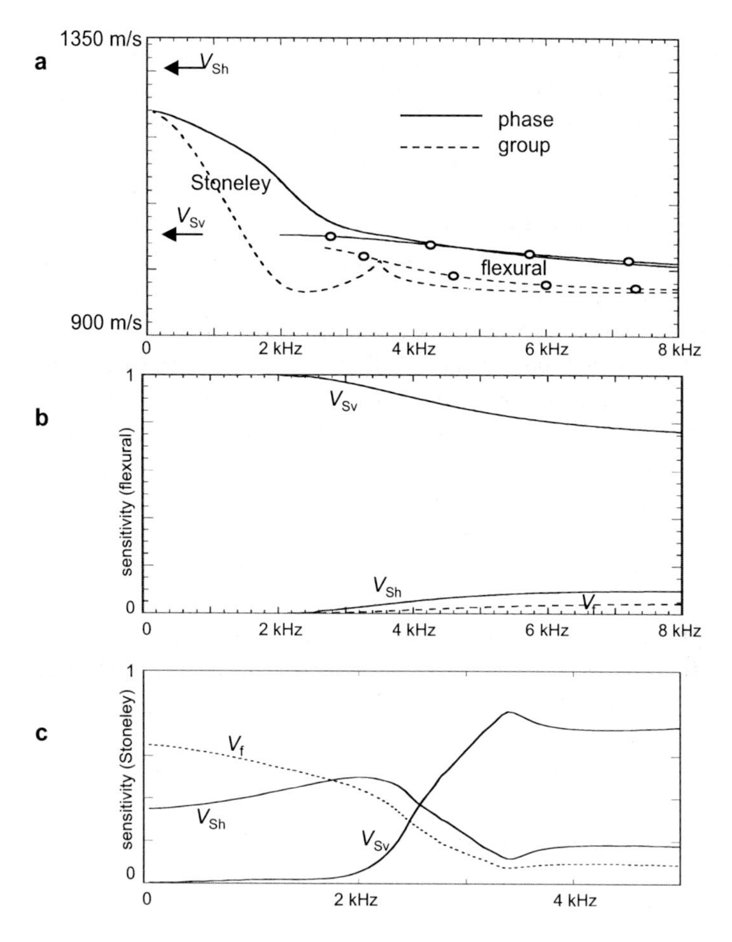

Figure 2.15. (a) Stoneley and flexural wave phase and group velocity dispersion curves in a TI formation. (b) and (c) show, respectively, the sensitivity of the flexural and Stoneley waves to the velocity of the horizontally propagating shear waves (V_{Sh}) and vertically propagating shear waves (V_{Sv}), as well as to the borehole fluid acoustic velocity (V_f).

wave TI property of the surrounding formation from borehole Stoneley waves, as discussed in Chapter 5.

Although the modeling results are shown for a slow formation, the conclusions derived from the modeling hold true in a fast formation situation. For this situation, two additional scenarios are worth mentioning: the propagation of the refracted shear wave, and the reduction of Stoneley-wave sensitivity to the formation. In a fast formation, the refracted shear wave can be generated at the borehole-formation interface. However, because this scenario is simply a shear wave propagating along the TI-symmetry axis, the wave travels at the velocity V_{Sv} of a vertically propagating S-wave. For the Stoneley wave, although the sensitivity of the wave to the TI formation is still dominated by c_{66} or V_{Sh}, the sensitivity becomes much smaller than in the slow-formation situation. This can be easily understood from the zero-frequency

Stoneley-wave velocity formula in equation (2.53). A fast formation is characterized by high shear rigidity, resulting in $c_{66} \gg K_f$ in equation (2.53). Consequently, the Stoneley-wave velocity is controlled mostly by V_f, causing a much-reduced sensitivity to c_{66}.

The theoretical modeling of wave propagation in a TI formation shows how the introduction of anisotropy can complicate the solution of the problem, even in the simple situation where the symmetry axis of the TI medium coincides with the borehole. When the symmetry axis is tilted from the borehole, an exact analytical solution is no longer available. In this situation, the solution is found either approximately/asymptotically using perturbation techniques (Ellefsen, 1990; Sinha et al., 1994), or numerically using finite difference or finite element techniques (Ellefsen, 1990; Cheng, 1994). A special situation is where the axis of TI-symmetry is tilted 90° from (or perpendicular to) the borehole. We will discuss this situation and its applications in Chapter 5.

CHAPTER 3: ELASTIC WAVE VELOCITY AND ATTENUATION ESTIMATION FROM ARRAY ACOUSTIC WAVEFORM DATA

3.1 Frequency Domain Methods

Modern acoustic logging tools contain an array of receivers for recording acoustic waveform data. Array processing methods can be used to calculate the velocity and attenuation of various acoustic wave modes in a borehole. Some borehole waves are dispersive guided wave modes, with velocity varying with frequency (e.g., flexural, pseudo-Rayleigh, and Stoneley waves). Velocity dispersion characteristics play an important role in acoustic logging. For example, by performing a dispersion analysis on an acoustic dipole data array one can compare the actual dispersion curve from data versus the theoretical curve. Such a comparison is essential for evaluating the design and performance of the dipole tool. Dipole-flexural wave dispersion is also used to diagnose formation alteration (Tang, 1996a) and stress-induced azimuthal anisotropy (Sinha and Kostek, 1995). The dispersion of Stoneley waves is used to analyze the effects of formation permeability (Winkler et al., 1989) and to estimate formation permeability. Frequency-domain array methods are commonly applied to estimate the dispersive velocity spectrum of the guided waves.

Two such methods will be described: Prony's method and a spectral weighted semblance method. Although the latter method is generally better than Prony's method, the theoretical basis of Prony's method, linear prediction theory, forms the foundation for developing a waveform inversion method in the time domain. As we will see later in this chapter, the waveform inversion method outperforms the semblance method in obtaining high-resolution slowness logs using short arrays.

3.1.1 Prony's Method

Assume that the array acoustic spectral data, obtained by Fourier-transforming the acoustic traces, consist of a number of propagating wave modes, as given by

$$\tilde{X}_n(\omega) = \sum_{k=1}^{p} h_k z_k^{n-1}, (n = 1, 2, ..., N), \tag{3.1}$$

where ω is angular frequency, $z_k = \exp(-i\omega s_k d)$ is a complex exponential, $s(\omega)$ is the slowness spectrum, and $h_k(\omega)$ is the amplitude of the kth wave mode. The total of the wave modes is p. The acoustic array consists of N receivers with an inter-receiver spacing d. The problem is now stated as: for each given frequency, find the slowness values $s_k(\omega)$ ($k = 1, ..., p$) for the wave modes in order to minimize the following least-squares error

$$\rho = \sum_{n=1}^{N} \left(X_n(\omega) - \tilde{X}(\omega; s_1, s_2, ..., s_p) \right)^2 . \tag{3.2}$$

The above is a difficult, nonlinear minimization problem, especially when each frequency needs to be estimated.

Prony (1795) recognized that the p exponentials z_k ($k = 1, ..., p$) are the roots of a characteristic polynomial equation

$$\prod_{k=1}^{p} (z - z_k) = \sum_{m=0}^{p} a_m z^{p-m} = a_0 z^p + a_1 z^{p-1} +, ..., + a_m z^0 = 0 . \tag{3.3}$$

where $a_0 = 1$. By multiplying both sides of (3.1) by a_m, shifting the index n by $-m$, and summing over m from 0 to p, then using equation (3.3), it can easily be shown that

$$\sum_{m=0}^{p} a_m X_{n-m}(\omega) = 0,$$

or, (3.4)

$$\tilde{X}_n(\omega) = -\sum_{m=1}^{p} a_m X_{n-m}(\omega); \quad (p + 1 \leq n \leq N),$$

where the ~ sign is used to denote the predicted receiver wave data. Equation (3.4) forms the basis of the linear prediction theory for the array data, which predicts the data of receiver n using a linear combination of data from other receivers. The actual algorithm based on Prony's method consists of three steps.

First, the coefficients a are computed using equation (3.4) by using spectral array data $X(\omega)$ in equation (3.4).

Second, the calculated coefficients are substituted into equation (3.3) to find the roots z_k ($k = 1, ..., p$), which yield the slowness values.

Finally, with the known slowness values, the complex amplitude h_k for each mode is calculated using the least-squares method to equation (3.1). The amplitude of a wave mode indicates the importance of the wave mode. A detailed calculation procedure will be discussed in Chapter 4.

3.1.2 WEIGHTED SPECTRAL SEMBLANCE METHOD

Prony's method described above has two drawbacks. The first is that it often generates spurious estimates (modes that have virtually zero amplitude). The second is that the method processes each frequency independently, which, for an acoustic array of a few receivers, uses only a few spatial samples in the processing. These drawbacks render the method unstable and inaccurate in the presence of noise. To overcome the drawbacks, Nolte and Huang (1997) developed a new method by weighting the semblance, or coherence function, of the array spectral data. This method processes a frequency by weighting the data over neighboring frequency points, significantly increasing the number of data points in the processing. The method searches the peak(s) of weighted semblance function over a range of slowness values to find the actual number of wave modes, eliminating the spurious estimates. The new method is generally better than Prony's method.

The spectral semblance, or coherence function, is defined as

$$\rho(\omega,s) = \frac{\left|\sum_{n=1}^{N} X_n^*(\omega) z^{n-1}\right|}{\sqrt{N \sum_{n=1}^{N} X_n^*(\omega) X_n(\omega)}} , \tag{3.5}$$

where $z = \exp(-i\omega sd)$ and * denotes taking the complex conjugate. Assume that the spectral data $X_n(\omega)$ correspond to an acoustic mode with a distinct slowness $s_k(\omega)$. The wave mode travels across each receiver in the array as:

$$X_n(\omega) = h_k z_k^{n-1} = h_k(\omega)\exp(-i\omega s_k (n-1)d) . \tag{3.6}$$

We see that if the slowness variable s attains the value of s_k, then the phase of $X_n^*(\omega)z^{n-1}$ in equation (3.5) is canceled and the semblance value is maximized (the value approaches 1 if the data are noise-free). Equation (3.5) is essentially a semblance/coherence stacking of the array data in the frequency domain (a time-domain stacking method will be described later).

To overcome the inadequate spatial sampling problem in Prony's method, a spectral weighting scheme is employed to enhance data information and reduce noise. To do so, the spectral data are first resampled to obtain denser data points. This can be done by padding zeros to time domain traces before taking the Fourier transform. The frequency in the more densely sampled data is denoted by ω_l. A weighted semblance function is defined as

$$F(\omega_l,s) = \sum_{j=l-m}^{l+m} W(\omega_j,\omega_m)\, \rho(\omega_m,s) , \tag{3.7}$$

where $W(\omega_j, \omega_m)$ is a weighting function. A Gaussian function is used to give the

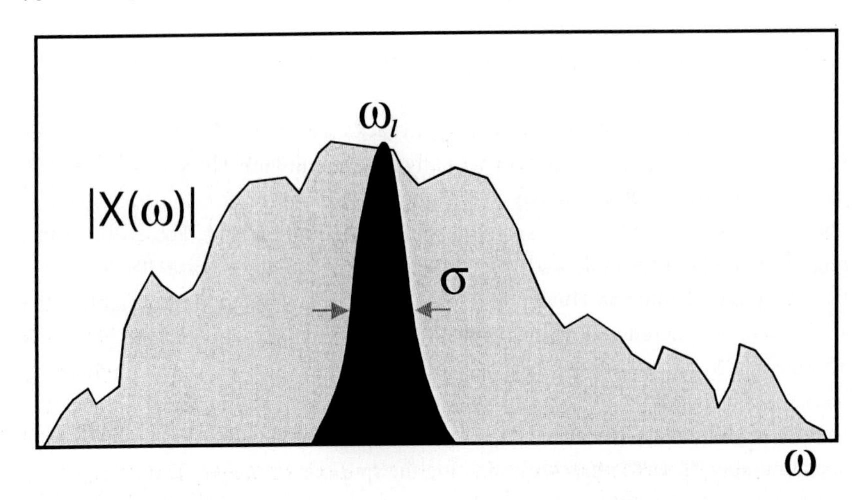

Figure 3.1 Weighting the spectral semblance computed from the data $X(\omega)$ with a Gaussian function centered at the l-th frequency ω_l. The width of the weighting function is controlled by σ.

maximum weight for the frequency to be estimated as

$$W(\omega_j,\omega_m) = \exp\left(-\frac{(\omega_m - \omega_j)^2}{2\sigma^2}\right).$$

This new function in equation (3.7) is the semblance function in equation (3.5) weighted by the nearby frequency points. This weighting scheme using neighboring data significantly increases the number of data samples used in the processing, thus reducing noise effects in the data and enhancing the robustness and accuracy of the processing. The number of neighboring points to be weighted over is controlled by σ, which specifies the width of the Gaussian function, as illustrated in Figure 3.1. For example, if one wants to use the nearby four data points in the weighting, one can specify the value of σ as $4\Delta\omega$, where $\Delta\omega$ is the increment of the re-sampled data.

The algorithm for performing the analysis using equation (3.7) is as follows.

1. Transform the acoustic array time series to obtain spectral data $X_n(\omega)$, ($n = 1, 2, \ldots, N$). Before the Fourier transform, pad zeros to the traces to obtain the desired upsampling. For example, an upsampling factor of 4 is recommended for a typical 512-sample trace sampled at 10 μs.
2. Specify the slowness range of interest ($s_{\min} \le s \le s_{\max}$). For a given s, calculate $\rho(\omega, s)$, and then weight it over a given frequency interval around ω_l using equation (3.7). For the specified slowness range, search the slowness values s_k ($k \ge 1$) that maximize the weighted spectral semblance function in equation (3.7).
3. Obtain $s_k(\omega)$ for all values of ω in the frequency range of interest. This will yield the slowness dispersion curve(s) as a function of frequency.

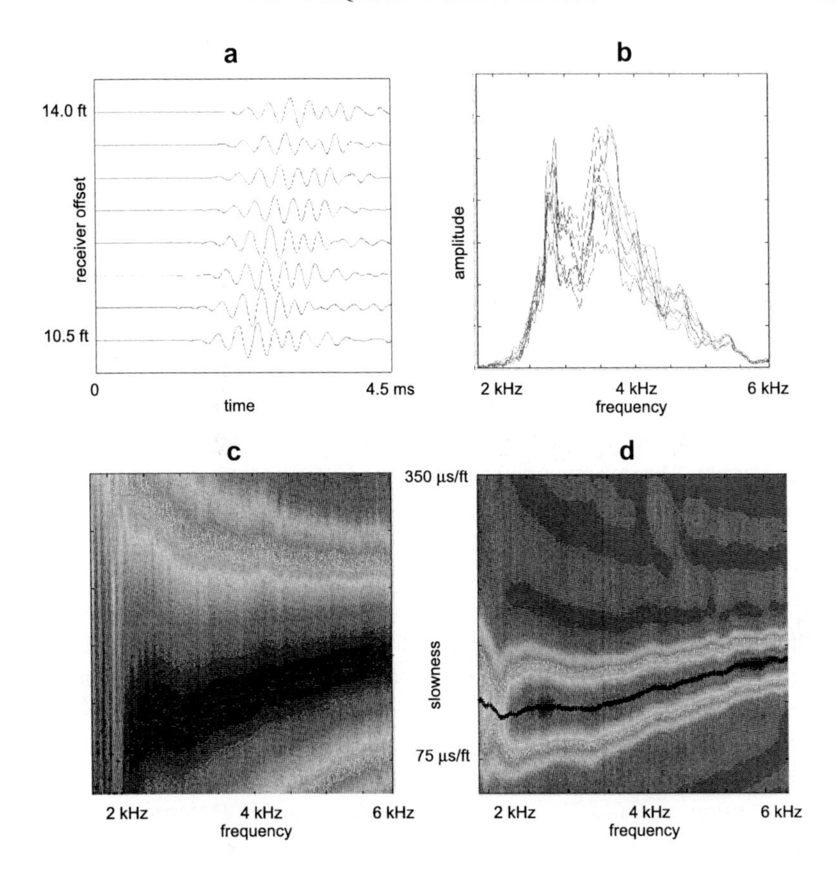

Figure 3.2. Comparison of processing results from Prony's method and from the weighted spectral semblance methods.
(a) Field array dipole acoustic data.
(b) Wave amplitude spectra.
(c) Slowness values (crosses) from Prony's method overlain with the fitness function (image).
(d) Slowness values (crosses) from the spectral semblance method overlain with the fitness function (image).

3.1.3 Comparison of Prony's Method With the Spectral Semblance Method

We now compare the performance of Prony's method and the "weighted spectral semblance" methods using a real data example. Figure 3.2a shows a flexural wave array data set from a field measurement. The array contains eight receivers with a 0.5 ft (0.152 m) receiver spacing. Figure 3.2b shows the wave amplitude spectra for all eight receivers. Figure 3.2c shows the dispersion analysis result from Prony's method. The points (crosses) represent the picked slowness values for the frequency range of the wave spectra (0–6 kHz). Overlain with the slowness picks are the values of a fitness function (this function is the modulus of 1 plus the characteristic equation (3.3), whose coefficients are found from equation (3.4) and the exponential z is let to vary with slowness). For a given frequency, a maximum function value along the slowness axis corresponds to a zero of the characteristic equation. The high values of the fitness function form a ridge along the frequency axis. Figure 3.2c shows that the ridge has quite a broad slowness range, especially towards low frequencies, indicating significant slowness picking errors. In comparison, Figure 3.2d shows the fitness function (i.e., the weighted semblance defined in equation (3.7)) and the slowness picks for the spectral semblance method. The ridge of the fitness function is narrow and sharp compared to that of the Prony's method (Figure 3.2c), indicating well-defined wave slowness values. The dispersion curve becomes flat towards the low-frequency range, exhibiting a typical flexural wave dispersion characteristic, as can be seen from the theoretical result in Figure 2.6. The flat portion of the dispersion curve from Prony's method is difficult to define because of relatively larger slowness picking errors. This example demonstrates the better performance of the spectral semblance method over Prony's method.

3.2 Time-Domain Methods

Routine processing of acoustic logging data needs fast and effective methods for estimating formation elastic wave velocity profiles. The frequency-domain methods described above have been used mainly as diagnostic tools because they are time and effort consuming. The time-domain methods described below are commonly used for routine processing. There are two types of such methods: waveform coherence stacking using semblance or *n*th-root calculation, and waveform matching or inversion based on linear prediction theory.

3.2.1 WAVEFORM COHERENCE STACKING METHODS

Waveform coherence stacking techniques step through a two-dimensional grid (time and slowness) and locate the appropriate wave arrival time and slowness values that maximize the coherence of the stacked data. The coherence refers to a windowed portion of the waveform data. For the semblance method, the coherence is defined as (Kimball and Marzetta, 1986)

$$\rho(s,T) = \frac{\int_T^{(T+T_w)} \left| \sum_{m=1}^{N} X_m(t + s(m-1)d) \right|^2 dt}{N \int_T^{(T+T_w)} \sum_{m=1}^{N} \left| X_m(t + s(m-1)d) \right|^2 dt}, \qquad (3.8)$$

where $X_m(t)$ represents the acoustic time signal at the *m*th receiver in the array of N receivers, with a receiver spacing d. The above semblance function consists of a denominator and a numerator. In the numerator, we back-propagate, or advance each wave trace, $X_m(t)$, $(1 < m \le N)$, to the first receiver position by applying a time shift $s(m-1)d$. Subsequently, we sum the N data values for the respective receiver

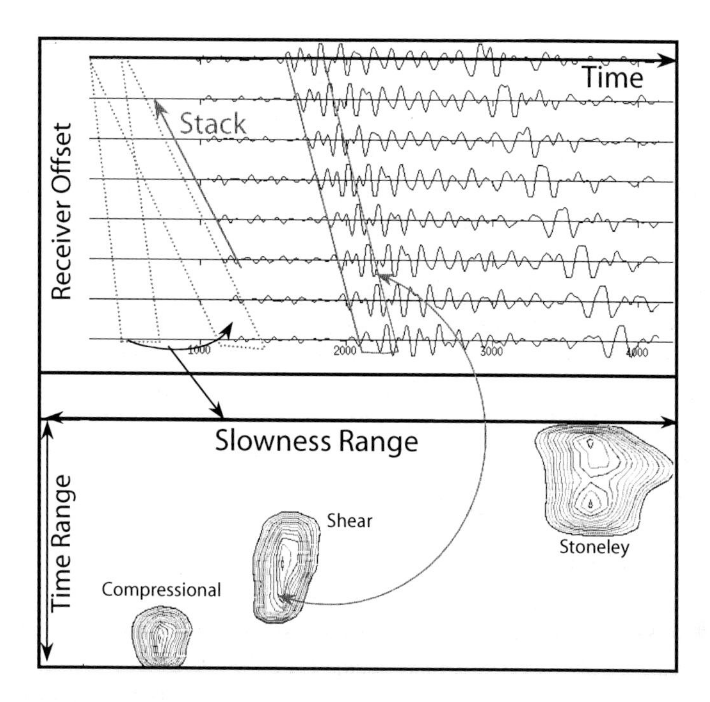

Figure 3.3 An illustration of the array coherence stacking or correlation process. A time window at the first receiver in an array steps through the waveform at successive positions. At each position, the window sweeps across the array at various slopes (slowness values). The data within the windowed corridor are stacked to compute a coherence value. Collecting the coherence value for all time positions and all slowness values give the correlogram (bottom), whose peak values correspond to the slowness and arrival time of a wave event in the array (the shear wave arrival is indicated in this figure).

waveforms (first receiver plus the $N - 1$ time-shifted waveforms). Then, we take the absolute value of the result and square it, then integrate it over a time window $[T, T_w]$. The time variable T is the center position of a time window of length T_w. The time window steps through the waveform at a certain time increment (in practice, T_w usually includes two or three wave cycles, and the stepping increment is half of T_w). In the denominator, the back propagation part is the same, except that we now take the absolute value of each waveform sample and square it, before summing it over the array.

Calculating the two-dimensional semblance function by stepping the time window T_w through a portion of – or the entire – waveform time span and for a specified slowness range of interest, we find some values of T and s, say (s_k, T_k) $(k \geq 1)$, which maximize the semblance coherence function. As illustrated in Figure 3.3, at each window position the slope of the window across the array, which is the slowness s, is varied, and the semblance value is calculated, through the given slowness range. By repeating the calculation for all window positions along the time axis, we obtain a semblance surface over the 2D time-slowness grid, called a "correlogram". This correlogram is shown as the contour plot in Figure 3.3. These peak semblance values mark the arrival time and the slowness of the acoustic wave mode(s) present in the array data. This scenario is illustrated for the shear wave arrivals in the array data shown in Figure 3.3. Therefore, by locating the peak values of equation (3.8), we can obtain the formation slowness value from the array data. With modern computing power, equation (3.8) can be implemented to perform a fast and effective processing of array acoustic data and obtain continuous slowness profiles.

Another measure of array coherence is the nth-root method. The nth-root coherence function (McFadden et al., 1986) is defined as

$$\rho(s,T) = \frac{\int_T^{(T+T_w)} \left| \sum_{m=1}^{N} \left| X_m[t+s(m-1)d] \right|^{1/n} \operatorname{sgn}\{X_m[t+s(m-1)d]\} \right|^n dt}{\int_T^{(T+T_w)} \left| \sum_{m=1}^{N} \left| X_m[t+s(m-1)d] \right|^{1/n} \right|^n dt}, \tag{3.9}$$

where sgn denotes the sign functionIn the nth-root calculation, the wave amplitudes are first scaled by an exponent $1/n$, then summed or stacked in the same way as in the semblance method. The stacked result is then raised to power n. The remaining steps are the same as those of the semblance method.

We now make some remarks on the semblance and the nth-root methods. Mathematically, semblance has a direct interpretation as being the power of the stacked wave data, while the nth-root stacking is mainly a mathematical manipulation. Both methods share the same property of $0 \le \rho \le 1$. The semblance method is considerably faster than the nth-root method, because the latter needs the nth root and power evaluations. The benefit of the latter method is, by raising the value of the exponent n, the nth-root method can enhance the quality of the coherence function (equation (3.9)). This is because taking the nth root of a waveform enhances its peaks (troughs) relative to other parts of the waveform. The subsequent correlation of the high-amplitude parts of the modified data sharpens the peak(s) of coherence function, thereby enhancing the resolution of the slowness estimation (see Smith et al., 1991). A value of $n = 4$ is commonly used in data processing.

3.2.2 QUALITY-CONTROL FOR SLOWNESS PROCESSING OF ARRAY WAVEFORM DATA

After obtaining the slowness from the array waveform data, it is important to provide quality-control indicator(s) for the validity of the processed results. One indicator is to check the overlay of the picked slowness value(s) with the maximum (maxima) of the coherence function (equation (3.8) or (3.9)). Errors in slowness picking or labeling often occur in zones of poor data quality or drastic slowness variations. The picking may also go astray from the desired wave mode (e.g., S) to other modes (e.g., P and Stoneley). Therefore, an indication for the quality of picking is needed. For each slowness value s, the maximum function value over all values of T is retained, yielding a 1D coherence function called the "combined correlogram", which is a function of slowness only. Then overlay the picked slowness value(s) with the 1D correlogram for the processed depth range. The deviation of the picked value(s) from the correlogram peak(s) indicates an erroneous picking and the need for correction.

Another effective and robust indicator is to check the overlay of the slowness-derived travel time with the waveform. The travel time curve versus depth is computed using

$$TT(z_w) = TT_f + \int_{source}^{receiver} s(z)dz \ , \tag{3.10}$$

where slowness s is integrated over source-to-receiver distance, which constitutes the primary contribution to the total wave travel time TT. Another (small) contribution TT_f is the travel time of the acoustic wave in the borehole fluid, which is approximately the fluid slowness times the difference of the borehole and tool diameters. The TT-curve is assigned to depth z_w at which the waveform acquisition is defined. For an acoustic tool, the acquisition point can be defined as the location of the source, receiver, or source-to-receiver middle, etc. The position of this point along the borehole is defined as z_w. The TT-curve can be displayed with the waveform of a single receiver (e.g., receiver 1) versus depth z_w. An example of this display has been shown in Figure 1.17. If the TT-curve tracks the wave event used to generate the slowness (hence TT), it means that the slowness profile is valid. Conversely, if the curve tracks away from the wave event, then the slowness profile at the corresponding depth is either inaccurate or invalid.

3.2.3 WAVEFORM INVERSION METHOD

In describing the frequency domain analyses, we derived a linear prediction theory (equation (3.4)), which states that the data of any receiver in an array can be expressed, or predicted, using a linear combination of the data of other receivers. Although Prony's method is based on the theory, the method becomes unstable in the presence of noise because it uses only a few spatial samples. However, if we apply the prediction theory to the time domain, we can greatly increase the data size and enhance processing accuracy and robustness, as this will enhance the redundancy of information and reduce the noise effect in the data.

Let us first introduce a modification to the prediction theory that doubles the data redundancy even in the frequency domain. For a wave mode propagating forwardly from the first to last receiver in the array, we can also view it as a wave propagating in the reverse direction from the last to first receiver. Mathematically, this means that the complex exponential for the wave propagation can be written as $z = \exp(\pm i\omega sd)$, where + (–) denotes forward (reverse) propagation. The prediction theory allowing for the forward and reverse propagation becomes

$$\tilde{X}_n(\omega) = -\sum_{m=1}^{p} a_m X_{n\pm m}(\omega), \quad (1 \le n \le N; 1 \le n \pm m \le N) \ . \tag{3.11}$$

For the wave spectra $X_{n\pm m}(\omega)$, $(m = 1, 2, \ldots, p)$ that are used for the prediction, the '–' sign represents forward prediction using receivers whose indices are smaller than n, and the '+' sign represents reverse prediction using receivers whose indices are greater than n. The prediction is valid only when all the receiver wave spectra summed in equation (3.11) are within the array. This is described by the condition $1 \leq (n \pm m) \leq N$ in equation (3.11). The summation coefficients a_m, $(m = 1, 2, \ldots, p)$ are related to the characteristic equation (equation (3.3)). Note that the propagation exponential in the equation is now $z = \exp(\pm i\omega\, s\, d)$.

We now take a new approach to the use of prediction theory by applying it directly in the time domain. Assuming that wave dispersion (frequency-dependence of slowness) is small across the array, we express the coefficients a_m, $(m = 1, 2, \ldots, p)$ using the roots z_m, $(m = 1, 2, \ldots, p)$, of equation (3.3), as

$$\begin{cases} a_1 = \sum_{k=1}^{p} z_k \\ a_2 = \sum_{\substack{k,j=1 \\ (k<j)}}^{p} z_k z_j \\ a_3 = -\sum_{\substack{k,j,m=1 \\ (k<j<m)}}^{p} z_k z_j z_m \\ \quad\vdots \\ a_p = (-1)^p z_1 z_2 \ldots z_p \end{cases} \quad . \tag{3.12}$$

Substituting equation (3.12) into equation (3.11), one sees that the predicted wave becomes a direct function of the p slowness values. Neglecting the dispersion (or frequency-dependence) of the slowness and Fourier-transforming equation (3.11) into the time domain and using the Fourier time-shift theorem (Bracewell, 1965), we obtain an expression for predicting a wave time series at receiver n,

$$\begin{aligned} X_n(t) = &\sum_{k=1}^{p} X_{n\pm 1}(t \pm s_k d) - \sum_{\substack{k,j=1 \\ (k<j)}}^{p} X_{n\pm 2}[t \pm (s_k + s_j)d] \\ &+ \sum_{\substack{k,j,m=1 \\ (k<j<m)}}^{p} X_{n\pm 2}[t \pm (s_k + s_j + s_m)d] +, \ldots, -(-1)^p X_{n\pm p}\,(t \pm \sum_{k=1}^{p} s_k d) \ . \end{aligned} \tag{3.13}$$

For example,

$$\begin{cases} p = 1: \quad X_n(t) = X_{n\pm 1}(t \pm s_1 d); \\ p = 2: \quad X_n(t) = X_{n\pm 1}(t \pm s_1 d) + X_{n\pm 1}(t \pm s_2 d) - X_{n\pm 2}[t \pm (s_1 + s_2)d]; \\ \quad\vdots \qquad\qquad \vdots \end{cases}$$

where the '–' sign corresponds to forward prediction and the '+' sign corresponds to

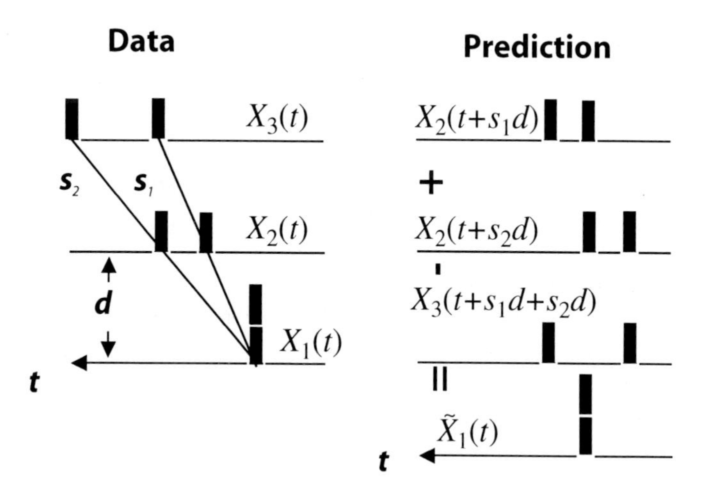

Figure 3.4. Graphic illustration of the operation of the time-domain wave prediction procedure ($p = 2$).

reverse prediction.

Although the general expression for waveform prediction (equation (3.13)) appears somewhat complicated, the basic prediction operation can be shown to comprise two major steps. The first step is time shifting according to the wave mode slowness value(s) across the array, and the other is the summation of the various shifted waveforms to form a wave packet for predicting the wave data at a given receiver. The prediction procedure is illustrated in Figure 3.4 using a simple example. As illustrated on the left-hand side of the figure, the array in this example consists of three receivers. There are two wave modes, each being a spike of unit height, propagating at slowness s_1 and s_2, respectively. The two spikes overlap at the first receiver to give a spike of two units height. Now let us see how the data at receivers 2 and 3 are used in the prediction theory to predict the wave data at receiver 1 (note this is a reverse prediction case). For $p = 2$, equation (3.13) reduces to only three terms, as is the $p = 2$ case in this equation (omit the '–' sign in the time-shifting). The first and second terms of the prediction equation represent back-shifting the data at receiver 2 in time by s_1d and s_2d, respectively. The third term represents back-shifting the data at receiver 3 by $(s_1 + s_2)d$. Finally, these shifted wave data are summed according to the prediction equation. This gives the predicted wave data at receiver 1, which is a spike of two units high. This result agrees with what one would expect from the simple data given in Figure 3.4. The shifted wave data and the summation results are also illustrated on the right-hand side of Figure 3.4. This simple example demonstrates that shifting and summation are the essence of the time-domain prediction theory.

By comparing the predicted waveform with the measured data at each receiver, an inversion procedure is formulated to estimate wave mode slowness. The objective function for this estimation is constructed as

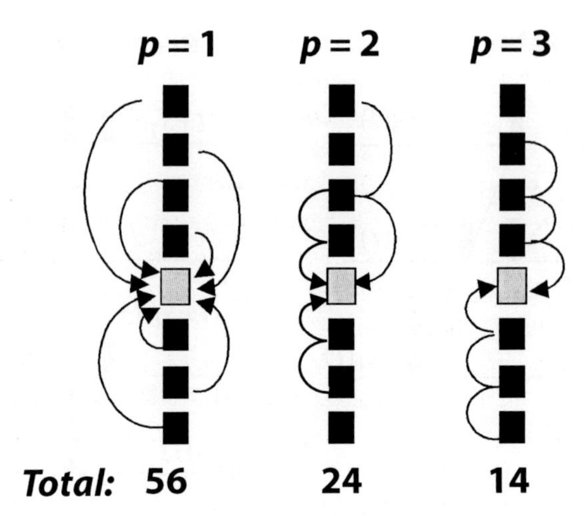

Figure 3.5. Wave prediction configurations for the one-, two-, and three-wave-mode situations (p=1, 2, and 3, respectively). The array contains eight receivers. The receiver data being predicted are those from receiver 4 (counted from bottom up). When the receiver whose data are being predicted is varied from one to eight, the total numbers of data combinations are 56, 24, and 14 for p =1, 2, and 3, respectively.

$$E(s_1,s_2,\ldots,s_p) = \sum_n \int_T \left[X_n(t) - \tilde{X}_n(t;\ s_1,s_2,\ldots,s_p\ )\right]^2 dt\ . \tag{3.14}$$

In equation (3.14), the integration over time is over a time window containing the wave modes of interest; the summation over receiver contains all possible combinations of forward and reverse prediction configurations, with various prediction distances d allowed by the array. Combining the variable prediction distance scheme with the constraint given in equation (3.11), one can obtain different wave prediction configurations depending on the number of wave modes in the data. The wave prediction configurations for p = 1, 2, and 3 are shown in Figure 3.5 for a typical eight-receiver array. As seen from equation (3.11), for p = 1, the data from any receiver in the array can be used to predict the data at another receiver (e.g., receiver 4 – counted from below – is being predicted in Figure 3.5), and there are seven configurations for the receiver being predicted. For all eight receivers, there are a total of 56 prediction configurations. For p = 2, wave data from two receivers are combined to perform a single prediction, as shown by equation (3.11). For receiver 4, there are three prediction configurations. Again, for all eight receivers, one gets 24 prediction configurations. Similarly, for p = 3, there are 14 prediction configurations.

The use of this variable prediction distance d allows maximum possible numbers of data combinations to be used in the slowness estimation, which effectively exploits the redundancy of information in the acoustic logging data, suppresses noise effects in the data, and improves the accuracy of the slowness estimates. For the objective function constructed by including all data combinations, the slowness values that minimize this function are taken as those of the wave modes across the array. The minimization can be performed effectively using the conventional Levenberg-Marquardt algorithm (Marquardt, 1963) or a simulated annealing algorithm (Chunduru and Tang, 1998). The minimization can be performed fast and efficiently when there are only a few (p = 1, 2, or 3) parameters to estimate. For the waveform inversion method, the overlap of the wave modes in time (or frequency) does not affect the estimation if the mode propagation characteristics (i.e., slowness values) are distinct. By applying the prediction theory in the time domain, the ability to estimate dispersion is sacrificed, but the amount of data that can be used for slowness estimation is increased by orders of magnitude as large numbers of wave time samples and all possible data combinations (Figure 3.5) are used. Thus, the noise effects are effectively suppressed, and the accuracy and robustness of the estimation are significantly enhanced compared to the frequency domain methods. Therefore, for non-dispersive or weakly dispersive waves (e.g., compressional, shear head waves, Stoneley, and low-frequency flexural waves), the waveform inversion method presents an effective tool for obtaining slowness profiles from these wave data. Worth special mention is that, by applying the method to a time window containing only one wave mode (p = 1), the waveform inversion becomes a pair-wise waveform matching scheme, which is particularly useful for enhancing the resolution of slowness estimation using short arrays. Resolution enhancement will be discussed in the following section.

3.3 Resolution Enhancement

In formation evaluation, it is often necessary to quantify the acoustic and petrophysical properties of laminated thin beds for better reserve estimation and reservoir characterization. Standard array acoustic processing yields a slowness log that tends to smooth, or average, the actual variations over the length of the receiver array (typically 3.5 ft or 1.067 m), obscuring the features that are smaller than the array aperture. This section describes techniques that can enhance the vertical (or borehole axial for deviated or horizontl wells) resolution of the acoustic slowness profiles.

Enhancing resolution of slowness estimates from an array acoustic tool is done through using overlapping sub-arrays across the same depth interval whose thickness is equal to the sub-array aperture. At some particular depth, the acoustic source on the tool is activated and an array (typically eight receivers) of waveform data is recorded. This procedure is repeated while the tool is pulled up a distance equal to one inter-receiver spacing (typically 0.5 ft or 0.152 m). Consequently, the receiver arrays at successive source locations are overlapped. Figure 3.6 shows all seven possible sub-array configurations for an eight-receiver array-acoustic tool. The apertures of the sub-arrays range from 3.5 ft–0.5 ft (1.067 m–0.152 m). One can therefore use the redundant information in overlapping arrays to improve both the vertical resolution and the accuracy of the formation acoustic slowness estimation. The processing techniques utilizing the overlapping arrays are described in the following sections.

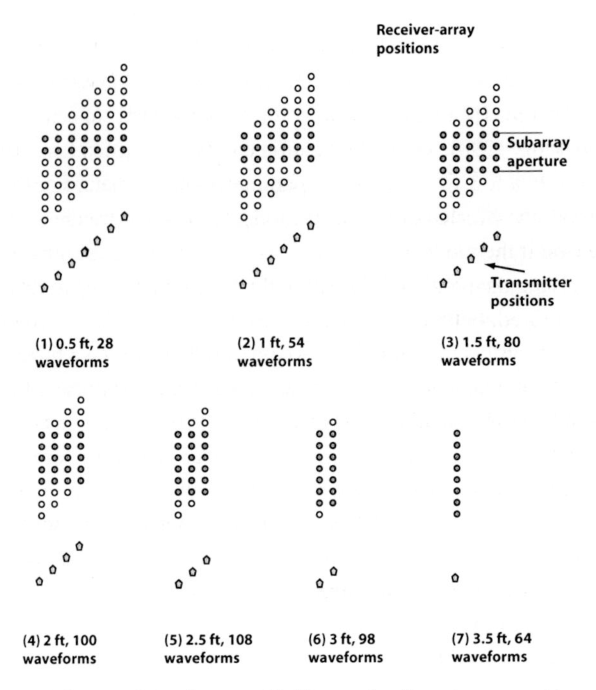

Figure 3.6. Various configurations provided by overlapping sub-arrays with apertures ranging from 3.5 ft–0.5 ft (1.067 m–0.152 m).

3.3.1 MULTI-SHOT SEMBLANCE

Signal processing techniques have been developed to enhance the vertical resolution of acoustic slowness logs. Hsu and Chang (1987) applied a multiple-shot semblance technique to process data from overlapping sub-arrays. The processing steps are as follows:

1. Select the sub-array aperture (i.e., number of sub-arrays that overlap the same depth interval).
2. Compute the semblance function for each sub-array. For each slowness, find the arrival time T_{max} that has the maximum semblance value, denoted by $\tilde{\rho}_k(s) = \rho_k(s, T_{max}),\ (k = 1, 2, \ldots, K)$, where K is the total number of sub-arrays.
3. Calculate the combined semblance function using either geometric or algebraic means

$$\bar{\rho}(s) = \left(\prod_{k=1}^{K} \tilde{\rho}_k(s)\right)^{1/K}, \quad \text{(geometric)}$$
$$\bar{\rho}(s) = \frac{1}{K}\sum_{k=1}^{K} \tilde{\rho}_k(s), \quad \text{(algebraic)}. \tag{3.15}$$

4. Estimate the wave slowness value(s) by finding the maximum (maxima) of equation (3.15).

Step 2 of the above process projects the 2D semblance function of each sub-array onto the slowness axis to become a 1D function of s only. This provides a common ground for combining the semblance statistics of all sub-arrays. Otherwise, it would be difficult to combine the 2D semblance function directly because the peak(s) of the 2D function has (have) a different arrival time T for different sub-arrays.

The multi-shot semblance technique can improve the accuracy and resolution for sub-arrays of moderate size (e.g., it is commonly applied to four-receiver sub-arrays). It should be pointed out that, while using shorter sub-array aperture is necessary to enhance resolution, it is more prone to noise contamination because less wave moveout and fewer data, and thus less slowness information and data redundancy, are utilized. One can therefore see the drawback of using array-stacking techniques such as semblance: for a small sub-array, noise severely degrades the quality of the stacking coherence (semblance) and increases the slowness estimation error. Thus it is difficult to apply the array stacking technique to the shortest sub-array (i.e., two receivers and 0.5 ft or 0.152 m aperture) that has the highest resolution.

3.3.2 PAIR-WISE WAVEFORM INVERSION

To obtain the highest resolution provided by the overlapping sub-arrays, the above-described waveform inversion method is adapted to the overlapping sub-array configurations. The key in obtaining a reliable, high-resolution acoustic slowness profile using short sub-arrays is to reduce the noise effect and to maximize the redundancy of information in the data. The waveform inversion method is particularly applicable to short arrays. If the waveform contains only one mode (p = 1), then there is only one slowness value to estimate, and any receiver data in the sub-array can be used to predict the data of another receiver. The general wave prediction formula in equation (3.13) is greatly simplified, as given by the case p = 1 in this equation. Using the case p = 1, it follows that for any given receiver n in the sub-array, the waveform at another receiver m can be shifted in time to substantially match with the waveform at receiver n as

$$W_m(t + s(m-n)d) \approx W_n(t) \ .$$

Therefore, by minimizing the difference between the above waveforms, we can formulate an inversion procedure to estimate the slowness s across the sub-array. The inversion essentially becomes a pair-wise waveform-matching scheme. The objective function for this estimation is

$$E(s) = \sum_{k=1}^{K} \sum_{m,n=1}^{N} \int_{T(m,n)} \{W_n^k(t) - W_m^k[t + s(m-n)d]\}^2 dt \ . \tag{3.16}$$

The superscript k denotes the sub-array index and K is the number of sub-arrays crossing the same depth interval. The integration is over the time window T in which the waveforms are matched. The placement of time window T is chosen to ensure

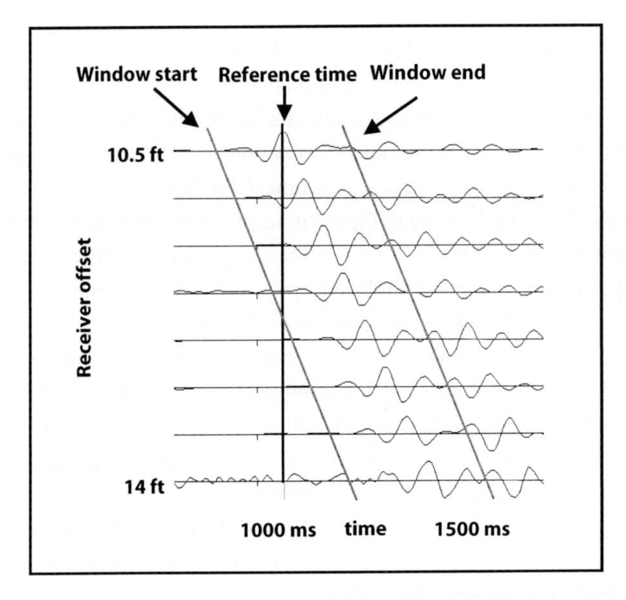

Figure 3.7. Example showing the isolation of a wave event of interest for waveform matching. A predetermined P-wave traveltime is used to place a time window for the P-wave event at the first receiver. The window position at subsequent receivers is along the wave's moveout across the array.

that it contains only one wave mode, so that the above analysis holds true.

Windowing the waveform data is done to isolate the most coherent portion of a wave event for processing, and to prevent noise and/or other wave events from adversely affecting the processing. Generally speaking, the wave onset, or first arrival portion of a wave event, has a high degree of coherence because of its shortest travel path from transmitter to receiver. In the present technique, one can first obtain a wave travel (or arrival) time curve as a reference time for the wave event. The travel time curve can be obtained by tracking the first portion of the wave event across depth, or by integrating the wave slowness curve, as could be obtained from the array processing method, over the transmitter-to-receiver distance. Figure 3.7 shows the placement of the time window to isolate the compressional wave event across the receiver array. The arrival time for the peak of the P-wave event was obtained from tracking the wave for the first receiver in the array. The start time of the window is placed ahead of the earliest wave arrival. The length, or time duration, of the window is chosen to include a few cycles of the wave event. The next step is to place the window for the rest of the receivers in the array. The wave event moves out, or propagates across the array according to its slowness. The placement of the window for each subsequent receiver is therefore along the wave moveout. The window start time at the nth receiver in the array is:

$$T_n = T_1 + (n-1)d / s_{av} , \tag{3.17}$$

where n is the receiver index; T_1 is the window start time at the first receiver; d is receiver spacing; and s_{av} is the average slowness across the array. We can use the predetermined slowness value (e.g., from an array processing method) for s_{av}, or we can simply use $T_1/trsp$ as an estimate for s_{av} where $trsp$ is transmitter-to-receiver spacing.

The slowness estimation using equation (3.16) uses all possible receiver combinations allowed by the sub-array of N receivers, in order to maximize the redundancy of information present in the waveform data. The receiver index m in the summation of equation (3.16) can be smaller (forward shift) or greater (reverse shift) than the index n. For example, for a four-receiver sub-array ($N = 4$), the waveform of any receiver in the sub-array can be shifted to match with the waveform of another receiver. There are three data combinations for one receiver being matched. For all four receivers, there are twelve data combinations with sixteen repeated usages of waveforms. Further, for all five sub-arrays across the same depth interval (see case 3 of Figure 3.6), there are 60 data combinations with a total of 80 waveforms used.

For the extreme case of a two-receiver sub-array ($N = 2$, case 1 of Figure 3.6), the waveform-matching technique has fourteen data combinations from performing forward and reverse waveform shifts. The total number of waveforms used is 28. The pair-wise waveform matching, therefore, utilizes the maximum possible number of waveforms to maximize the redundancy of information for all sub-array configurations of Figure 3.6. For each sub-array configuration, the number of waveforms utilized in the waveform inversion analysis is indicated in Figure 3.6.

We can now discuss the validity condition of the above resolution enhancement method using sub-arrays of reduced aperture. Obviously, this method cannot be applied to waves with very long wavelengths. There must be a limit beyond which the method will become invalid. The validity condition is that the processing aperture should be no less than a quarter of the wavelength for the acoustic wave used (e.g., compressional or dipole shear waves). A physical explanation for this condition is that, for the wave to be still recognizable as an oscillating wave across an observation interval (i.e., the processing aperture), the interval must be greater than a quarter of the wavelength. For the 0.5 ft-aperture sub-array (case 1 of Figure 3.6), this condition is satisfied for most logging situations, especially for P-wave data. For example, for a 10 kHz P-wave with a velocity of 10 000 ft/s, the quarter wavelength is 0.25 ft, which is smaller than 0.5 ft. This condition may occasionally be violated with low-frequency dipole waves when the 0.5 ft aperture configuration is used. In this case, the 1 ft aperture configuration, as in case 2 of Figure 3.6, should be used.

3.3.2.1 Example of Resolution Enhancement

Figure 3.8 demonstrates resolution enhancement from the waveform inversion method. Track 1 of this figure shows the compressional- (P-) wave portion of the acoustic log data across a depth segment of 100 ft. Only data from receiver 1 of an eight-receiver array are displayed. Track 1 also shows the P-wave traveltime

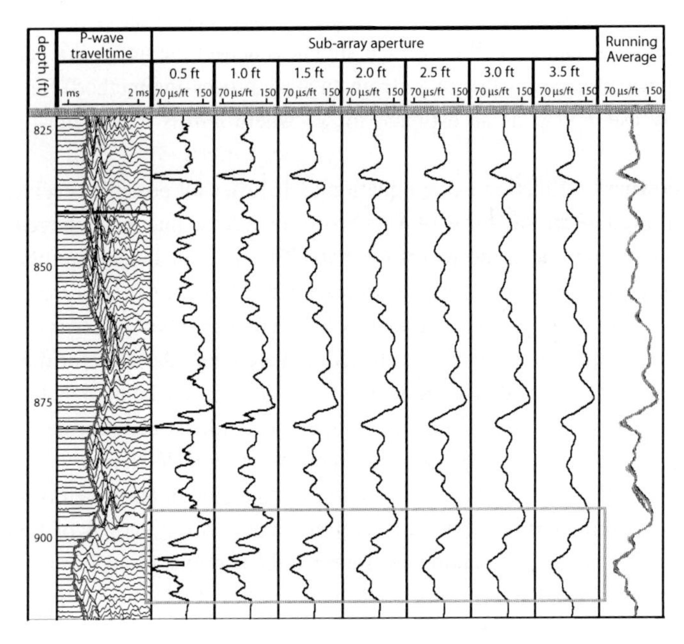

Figure 3.8. P-wave slowness curves estimated for various configurations (apertures) of Figure 3.6. Note the increasing resolution of the curves with decreasing sub-array aperture (e.g., the depth section indicated by a box). Track 9 is a consistency check by averaging the curves to 3.5 ft aperture and overlaying the results.

curve that is used to place the time window for processing. Tracks 2 through 8 show slowness profiles obtained for various resolutions provided by the sub-array configurations in Figure 3.6. It is clear that the resolution of formation features is increasingly enhanced when the sub-array aperture decreases from 3.5 ft–0.5 ft (from track 8 to 2). Features that are obscure on the conventional profile (track 8, 3.5 ft aperture) are clearly identified on the profile with the highest resolution (track 2, 0.5 ft aperture). For example, the 0.5 ft aperture profile in track 2 reveals a layered formation between 890 ft and 915 ft, while the layering cannot be seen on the 3.5 ft aperture profile (track 8). Track 9 provides a simple consistency check of the processing results by overlaying the running average of each slowness profile from track 2–7 with the conventional curve (track 8). The length of the average is 3.5 ft for track 2, 3.0 ft for track 3, ..., and 0.5 ft for track 7. The different averaging lengths are used to average the curves of different resolutions, so as to match with the resolution of the conventional (3.5 ft aperture) curve. The various averaged curves overlay with the conventional curve very well with only small differences. This comparison demonstrates that these curves of enhanced resolution are inherently consistent with one another, although the magnitude of variations may be quite different on curves with different resolutions. The consistency of the curves shows that the waveform matching can determine formation acoustic slowness with various resolutions by using different measurement apertures ranging from 3.5–0.5 ft.

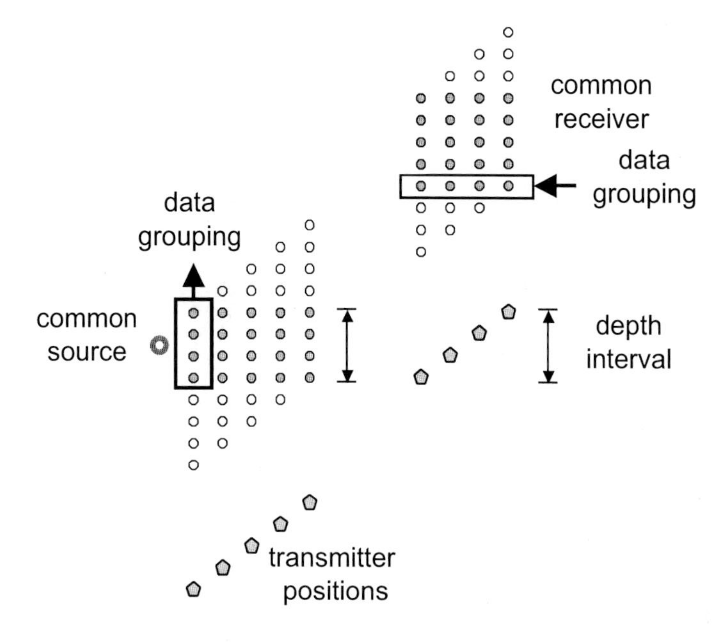

Figure 3.9. Diagram illustrating the gather of array acoustic data into either common-source subarrays or common-receiver subarrays. These sub-arrays cover the same depth interval. Data grouping (or gathering) is in the vertical direction for the source gather, and in the horizontal direction for the receiver gather.

3.4 Borehole Compensation

The change of borehole size often affects the acoustic travel time in the borehole fluid, resulting in inaccurate slowness estimates. The effects of borehole changes can be compensated by averaging the slowness values from two types of acoustic data arrays. These arrays are, respectively, gathered according to successive receiver or transmitter depth locations.

The various sub-array configurations shown in Figure 3.6 are formed for successive transmitter locations. They are called common-source gathers (this gather of data is called the receiver array). Analogous to the common-source gathers, we can form various common-receiver sub-array configurations (the acoustic waveform array formed from the common-receiver gather is called the transmitter array). Figure 3.9 shows the configuration of a four-receiver sub-array for the common-receiver gather (i.e., gathered for a common receiver location). As the acoustic tool is raised during logging, successive transmitter locations eventually cross the same depth interval spanned by the common-source sub-arrays. For an array of eight receivers, there are five common-receiver sub-arrays covering the same depth interval. Data grouping (or gathering) is in the vertical direction for the receiver array, and in the horizontal direction for the transmitter array. Similar to the common-source configurations shown in Figure 3.6, this construction of common-receiver sub-arrays can be

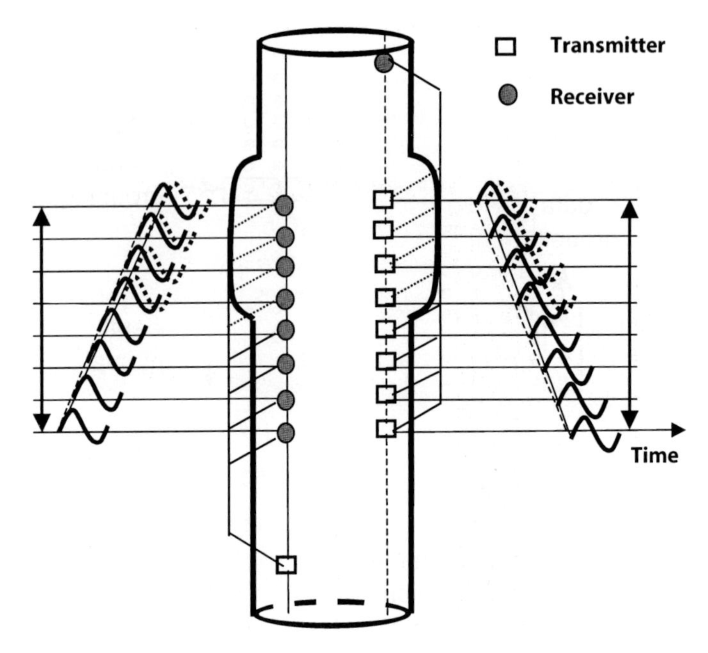

Figure 3.10. Diagram illustrating the acoustic wave time delay in an enlarged bore section. For the receiver array gathered by the source, this time delay causes an over-estimation of the wave slowness. In comparison, for the transmitter array gathered by the common receiver location, this delay causes an underestimation of the wave slowness. Consequently, averaging the two results tends to compensate for the effect of borehole enlargement.

made for various sub-array apertures ranging from two to seven receiver spacings. The above-described analysis can be applied to the common-receiver sub-arrays to determine the slowness value of a wave mode.

Figure 3.10 illustrates why averaging the receiver- and transmitter-array results can compensate the borehole changes. This figure uses an eight-receiver array as an example. The upper part of figure illustrates an enlargement of the borehole. The acoustic array tool is pulled up from below the enlargement. When some receivers of the array are in the enlarged borehole, the wave arrival time at these receivers increases because of the additional time delay in the borehole fluid (assuming the fluid is acoustically slower than the formation). Consequently, the estimated slowness (slope of the solid line in the figure) increases. An opposite situation will occur for the transmitter array gathered from successive transmitter locations. The same additional time delay in the fluid causes a steeper slope along the wave moveout in the transmitter array, decreasing the estimated slowness (obviously, a shrinking of the borehole will decrease (increase) the estimated slowness from the receiver (transmitter) array). Thus, averaging the slowness values from the two different directions eliminates, or reduces, the effect of borehole changes on slowness values.

The transmitter-array result can also enhance or make up for the missing estimates in the receiver arrays. For example, if there is a wave propagation obstacle (e.g., frac-

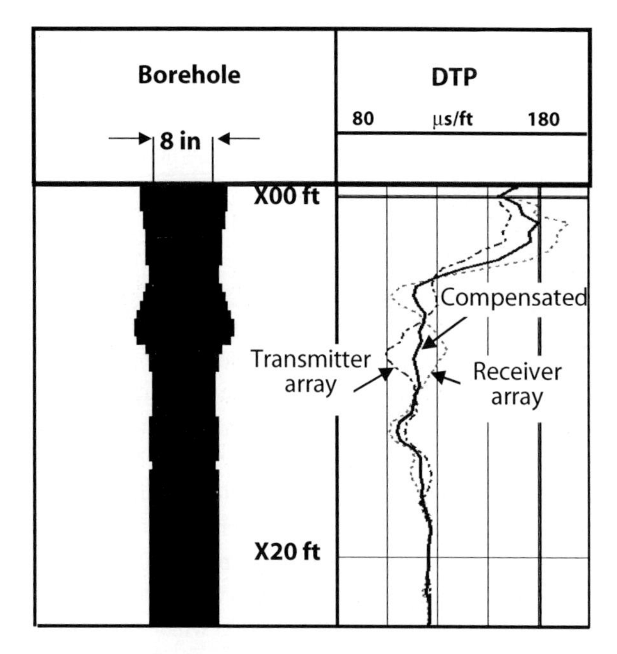

Figure 3.11. An example of borehole compensation using respective compressional-wave slowness (DTP, track 2) curves from transmitter and receiver arrays. An increase (decrease) in the borehole (shaded area in track 1) diameter causes the slowness from the receiver array to increase (decrease) and the slowness from the transmitter array to decrease (increase). Consequently, averaging the two slowness curves compensates the effect caused by borehole diameter changes.

tures) somewhere between the transmitter and receiver, the wave events received by the receiver array may have poor quality or even disappear, resulting in poor or missing estimates of the wave slowness values. However, the common-receiver data gathered for transmitter positions above the obstacle will not be affected. Since this gather covers the same depth interval as does the source gather (Figure 3.9 or 3.10), the poor or missing result of the source gather can be recovered from the receiver gather data.

Figure 3.11 shows an example of borehole-change-induced slowness variation and the borehole compensation result. The shaded area in track 1 illustrates the borehole with significant diameter changes. The compressional-wave slowness (DTP) curves shown in track 2 are obtained from the logging of an eight-receiver array tool (source below receiver). The logging is in the upward direction. When the borehole diameter is almost constant, as shown in the lower portion of the figure, the DTP curves from the receiver array (short dashed curve) and transmitter array (long dashed curve) merge into one curve. However, they differ significantly when there are significant borehole changes. A good example is the behavior of the slowness curves around X10 ft. When the receivers of the tool begin to enter the enlarged borehole, the slowness from the receiver array starts to increase, while the slowness from the transmitter array starts to decrease. The situation is reversed as the receivers leave the enlargement encountering a decreasing borehole diameter. The trend of the slowness variation caused by borehole changes agrees well with the illustration and explanation of Figure 3.10. Consequently, we can compensate for the effect by averaging the two slowness curves to produce a borehole-compensated slowness curve (solid curve in track 2).

3.5 Dispersion Effects and Correction

In this section, we analyze the effects of dispersion on the slowness estimation obtained from time-domain array processing techniques. In particular, we present a useful, general theoretical result that relates the time-domain moveout of a dispersive wave to the wave's dispersion characteristics. This theoretical result can be used to provide a simple method for correcting the dispersion effect in the dipole-shear slowness measurement. It can also be used to estimate formation anisotropy from the measured Stoneley-wave slowness (see Chapter 5).

The above-described time-domain processing methods, such as semblance and waveform inversion methods, process the array acoustic data assuming non-dispersive waves (waveforms within the processing time window propagate at the same velocity). The actual waveform data, especially the dipole acoustic data, are a dispersive wavetrain. An important question then arises: what velocity do we measure when we process a dispersive wave using the non-dispersive methods? The following theoretical analysis answers this question.

In the time-domain methods, such as semblance (equation (3.8)) and waveform inversion (equations (3.14) and (3.16)), the energy of the waveform data (semblance) or the waveform mismatch (waveform inversion) is integrated over a time window T. For the following analysis, we let this window be long enough to include the entire dispersive wavetrain (this is especially the case for the waveform inversion method that usually uses a longer window than the semblance method). Mathematically, this means that the time interval for the integration extends to infinity. As a result, we can readily apply Parseval's theorem (Bracewell, 1965):

$$\int_{-\infty}^{+\infty} |X(t)|^2 dt = \frac{1}{2\pi} \int_{-\infty}^{+\infty} |X(\omega)|^2 d\omega \ , \tag{3.18}$$

where $X(\omega)$ is the Fourier transform of the wave data $X(t)$ recorded by a receiver in the acoustic array. In the time-domain formulations, the waveform data are shifted from one receiver to another using a slowness value S^*. In the Fourier domain, this is equivalent to phase-shifting the data across the receivers. For example, in an array of N receivers, the time shifting and frequency phase-shifting of the data of receiver n to the position of receiver m are related by the following Fourier transform pairs:

$$X_n(t + S^*(n-m)d) \Leftrightarrow X_n(\omega)\, e^{-i\omega S^*(n-m)d} \ . \tag{3.19}$$

For the semblance method, the array data are shifted to receiver 1 (m = 1); for waveform inversion, the data can be shifted to any receiver position (m = 1, ..., N). Notice that the shifting is made with the frequency-independent slowness S^* (here we consider only one wave mode, p = 1). The actual acoustic wave data, however, propagate across the receiver array with a frequency-dependent slowness function $S(\omega)$. The actual time shifting of the wave corresponds to the phase shifting in the Fourier domain as:

$$\text{time-shifting of } X_n(t) \text{ to receiver } m \Leftrightarrow X_n(\omega)\, e^{-i\omega S(\omega)(n-m)d} \quad . \tag{3.20}$$

Combining equations (3.19) and (3.20) and applying Parseval's theorem in equation (3.18) respectively to the semblance and waveform inversion formulations (equations (3.8) and (3.14); note p = 1 in the latter equation), we obtain:

$$\left.\begin{array}{l}\rho(S^*)\\ E(S^*)\end{array}\right\} \propto \left\{\begin{array}{l}\int_{-\infty}^{+\infty} A^2(\omega)\, W(u(\omega, S, S^*))\, d\omega \\ \int_{-\infty}^{+\infty} A^2(\omega)\big(1 - W(u(\omega,S,S^*))\big)\, d\omega\end{array}\right. \quad . \tag{3.21}$$

The above results show that the semblance is a weighted average of thc power spcctrum, weighted by a dimensionless function W, and the waveform misfit function is the total power minus the semblance. Thus maximizing the semblance is equivalent to minimizing the waveform misfit. The weight W is a function of the dimensionless variable u, where

$$u(\omega,S,S^*) = \frac{\omega(S(\omega) - S^*)d}{2} \quad . \tag{3.22}$$

For the semblance method, it is relatively straightforward to show that (Kimball, 1998)

$$W(u) = \frac{\sin^2(Nu)}{N^2 \sin^2 u} \quad . \tag{3.23}$$

For waveform inversion, the same expression would be obtained if we shift other receiver data to match with the first receiver data, as in the semblance case. However, if all data combinations (Figure 3.5, case p = 1) are to be included, the expression for $W(u)$ becomes quite lengthy, as given by the following expression,

$$\begin{aligned} W(u) = &\frac{(2N-1)(2\sin^2 u - \cos u + \cos(3u))}{2N(N-1)(4\sin^2 u - \cos u + \cos(3u))} \\ &+ \frac{\cos((2N-3)u] - \cos((2N-2)u) - \cos((2N-1)u) + \cos(2Nu)}{2N(N-1)(4\sin^2 u - \cos u + \cos(3u))} \quad . \end{aligned} \tag{3.24}$$

It can be shown that the two functions have the same maximum value at $u = 0$ ($W(0) = 1$) and a quite similar behavior in the vicinity of $u = 0$. Both equations (3.23) and (3.24) are sharply peaked at $u = 0$, indicating that ρ or E in equations (3.21) is dominated by the contribution of $W(u)$ in the vicinity of $u = 0$.

The key in the semblance or waveform inversion method is to find the value of the slowness S^* that either maximizes the semblance (equation (3.8)) or minimizes the wave misfit function (equation (3.14)). The mathematical condition for S^* to maximize the semblance function or minimize the waveform misfit function is: $\partial\rho/\partial S^* = 0$ (semblance) or $\partial E/\partial S^* = 0$ (inversion). Taking the derivative of either ρ or E in equations (3.21) with respect to S^* involves $W'(u)$, the derivative of $W(u)$ with respect to u. Since ρ or E in equations (3.21) is dominated by the contribution of $W(u)$ in the vicinity of $u = 0$, we only need to retain the leading order of $W'(u)$ around $u = 0$, which is $W'(u) \sim u$. With this consideration, the condition $\partial\rho/\partial S^* = 0$ (semblance) or $\partial E/\partial S^* = 0$ (inversion) leads to the following result:

$$\int_{-\infty}^{+\infty} \omega^2 A^2(\omega)\left(S(\omega) - S^*\right)d\omega = 0,$$

or, (3.25)

$$S^* = \frac{\int_{-\infty}^{+\infty} S(\omega)\omega^2 A^2(\omega)\,d\omega}{\int_{-\infty}^{+\infty} \omega^2 A^2(\omega)\,d\omega}.$$

This result can be called the Weighted Spectral Average Slowness theorem. The theorem states that the slowness (or velocity) resulting from a non-dispersive time domain array processing method is a weighted spectral average of the wave's slowness dispersion curve over the frequency range of the wave spectrum. The weighting function is given by $\omega^2 A^2(\omega)$. This theorem gives the answer to the previous question concerning the meaning of the slowness value S^*.

3.5.1 A Numerical Test of the Weighted Spectral Average Slowness Theorem

We use a numerical example to demonstrate the validity of the important result in equation (3.25). Although this result is aimed at correcting the dispersion effect for dipole-flexural waves, we apply it to a synthetic Stoneley-wave data set in order to demonstrate the universal applicability of this result to dispersive waves. Figure 3.12a shows an array of synthetic Stoneley waves for a slow TI formation (see model parameters given in Table 2.3). Figure 3.12b shows the corresponding wave amplitude spectrum and the wave slowness dispersion curve. Both the waveform and dispersion curve show that the wave is quite dispersive toward the low frequency range, which is needed to test the theoretical result in equation (3.25). As illustrated in Figure 3.12b, a band-pass filter of 1.5 kHz width is used to pass the wave

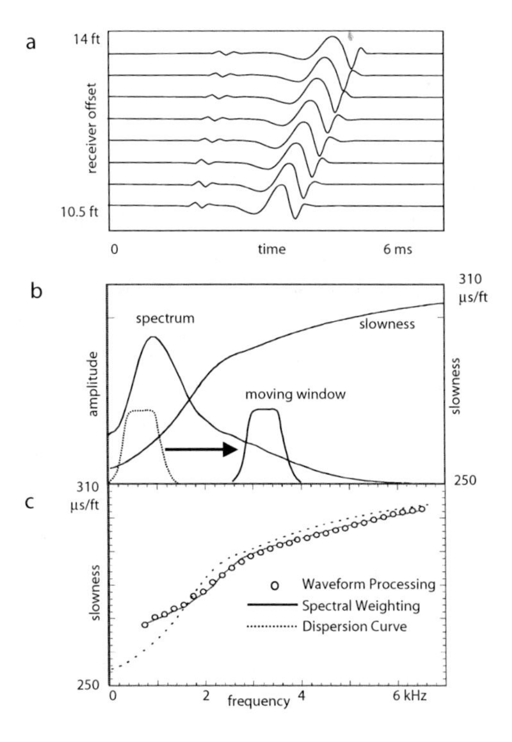

Figure 3.12. A numerical test of the weighted spectral average theorem.
(a) Synthetic array Stoneley-wave data.
(b) Wave spectrum and dispersion curve. The wave data are filtered with a band-pass filter moving from low to high frequencies. At each filter position, the filtered waveform data are processed to give a slowness value in (open circles in panerl c) and the weighted spectral average slowness (solid curve in panel c) is calculated for the filter's frequency band.
(c) Comparison of the two results.

frequency components within the window. The filtered wave data are then processed using the waveform inversion method (equation (3.14) for the case $p = 1$) to yield a wave slowness value for S^*. This slowness value is then assigned to the frequency centered by the band-pass filter. In the frequency domain, the amplitude spectrum of the filter is multiplied with that of the Stoneley wave, as shown in the same figure, to form the filtered wave amplitude spectrum $A(\omega)$. The wave spectrum is then used in equation (3.25) to weight the wave dispersion curve and produce the weighted slowness value. These calculations are repeated for a range of the filter center frequency from 0.75 kHz–6 kHz, at an increment of 0.2 kHz. Figure (3.12c) shows the comparison of the waveform-processed slowness (open circles) and the weighted spectral average slowness (solid curve) values, together with the wave's dispersion curve (dashed curve). As predicted by the theory (equation (3.25)), the waveform-processed slowness does not fall onto the wave's phase dispersion curve. Instead, it agrees excellently with the weighted spectral average slowness value given by equation (3.25). This numerical study proves this general theoretical result.

3.5.2 APPLICATION TO DIPOLE-FLEXURAL WAVE DISPERSION CORRECTION

Equation (3.25) can be used to provide a simple and effective correction for the dispersion effect of shear slowness from dipole acoustic logging data. Dipole-wave data are usually processed using a non-dispersive method (semblance or waveform inversion). The obtained slowness value, as shown by equation (3.25), is a weighted average of the slowness dispersion curve. The flexural wave dispersion curve $S(\omega)$ is parameterized by the formation shear slowness S_S as $S(\omega, S_S)$ (assume that other parameters, i.e., borehole diameter, formation density and compressional-wave slowness, are available from log data). With this parameterization, equation (3.25) becomes

$$S^* = \frac{\int_{-\infty}^{+\infty} S(\omega, S_s)\omega^2 A^2(\omega)d\omega}{\int_{-\infty}^{+\infty} \omega^2 A^2(\omega)d\omega} . \quad (3.26)$$

If there is no dispersion ($S(\omega, S_S) = S_S$), then S^* should be the desired shear slowness S_S. For a flexural wave, the slowness $S(\omega, S_S)$ increases with frequency; thus S^* is always greater than S_S. This difference is known as the dispersion effect on the estimated dipole-shear slowness data. The dispersion effect can be corrected using equation (3.26) by the following procedure.

1. Process array dipole acoustic data to obtain S^*.
2. Compute the acoustic wave spectrum $A(\omega)$. Integrate $\omega^2A^2(\omega)$ over the frequency range of $A(\omega)$ to calculate the denominator of equation (3.26).
3. For a trial shear slowness S_S, calculate the dispersion curve $S(\omega, S_S)$.
8. Weight $S(\omega, S_S)$with the weighting function $\omega^2A^2(\omega)$ and integrate the weighted slowness curve over the frequency range of $A(\omega)$. Divide the integral value by the denominator from step 2, then equate the result to S^*.
5. Repeat steps 3 and 4 until equation (3.26) is satisfied. Then, output S_S as the true formation shear slowness.

The importance of equation (3.25) – or (3.26) – is that it demonstrates that the moveout slowness (or velocity) of a dispersive wave, as obtained from a non-dispersive array processing method (semblance or inversion), is a weighted spectral average of the wave's dispersion curve. Although this result is obtained assuming an infinitely long time window, the same spectral weighting concept, as stated in this equation, can be extended to situations where a finite time window length is used. In the actual processing, especially when the semblance method is used, the window length includes only one to three cycles of the waveform data (waveform inversion may take more cycles). As a result, the assumption of the infinite window length, as is required to apply Parseval's theorem, does not hold and the theoretical result (equation (3.25) or (3.26)) may not be directly applicable. However, the same weighted spectral averaging concept as equation (3.26) can still be used. Taking the

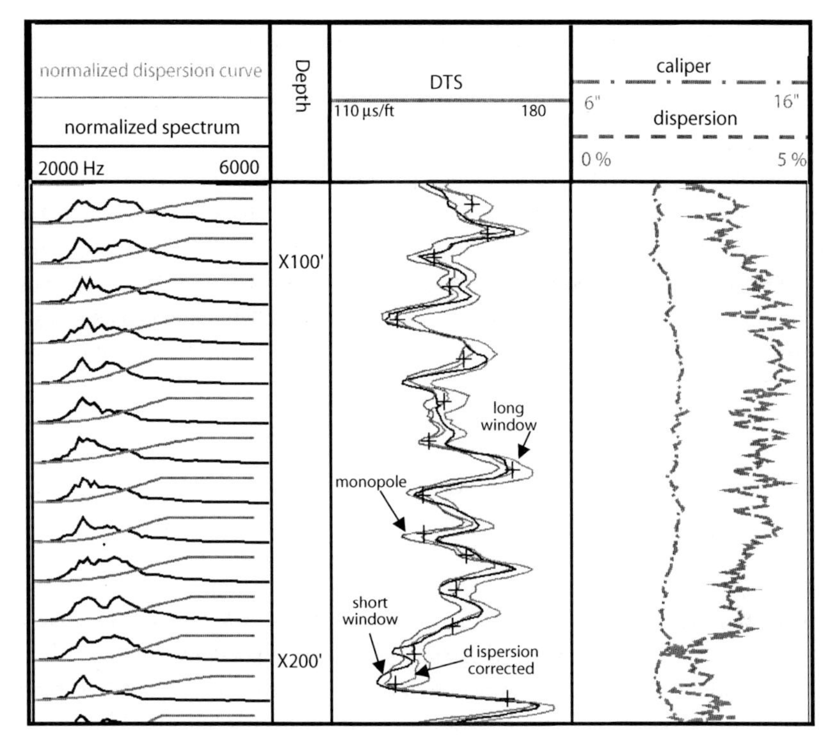

Figure 3.13. An example of dispersion correction for dipole-shear wave data. Track 1 shows the normalized wave spectrum and dispersion curve for every 10 ft. Track 2 shows various slowness curves: monopole shear, dipole shear processed with large and small windows, and the dispersion-corrected shear (marked by crosses) for the large window result. The percentage of correction is shown in track 3 together with the borehole caliper curve.

entire dispersive wavetrain, as in the derivation of equation (3.26), results in weighting $S(\omega, S_S)$ towards higher frequency, as can be seen from the weighting function $\omega^2A^2(\omega)$. If we window the first arrival part of the dipole wave data, which usually has lower frequency and higher velocity, an 'effective' weighting function can be used to replace $\omega^2A^2(\omega)$. This function should weight $S(\omega, S_S)$ towards the lower frequency range. For example, we can use $A(\omega)$ instead of $\omega^2A^2(\omega)$, or some other empirical forms that fit the data.

Figure 3.13 demonstrates a field data example of dispersion correction using equation (3.26). Track 1 shows the wave spectrum and calculated dispersion curve. The dispersion curve is displayed as the percentage difference relative to the formation shear slowness (maximum 10 %). If a portion of the spectrum overlays with the dispersion curve with significant curvature (i.e., dispersion), then a significant amount of dispersion correction is expected. Track 4 displays the amount of dispersion correction in percentage, together with the borehole caliper. The dispersion correction correlates well with borehole caliper, showing that a larger borehole causes higher dispersion. As proof of the correction result, track 3 plots four shear slowness curves obtained respectively from: 1) refracted monopole shear wave data (presumably free from dispersion); 2) correcting the dipole-wave slowness processed using a long time window (including 3–5 cycles of the waveform); 3) dipole-shear slowness processed using a short time window (including the first cycle of the wave); and 4) dipole-shear slowness processed with a large window (the right-most curve). As expected, the slowness curve from the short window is quite close to the monopole shear curve, showing minimal dispersion effect. However, the slowness using the long window is significantly higher than the monopole slowness. After correction using equation (3.26), the corrected slowness (curve marked by crosses) is quite close to the monopole shear, showing the effectiveness of the dispersion correction.

3.6 Wave Attenuation Estimation

Elastic wave attenuation estimation from borehole acoustic logs has always been a difficult problem. On one hand, the attenuation of wave amplitude is sensitive to formation rock and fluid properties and should be utilized as a means of formation evaluation. On the other hand, however, many factors significantly affect wave attenuation. These factors, to mention just a few, are wave scattering due to borehole and formation changes, geometric spreading, receiver mismatch, etc.

A conventional method for estimating attenuation is the spectral-ratio method. This method derives attenuation by taking the ratio of the amplitude spectra at two receiver locations, as given by the following formula:

$$\ln\frac{X(\omega,z_2)}{X(\omega,z_1)} = (z_2 - z_1)\frac{\pi f}{QV} + \ln\frac{G(z_2)}{G(z_1)} , \tag{3.27}$$

where X is the amplitude spectrum of the wave, f is frequency, and V is wave velocity. The geometric spreading G is commonly assumed to be independent of frequency. Thus, by linearly fitting the natural logarithm of the spectral ratio versus frequency, the attenuation Q^{-1}, in terms of the inverse of the quality factor Q, can be estimated from the slope of the fitted line. In the field application of the above method, however, two problems seriously limit the reliability of the method. The first is that the wave spectrum has significant local variations along the borehole, causing large variations in the spectral ratio values. The second is that the receivers may not be in the far field of the source such that G can depend strongly on frequency. Below we describe a newly developed method that overcomes the two problems and appears to give a robust estimation of attenuation.

For a wave recorded by a receiver of an acoustic tool, the wave's amplitude spectrum, recorded at depth z, can be written as

$$X(\omega,z) = S(\omega)G(\omega,z)R(\omega)\exp\left(-\frac{\omega T(z)}{2Q(z)}\right), \tag{3.28}$$

where S and R now denote the amplitude spectrum of the source and receiver response, respectively; G is Green's function that governs the wave's geometric spreading from source to receiver (in a borehole environment, the geometric spreading is a complicated function of ω and z); T is the wave travel time, and Q^{-1}, the inverse of quality factor Q, is the attenuation to be estimated.

Rearranging equation (3.28) and taking its natural logarithm result in

$$\frac{2\ln X}{\omega} = -Q^{-1}T + \frac{2\ln(SRG)}{\omega} \quad . \tag{3.29}$$

The first term on the right-hand-side depends only on z (assuming Q is independent of ω in the frequency range of acoustic logging). The second term depends on both ω and z. If we can make the second term a function of ω only, then we can apply a statistical average/median over the ω- and z-domain, respectively, to yield a robust estimate of Q^{-1}.

Synthetic modeling is used to remove the z-dependence of the second term. By modeling equation (3.29) without attenuation ($Q^{-1} = 0$) and taking the difference between this equation and its modeled counterpart, we get

$$\Delta\Phi(z,\omega) = \phi(z) + \Delta A(\omega) \ , \tag{3.30}$$

where

$$\begin{cases} \phi = -Q^{-1}T \ , \\ \Delta\Phi = \dfrac{2\ln X}{\omega} - \dfrac{2\ln X_{syn}}{\omega} \ , \\ \Delta A = \dfrac{2\ln(SRG)}{\omega} - \dfrac{2\ln(SRG)_{syn}}{\omega} \quad . \end{cases} \tag{3.31}$$

If the synthetic Green's function G correctly accounts for the z-dependence of the geometrical spreading function, then the second term in equation (3.30) is independent of z. The synthetic and the actual tool's source-receiver functions need not be the same because they do not depend on z. The only requirement is that they should be approximately in the same frequency range of interest.

The separation of ω- and z-dependences in equation (3.30) is advantageous. Because the ω-dependence term $\Delta A(\omega)$ is common for all depths, it can be estimated reliably using data from all depths. In practice, $\Delta\Phi(\omega, z)$ is a matrix with m columns and n rows, where $\omega = \omega_1, \ldots, \omega_m$; $z = z_1, \ldots, z_n$:

$$\Delta\Phi(\omega,z) = \begin{pmatrix} \Delta\Phi_{11} & \ldots & \Delta\Phi_{1m} \\ \vdots & \ddots & \vdots \\ \Delta\Phi_{n1} & \ldots & \Delta\Phi_{nm} \end{pmatrix} \quad . \tag{3.32}$$

The novel part of the method is to use the statistical averaging method over the rows and columns of the data matrix to yield a robust estimation of attenuation. First, take the average equation (3.30) over ω for each z, as

$$\Delta\tilde{\Phi}(z) = mean_{\omega}\{\Delta\Phi(\omega,z)\} \; , \tag{3.33}$$

where $mean_{\omega}$ denotes the mean value over ω. We can see that $\Delta\tilde{\Phi}(z)$ = $(z) + \Delta\bar{A}$, where the last term $\Delta\bar{A}$ is now a constant that does not depend on either ω or z. Subtracting this equation from equation (3.30) to get a difference and calculating the median of the difference over z, we can estimate the frequency-dependent portion of $\Delta A(\omega)$, as in the following expression:

$$\delta\left(\Delta\tilde{\Phi}(\omega)\right) = \Delta A(\omega) - \Delta\bar{A} = median_{z}\left\{\Delta\Phi(\omega,z) - \Delta\tilde{\Phi}(z)\right\} \; , \tag{3.34}$$

where $median_{z}$ denotes taking the median value over z. It is clear now that if we subtract $\delta(\Delta\tilde{\Phi}(\omega))$ from equation (3.30) and remove its ω-dependence by taking the median over ω, as

$$\hat{\Phi}(z) = median_{\omega}\left\{\Delta\Phi(\omega,z) - \delta\left(\Delta\tilde{\Phi}(\omega)\right)\right\} \; , \tag{3.35}$$

then we obtain an estimate of $Q^{-1}T$ plus an unknown constant:

$$\hat{\Phi}(z) = -Q^{-1}T + \Delta\hat{A} \; . \tag{3.36}$$

Finally, if we know the value of Q at some reference depth z_0, (e.g., $Q(z_0) = 100$), we can calculate the absolute attenuation using

$$Q^{-1} = \frac{\Delta\hat{A} - \hat{\Phi}(z)}{T(z)} = \frac{\hat{\Phi}(z_0) - \hat{\Phi}(z) + Q^{-1}(z_0)\,T(z_0)}{T(z)} \; . \tag{3.37}$$

Otherwise, the attenuation profile is regarded as a relative curve with valid variability but an uncertain baseline shift.

Synthetic modeling is required in the above method (equation (3.30)), which is, however, quite time-consuming. In the practical implementation of the method, the synthetic data can be calculated as tabulated values and stored in the computer disk as a lookup table. Because the ω-dependence of $\Delta\Phi(\omega, z)$ is removed by taking the mean value over ω (see equation (3.33)), we can simplify the data as

$$\Delta\Phi(\omega,z) = \frac{2\ln X(\omega,z)}{\omega} - \bar{\Phi}_{syn}(z) \; , \tag{3.38}$$

where

$$\bar{\Phi}_{syn}(z) = mean_{\omega}\left\{\frac{2\ln X_{syn}(\omega,z)}{\omega}\right\} ,$$

This means that we can average the synthetic data first and store the mean value $\bar{\Phi}_{syn}$ in the lookup table. The table contains the value of $\bar{\Phi}_{syn}$ for various formation

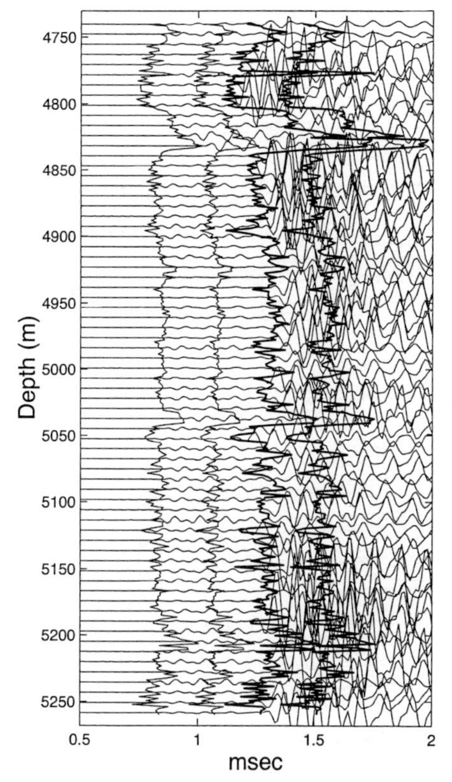

Figure 3.14. Synthetic waveforms for testing the attenuation estimation method. The receiver offset is 3.2 m. A time window of 240 ms is applied to both P- and S-wavetrains. The waveforms were calculated at 0.152 m depth increment and plotted with a decimation of 50.

density-, compressional, and shear velocity values. The value of $\overline{\Phi}_{syn}$ at depth z is given by finding the $\overline{\Phi}_{syn}$ value in the table corresponding to the formation parameter values at z.

3.6.1 EXAMPLES OF ATTENUATION ESTIMATION

Synthetic and field data examples are used to demonstrate the application of the new method. Figure 3.14 shows the synthetic single-receiver full-waveform acoustic data across a depth range of 500 m. The data are computed by inputting the actual P- and S-wave slowness, borehole caliper (diameter), and density logs into a wave modeling program, assuming a wave source center frequency appropriate for the field waveform logging data. The synthetic waveforms are calculated with attenuation by inputting an attenuation profile obtained from the field-measured acoustic logging data. The goal is to test the ability of the method to recover the input attenuation profile. The P- and S-waves of the data are first windowed using a floating window guided by the travel time of the waves. The windowed wave data are then used to

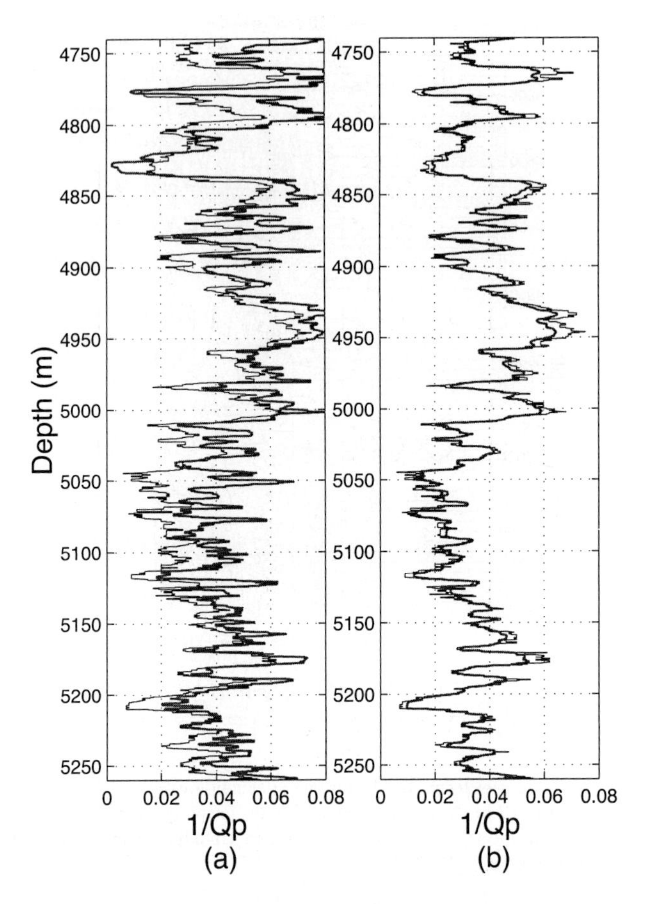

Figure 3.15. Attenuation extracted from synthetic data. Thick line is calculated attenuation; thin line is attenuation input into the model. Left track shows the result without geometric spreading correction. Right track shows the result with the correction.

compute the Fourier spectra for the two waves and used in equations (3.30) through (3.38) to estimate the attenuation profile.

To test the role of synthetic modeling in equations (3.30) and (3.38), the estimation is made with and without synthetic modeling. In the latter case the synthetic terms in equations (3.31) are dropped, while in the former case the synthetic calculations are made using the data input from the actual logs. The results are shown in Figure 3.15, along with the input attenuation (Q_P^{-1}) profile. From the first column, we see that attenuation values estimated without compensation for geometric spreading in Green's function G are biased from input values, which is caused mainly by the variation of the term ΔA in equation (3.30) with depth. With synthetic geometric spreading correction, the agreement between the estimated and input Q_P^{-1} values is significantly improved (right column). This example demonstrates the necessity of correcting for the geometric spreading effect in the attenuation estimation using

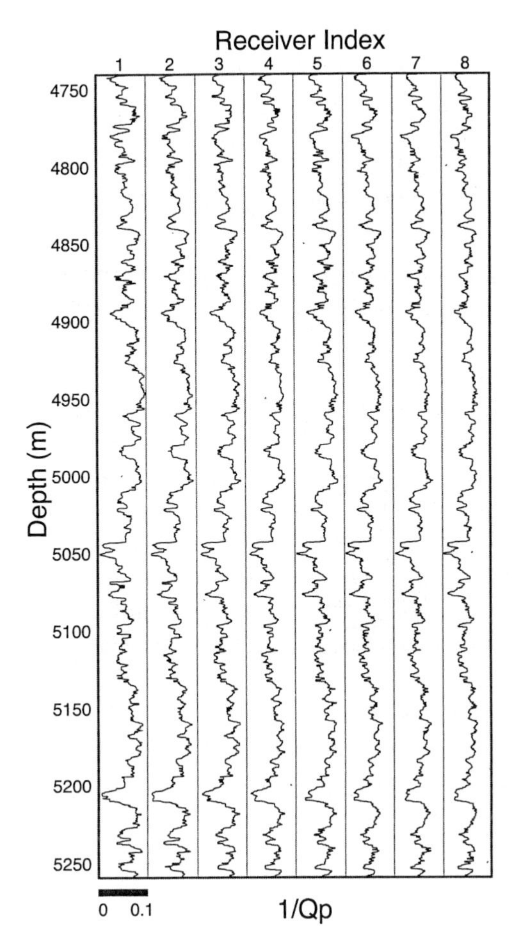

Figure 3.16. Comparison of attenuation profiles for the eight receivers in the array. Each is calculated independently with the same processing procedure.

acoustic logging waveform data.

The consistency and repeatability of the estimation method are checked using all (eight) receivers of the acoustic array. Because the method uses the single receiver data for attenuation estimation, the results from different receivers can be used to provide a measure of the repeatability and reliability of the method. The eight P-wave attenuation (Q_P^{-1}) profiles shown in Figure 3.16 are calculated independently using the respective receiver waveform data. The profiles are in generally good agreement except for some small differences. The correlation between any two of them is better than 0.9. The attenuation log for the first receiver shows more variations and higher quality than the log for the last receiver. This is because attenuation causes the last receiver waveform to have lower amplitude and therefore a lower S/N ratio compared to the first receiver because of their different offsets (4.27 m versus 3.2 m). Another important reason is that the attenuation value calculated using equation (3.30) is an average over the source-receiver offset. As a result, the attenuation profile from different offsets gets different depth resolution. The shorter the offset, the higher the depth resolution of the attenuation profile. Therefore, the attenuation log from the first receiver wave data is usually used as the estimation results.

3.6.2 REMARK ON THE ATTENUATION ESTIMATION METHOD

The above-described method has two key important contributions to attenuation estimation from acoustic logging data. The first is using synthetic modeling to separate the frequency and depth dependencies of acoustic data. The second is the use of statistical averaging methods to estimate the frequency-dependence of the data and remove it from the attenuation estimation. This greatly enhances the robustness and reliability of the attenuation estimates. In the complicated borehole environment, the acoustic logging wave spectrum is affected by local borehole conditions. Because of these local (or random like) variations of the wave (spectral) data, attenuation estimation using a depth-by-depth approach (e.g., spectral ratio method, etc.) is inevitably affected by local variations. In the new method, the functional form of equation (3.30) makes the frequency-dependence of the spectral data a common term for all depths. Consequently, estimation of the common term using statistical averaging removes the local, random effects and yields robust estimation results. A shortcoming of this method is that it estimates the attenuation over the span from transmitter to receiver (this is also why it gives a better result than other methods). The resolution of the attenuation profile is lower (~10 ft or 3 m) than the array aperture (3.5 ft or 1.07 m).

CHAPTER 4: PERMEABILITY ESTIMATION – THEORY, METHODS, AND FIELD EXAMPLES

4.1 Theory for Acoustic Propagation Along a Permeable Porous Borehole

4.1.1 Introduction

In the past few decades, the estimation of rock permeability from borehole acoustic measurements has become a topic of intense research. Several theoretical models have been developed to study the relationship between permeability and acoustic waves. Rosenbaum (1974) was the first to use Biot's (1956a,b; 1962) poroelastic wave theory to model acoustic logging in a porous formation. This model is referred to as the Biot-Rosenbaum theory. Later D. Schmitt generalized this model to radially layered isotropic or transversely isotropic formations, taking into account monopole, dipole, and quadrupole excitations (Schmitt 1988a, b, 1989, 1993). In this chapter we will mainly concentrate on the original Biot-Rosenbaum formulation for monopole excitation in an open borehole with a porous formation since we will be discussing mainly the effect of formation permeability on Stoneley waves.

In Biot's theory, there are three types of waves that can propagate in a fluid-saturated porous solid (Biot, 1956a, b, 1962): the fast compressional wave, the shear wave, and the slow compressional wave. The fast waves are primarily associated with motion of the solid matrix, but modified by the presence of the pore fluid. The slow wave is primarily associated with motion of the pore fluid, but modified by the presence of the solid matrix. The prediction of slow waves in a porous solid is a unique contribution of the Biot theory, even though recent work of Dvorkin and Nur (1992) found that this theory is inadequate to describe the significant fast-wave attenuation and dispersion in real rocks. The interaction of formation permeability with borehole acoustic waves, particularly the Stoneley wave, is through the excitation and propagation of the slow wave in the formation pore fluid. This mechanism is the basis for permeability estimation from borehole Stoneley waves.

During the propagation of a Stoneley wave along a porous borehole, the wave excites all three types of waves in the formation. The Biot-Rosenbaum theory deals with the interaction between the borehole wave and the three formation waves by rigorously solving the Biot's poroelastic wave equation in connection with the boundary conditions at the borehole wall. Although such an approach is complete and accurate within the context of the Biot theory, the mathematics and computation involved make the theory quite complicated, particularly in an inverse problem to extract formation permeability from Stoneley-wave measurements. This approach is quite time consuming. To facilitate the practical application of the Biot-Rosenbaum theory, a simplified model theory has been developed (Tang et al., 1991a). Because the effect of permeability on a borehole Stoneley wave is primarily through the slow-wave mechanism, the slow wave can be treated separately from the fast waves. Specifically, the interaction between the Stoneley wave and the porous formation is decomposed into two parts. The first is the interaction of the Stoneley wave with the formation fast compressional and shear waves in the absence of the slow wave. This problem is equivalent to the one with a formation having effective elastic moduli corresponding to the fast waves, for which the solution is known. On the basis of the first part, the second part is the interaction between the Stoneley wave and the formation slow waves. This second interaction describes the loss of Stoneley-wave energy that is carried away by the slow wave into the formation. Since this latter interaction is largely responsible for the effect of permeability on the Stoneley wave, the simplified theory is expected to be consistent with the full Biot-Rosenbaum theory.

The consistency between the full and simplified theories will not only verify the slow wave mechanism in for Stoneley-wave propagation in porous boreholes, but also provide a much simplified useful model for forward and inverse problems concerning permeability logging using Stoneley waves. The decomposition of the problem into elastic and porous problems is particularly advantageous in the inverse problem where iterative calculations of the forward model are used. Because of the decomposition, the elastic problem needs only to be calculated once; the iterations adjust only the fluid-flow property, which greatly speeds up the inversion process. The complete and simplified theories will be described respectively in the following two sections.

4.1.2 Biot-Rosenbaum Theory

We describe the approach for modeling acoustic wave propagation in porous boreholes using Biot's poroelastic wave theory. Although the notations and equations are modified from those originally used by Rosenbaum (1974), the result obtained here is not fundamentally different from his. We therefore still refer to the model as the Biot-Rosenbaum model. For the Biot theory formulations, we will use a modified version by Auriault et al. (1985), as the notations are more concise than those of the original theory.

Biot (1962) dealt with the wave propagation in a porous solid by connecting the elastic motion of the solid matrix with that of the pore fluid flow (called relative motion between pore fluid and solid, denoted by $\mathbf{w}$) through viscous coupling of the

fluid with solid grains. He made three major assumptions: (1) both the pore and grain sizes are much smaller than a wavelength; (2) the solid grains are all connected; and (3) the pore space is connected. Based on the assumptions, one can derive the following stress-displacement constitutive relationships (e.g., Auriault et al., 1985):
Stress in the solid

$$\sigma_{ij} = 2\mu e_{ij} + (\lambda e - \alpha p)\delta_{ij} \ . \tag{4.1}$$

Stress in the pore fluid

$$p = -(\alpha \nabla \cdot \mathbf{u} + \nabla \cdot \mathbf{w})/\beta \ . \tag{4.2}$$

Dynamic (frequency-dependent) Darcy's law

$$\mathbf{w} = \phi(\mathbf{U} - \mathbf{u}) = \theta\left(\nabla p - \rho_{pf}\omega^2 \mathbf{u}\right) , \tag{4.3}$$

with

$$\begin{array}{lll} e = \nabla \cdot \mathbf{u} = \text{volumetric strain} & \phi = \text{porosity} & K_d = (1-\phi)(\lambda_s + 2\mu_s/3) \ , \\ \mathbf{u} = \text{solid displacement} & \lambda = (1-\phi)\lambda_s & \rho = \rho_s(1-\phi) + \rho_{pf}\phi \ , \\ \mathbf{U} = \text{fluid displacement} & \mu = (1-\phi)\mu_s & \beta = (\alpha - \phi)/K_s + \phi/K_{pf} \ , \\ K_{pf} = \text{pore fluid bulk modulus} & \alpha = 1 - K_d/K_s & \theta = i\kappa(\omega)/\eta\omega \ . \end{array} \tag{4.4}$$

where η is fluid viscosity. The parameters λ_s and μ_s are the Lamé constants of the porous skeleton or matrix; ρ with subscripts s and pf denotes the density of the pore skeleton and pore fluid, respectively. Worth special mention is the dynamic permeability parameter, which, according to Johnson et al. (1987), is given by

$$\kappa(\omega) = \frac{\kappa_0}{\left(1 - 4i\tau^2\kappa_0^2\rho_{pf}\omega / (\eta\Lambda^2\phi^2)\right)^{1/2} - i\tau\kappa_0\rho_{pf}\omega / (\eta\phi)} , \tag{4.5}$$

where τ is tortuosity, which describes the tortuous, winding pore spaces (e.g., $\tau = 1$ for fractures, $\tau = 3$ for round pores, etc.); Λ is a measure of pore size, which can be approximately given below (see Johnson et al., 1987)

$$\Lambda = \left(\frac{8\tau\kappa_0}{\phi}\right)^{1/2} . \tag{4.6}$$

The dynamic permeability is a measure of the fluid transport property of a porous medium when the medium is subject to dynamic wave excitation. The low- and high-frequency behavior of $\kappa(\omega)$ can be readily shown. The low-frequency regime refers to the frequency range below Biot's critical frequency. This frequency is defined by

$$\omega_c = \frac{\eta\phi}{\tau\rho_{pf}\kappa_0} \quad . \tag{4.7}$$

At low frequencies ($\omega \ll \omega_c$), $\kappa(\omega) \to \kappa_0$, where κ_0 is the static Darcy permeability; at high frequencies ($\omega \gg \omega_0$), $\kappa(\omega) \to i\eta\phi/(\tau\rho_{pf}\omega)$, varying inversely proportional to frequency with no dependence on κ_0.

Combining equation (4.1) through (4.3) leads to two coupled partial differential equations for the solid displacement and fluid pressure:

$$(\lambda + \mu)\nabla(\nabla \cdot \mathbf{u}) + \mu\nabla^2\mathbf{u} + \hat{\rho}\omega^2\mathbf{u} - \hat{\alpha}\,\nabla p = 0 \ , \tag{4.8}$$

and

$$\theta\nabla^2 p - \beta\, p - \hat{\alpha}\,\nabla \cdot \mathbf{u} = 0 \ , \tag{4.9}$$

with

$$\hat{\alpha} = \alpha + \rho_{pf}\omega^2\theta; \quad \hat{\rho} = \rho + \rho_{pf}^2\omega^2\theta.$$

4.1.2.1 *Biot's fast and slow wave characteristics*

Equation (4.8) resembles an elastic wave equation (e.g., see equation (2.9)), except for a term related to pore-pressure gradient. The elastic parameters involved in equation (4.8) are those of the fluid-saturated porous solid. This equation describes mostly the fast compressional and shear waves in the porous solid, even though the waves are coupled with the slow wave through the pore-pressure gradient.

Equation (4.9) governs the fluid pressure associated with the slow wave, which is coupled with the porous solid displacement. The slow wave will decouple from the fast waves when the solid is hard or rigid relative to the fluid. In fact, in the limit as $K_s \to \infty$ (and hence $\Delta \cdot \mathbf{u} \to 0$), equation (4.9) becomes

$$\nabla^2 p + \frac{i\omega}{D}\, p = 0 \ . \tag{4.10}$$

where $D = \kappa(\omega)K_{pf}/\phi\eta$, which, in the limit as $\omega \to 0$, is the diffusivity of a viscous fluid. The corresponding slow-wave motion is therefore that of a viscous fluid flow. At high frequencies as $\omega \to \infty$ and $\kappa(\omega) \to i\eta\phi/(\tau\rho_{pf}\omega)$, equation (4.10) describes a propagating wave, with a wavenumber $k_{slow}^2 = i\omega/D = \omega^2/(K_{pf}/\tau\rho_{pf})$ for the slow wave.

For a porous solid of finite rigidity, the effect of solid elasticity should be included. As will be described later, this effect can be approximately modeled by modifying the diffusivity to correct for the medium elasticity, as

$$D = \frac{\kappa(\omega)K_{pf}}{\phi\eta} \bigg/ \left\{ 1 + \frac{K_{pf}}{\phi(\lambda + 2\mu)} \left(1 + \frac{4\alpha\mu/3 - K_d - \phi(\lambda + 2\mu)}{K_s} \right) \right\} . \tag{4.11}$$

Equations (4.10) and (4.11) will be used to derive a simplified model for Stoneley-wave propagation in a porous borehole.

To get a complete description of the fast- and slow-wave propagation characteristics, we need to solve equations (4.8) and (4.9) simultaneously. We first solve for the shear wave. The shear wave is not coupled with the fluid pressure, because equation (4.9) involves only the divergence of the displacement. It is the curl of the displacement vector that gives rise to the shear motion of the porous solid. By taking the curl of equation (4.8), one gets

$$(\nabla \times \mathbf{u}) + \hat{\rho}\omega^2 / \mu (\nabla \times \mathbf{u}) = 0 . \tag{4.12}$$

This is a wave equation for a shear wave with a wavenumber given by

$$k_{\text{shear}}^2 = \omega^2 / (\mu / \hat{\rho}) . \tag{4.13}$$

It can thus be seen that the shear-wave propagation is only weakly affected by the pore-fluid transport property through a complex density $\hat{\rho}$.

We now solve for the fast and slow compressional waves. Taking the divergence of equation (4.8), we get

$$(\lambda + 2\mu)\, \nabla^2 (\nabla \cdot \mathbf{u}) + \hat{\rho}\,\omega^2 (\nabla \cdot \mathbf{u}) - \hat{\alpha}\, \nabla^2 p = 0 . \tag{4.14}$$

To understand the wave propagation characteristics in an unbounded porous medium, we solve this equation and equation (4.9) simultaneously using a plane wave solution. For simplicity, we assume that wave propagation direction is along the x-axis. We also use a displacement potential $a\exp(ikx)$, where a is the amplitude coefficient of the potential. This results in

$$\begin{Bmatrix} \nabla \cdot \mathbf{u} \\ p \end{Bmatrix} = \begin{Bmatrix} -k^2 \\ p_0 \end{Bmatrix} a e^{ikx} , \tag{4.15}$$

where p_0 is a to-be-determined multiplying coefficient related to pressure amplitude. Substitution of equations (4.15) into equations (4.9) and (4.14) leads to two linear equations for coefficients a and ap_0. A non-trivial solution of a and ap_0 requires that the determinant of the 2×2 system vanishes. This leads to the following bi-quadratic equation to determine the wavenumber of the fast and slow waves:

$$\theta(\lambda + 2\mu) k_{fast,\,slow}^4 + (\beta(\lambda + 2\mu) - (\hat{\rho}\omega^2\theta - \hat{\alpha}^2))\, k_{fast,\,slow}^2 - \hat{\rho}\omega^2 \beta = 0 , \tag{4.16}$$

where the subscripts *fast* and *slow* of the wavenumber k, i.e., k_{fast} and k_{slow}, denote the wavenumber (root of the above equation) associated with the fast and slow compressional waves, respectively. The wavenumbers can be calculated using the equations

$$k_{fast} = k_{p0}\sqrt{\frac{1+b_{fast}\rho_{pf}/\rho}{1-b_{fast}/b_0}}, \qquad k_{slow} = k_{p0}\sqrt{\frac{1+b_{slow}\rho_{pf}/\rho}{1-b_{slow}/b_0}}, \tag{4.17}$$

with

$$\left.\begin{matrix} b_{fast} \\ b_{slow} \end{matrix}\right\} = \frac{1}{2} b_0 \left\{ c \mp \sqrt{c^2 - 4\alpha(1-c)/b_0} \right\}, \quad c = \frac{\alpha - b_s \rho/(\rho_{pf} b_0)}{\alpha + b_s},$$

$$b_0 = -\frac{\beta(\lambda + 2\mu + \alpha^2/\beta)}{\alpha}, \quad k_{p0} = \frac{\omega}{\sqrt{(\lambda + 2\mu + \alpha^2/\beta)/\rho}}, \quad b_s = \frac{\hat{\rho} - \rho}{\rho_{pf}} = \rho_{pf}\theta\omega^2 .$$

From the expressions in (4.17), the low-frequency property of the fast and slow waves can be readily derived. In the limit as $\omega \to 0$ (or $b_s \to 0$), the fast compressional-wave velocity is given by $((\lambda + 2\mu + \alpha^2/\beta)/\rho)^{1/2}$, in agreement with Gassmann's results given in equations (1.2) and (1.3). At low frequencies, the slow-wave motion is dominated by viscous diffusion. Thus the slow wavenumber $k^2_{slow} = i\omega/D$ in equation (4.10) should well characterize the wave motion. By comparing this wavenumber with that of equation (4.17) for the low-frequency condition, we can derive the expression for D in equation (4.11). In fact, this approximate slow wavenumber solution is in complete agreement with the exact solution in the low-frequency regime. In the high-frequency regime, there is a discrepancy between the approximate and exact solutions, which tends to increase (decrease) with the softness (rigidity) of the porous solid (Zhao, 1994). This rigidity-dependent accuracy of the approximate solution is easily understood since the approximate solution becomes exact in the limit as $K_s \to \infty$, regardless of the frequency range involved (see the condition leading to equation (4.10)).

Because of the existence of fast and slow compressional waves, the wavefield quantities such as solid- and fluid-displacement components, stress, and pore-fluid pressure, involve contributions from both waves. For example, the one-dimensional (1-D) plane wave $a\exp(ikx)$ in equation (4.15), when the two eigenvalues for k are included, should be written as $a_{fast}\exp(ik_{fast}x) + a_{slow}\exp(ik_{slow}x)$. The corresponding solution, after solving p_0 for both fast and slow waves, becomes

$$\begin{Bmatrix} \nabla\cdot\mathbf{u} \\ p \end{Bmatrix} = k^2_{fast}\begin{Bmatrix} -1 \\ (\alpha + b_{fast})/\beta \end{Bmatrix} a_{fast} e^{ik_{fast}x} + k^2_{slow}\begin{Bmatrix} -1 \\ (\alpha + b_{slow})/\beta \end{Bmatrix} a_{slow} e^{ik_{slow}x}, \tag{4.18}$$

where the coefficients a_{fast} and a_{slow} are determined by the source excitation (unbounded medium) or the boundary condition (bounded medium) of a specific problem. In a general problem of wave propagation in a porous medium, the shear and the fast- and slow-compressional waves are combined to yield the solution to the problem. Acoustic logging in porous formations is one such problem.

4.1.2.2 Application to borehole in a porous formation

We now return to the logging problem to describe the modeling of acoustic wave propagation in a porous borehole. We consider only the monopole case (see Schmitt et al., 1988, and Schmitt 1988b, 1989 for the dipole case). In the monopole case, there are three types of waves in the formation: fast compressional wave, slow compressional wave, and the shear wave (SV type). The displacement potentials for the three formation waves are given by the following expressions:

$$\begin{Bmatrix} \Phi_{fast} \\ \Phi_{slow} \\ \Psi \end{Bmatrix} = \begin{Bmatrix} B_{fast} K_0(q_{fast} r) \\ B_{slow} K_0(q_{slow} r) \\ B_s K_1(q_s r) \end{Bmatrix} \exp(ikz) \; , \tag{4.19}$$

where the symbol B with subscripts *fast*, *slow*, and s represents the amplitude coefficient for the fast compressional wave, slow compressional wave, and shear wave, respectively; the radial wavenumbers are given by

$$q_{fast} = \sqrt{k^2 - k_{fast}^2} \; ; \quad q_{slow} = \sqrt{k^2 - k_{slow}^2} \; ; \quad q_s = \sqrt{k^2 - k_{shear}^2} \; ,$$

where k now denotes the wavenumber along the borehole axial direction; the wavenumber of the respective formation waves is given in equations (4.13) and (4.17). In terms of the potentials, the formation displacement components and stress elements are calculated using

$$\begin{cases} u_r = \dfrac{\partial \Phi_{fast}}{\partial r} + \dfrac{\partial \Phi_{slow}}{\partial r} - \dfrac{\partial \Psi}{\partial z} \; , \\ w_r = b_{fast} \dfrac{\partial \Phi_{fast}}{\partial r} + b_{slow} \dfrac{\partial \Phi_{slow}}{\partial r} - b_s \dfrac{\partial \Psi}{\partial z} \; , \\ \sigma_{rr} = 2\mu \left(\dfrac{\partial^2 \Phi_{fast}}{\partial r^2} + \dfrac{\partial^2 \Phi_{slow}}{\partial r^2} - \dfrac{\partial^2 \Psi}{\partial r \partial z} \right) - \\ \qquad k_{fast}^2 (\lambda + \dfrac{\alpha}{\beta}(\alpha + b_{fast})) \Phi_{fast} - k_{slow}^2 (\lambda + \dfrac{\alpha}{\beta}(\alpha + b_{slow})) \Phi_{slow} \; , \\ p = k_{fast}^2 ((\alpha + b_{fast}) / \beta) \Phi_{fast} + k_{slow}^2 ((\alpha + b_{slow}) / \beta) \Phi_{slow} \; , \\ \sigma_{rz} = -k_{shear}^2 \mu \Psi + 2\mu \left(\dfrac{\partial^2 \Phi_{fast}}{\partial r \partial z} + \dfrac{\partial^2 \Phi_{slow}}{\partial r \partial z} - \dfrac{\partial^2 \Psi}{\partial z^2} \right) . \end{cases} \tag{4.20}$$

We now specify the boundary conditions at the borehole wall ($r = R$). We still use the notations in Chapter 2 to describe the borehole fluid stress and displacement field. There are four boundary conditions. The first three are continuity of averaged normal stress, continuity of tangential stress, and continuity of averaged normal displacement, as respectively given by

$$\begin{cases} \sigma_{rrf} = \sigma_{rr} \\ 0 = \sigma_{rz} \\ u_{rf} = u_r + w_r \end{cases} , \quad (\text{at } r = R) \; . \tag{4.21}$$

The last boundary condition concerns pressure communication, or hydraulic exchange, between borehole and formation fluid. It has long been thought that the presence of borehole mud cakes may affect the pressure communication. Rosenbaum (1974) used a flow impedance to model the effect of the mud cake. In our notation, we use T to denote the impedance factor. The impedance effect due to the mud cake reduces the flow between the borehole and formation. The pressure communication condition is written as

$$p_f - p = T\phi(U_r - u_r) = T\, w_r \ , \tag{4.22}$$

where p_f is the borehole fluid pressure. However, the value of T for a mud cake was unknown. Rosenbaum therefore assumed two extreme cases. The first case is $T = 0$. For this case, the borehole pressure equals the pore fluid pressure and there is a free hydraulic exchange between borehole and formation. This case is referred to as the open-pore case. The second case corresponds to $T \to \infty$. In this case there is no hydraulic communication between borehole and formation. This is referred to as the "sealed-pore case". A finite value of T should correspond to a realistic mud cake that allows partial communication with a pressure discontinuity at the borehole wall. A mud cake model will be discussed later in this chapter. In this section, we will discuss the open-pore case with $T = 0$.

Substitution of the stresses and displacements of equations (4.20) into the boundary conditions given in equations (4.21) and (4.22) (note $T = 0$ for the present situation) results in

$$\begin{pmatrix} N_{11} & N_{12} & N_{13} & N_{14} \\ N_{21} & N_{22} & N_{23} & N_{24} \\ N_{31} & N_{32} & N_{33} & N_{34} \\ N_{41} & N_{42} & N_{43} & N_{44} \end{pmatrix} \begin{pmatrix} A_0' \\ B_{fast} \\ B_{slow} \\ B_s \end{pmatrix} = \begin{pmatrix} u_f^d \\ \sigma_{rrf}^d \\ 0 \\ 0 \end{pmatrix} , \tag{4.23}$$

where A_0' is the reflected wave amplitude coefficient in the borehole fluid; the subscript 0 means monopole. The borehole-fluid displacement and stress at the right-hand side are given in equation (2.17), which belong to the direct wavefield radiated from a point source at the center of the borehole. After A_0' is found from the above equation, the procedure for calculating the synthetic micro-seismogram is the same as in the elastic case.

The compressional and shear waves along a borehole, as modeled using the Biot-Rosenbaum theory, have quite similar wave propagation characteristics compared to their counterparts in an unbounded porous medium (see Schmitt et al., 1988, for modeling examples). In other words, we can directly measure the fast compressional and the shear waves from acoustic logging. The slow compressional wave, however, cannot be directly measured in the borehole because it is not refracted along the borehole since its velocity is always below that of the borehole fluid. Nevertheless, the slow wave effect is manifested through the pressure communication between borehole and formation pore fluid. The borehole Stoneley wave, being a fluid-borne low-frequency pressure wave, is most sensitive to this pressure communication at the borehole interface, and therefore, is used to relate the borehole acoustic measurement to formation fluid-transport properties.

The calculation of wave-mode dispersion involves finding the root of the period equation, which is obtained by setting the determinant of the above matrix to zero, i.e.,

$$D(k,\omega) = \det \mathbf{N} = 0 \ . \tag{4.24}$$

The elements of the matrix $\mathbf{N}$ are:

$$\begin{aligned}
N_{11} &= -fI_1(fR) \\
N_{12} &= -q_{fast}(1+b_{fast})K_1(q_{fast}R) \\
N_{13} &= -q_{slow}(1+b_{slow})K_1(q_{slow}R) \\
N_{14} &= -ik(1+b_s)K_1(q_sR) \\
\\
N_{21} &= \rho_f\omega^2 I_0(fR) \\
N_{22} &= -2\mu\left(q_{fast}^2 K_0(q_{fast}R) + q_{fast}K_1(q_{fast}R)/R\right) + k_{fast}^2(\lambda + \frac{\alpha}{\beta}(\alpha+b_{fast}))K_0(q_{fast}R) \\
N_{23} &= -2\mu\left(q_{slow}^2 K_0(q_{slow}R) + q_{slow}K_1(q_{slow}R)/R\right) + k_{slow}^2(\lambda + \frac{\alpha}{\beta}(\alpha+b_{slow}))K_0(q_{slow}R) \\
N_{24} &= 2i\mu k\left(q_s^2 K_0(q_sR) + q_sK_1(q_sR)/R\right)
\end{aligned} \tag{4.25}$$

$$\begin{aligned}
N_{31} &= 0 \\
N_{32} &= -2ik\mu\, q_{fast}K_1(q_{fast}R) \\
N_{33} &= -2ik\mu\, q_{slow}K_1(q_{slow}R) \\
N_{34} &= \mu(2k^2 - k_{shear}^2)q_sK_1(q_sR) \\
\\
N_{41} &= -\rho_f\omega^2 I_0(fR) \\
N_{42} &= k_{fast}^2((\alpha+b_{fast})/\beta)K_0(q_{fast}R) \\
N_{43} &= k_{slow}^2((\alpha+b_{slow})/\beta)K_0(q_{slow}R) \\
N_{44} &= 0
\end{aligned}$$

Solving the dispersion equation to find the wavenumber k for the Stoneley-wave mode, we can calculate the Stoneley-wave phase velocity and attenuation using

$$\begin{cases} V_{ST} = \omega / \mathrm{Re}(k)\ , \\ Q^{-1} = 2\,\mathrm{Im}(k)/\mathrm{Re}(k)\ . \end{cases} \tag{4.26}$$

As can be seen from the above matrix elements, the elastic property of the porous rock is intimately coupled with the flow parameters (i.e., permeability and pore-fluid parameters) in the Biot-Rosenbaum model. This makes the calculation of the model result somewhat involved. Particularly, in an inverse problem to estimate the flow parameters, in each iteration step of the inversion process one has to calculate both the elastic and the flow effects, which is quite time-consuming and inefficient. The modeling efficiency can be greatly enhanced by decoupling the flow and elastic effects in the Biot-Rosenbaum model, which will be described in the following section.

4.1.3 SIMPLIFIED BIOT-ROSENBAUM THEORY

We now describe a simple theory to account for the effects of permeability on the borehole Stoneley-wave propagation. The Stoneley wave is an axi-symmetric mode with no θ-dependence, whose associated borehole-fluid displacement potential satisfies the following wave equation:

$$\nabla_t^2 \Phi_f + \frac{\partial^2 \Phi_f}{\partial z^2} + k_f^2 \Phi_f = 0 , \tag{4.27}$$

where the 2-D Laplace operator is

$$\nabla_t^2 = \frac{\partial^2}{\partial r^2} + \frac{1}{r}\frac{\partial}{\partial r} .$$

In terms of the potential, the borehole fluid pressure and displacement are given by

$$\begin{cases} p_f = \rho_f \omega^2 \Phi_f , \\ u_f = \dfrac{\partial \Phi_f}{\partial r} . \end{cases} \tag{4.28}$$

For Stoneley-wave propagation along the borehole, we can separate the axial and radial dependences, as

$$\begin{cases} \Phi_f(r,z) = \varphi(r)\exp(ikz) \\ p_f(r,z) = p'(r)\exp(ikz) , \quad (r \le R) . \\ u_f(r,z) = u'(r)\exp(ikz) \end{cases} \tag{4.29}$$

Using equations (4.28) and (4.29), we can relate the (radial dependence of) fluid displacement potential φ and its radial derivative through the equation

$$\frac{\partial \varphi}{\partial r} - (\rho_f \omega^2 \frac{u'}{p'})\varphi = 0 , \tag{4.30}$$

which, when evaluated at the borehole boundary $r = R$, becomes the boundary condition for φ. Consequently, φ is given by the following boundary-value problem

$$\begin{cases} \nabla_t^2 \varphi + \nu^2 \varphi = 0 , \quad (\text{with } \nu^2 = k_f^2 - k^2) , \\ \partial_r \varphi = \rho_f \omega^2 (u' / p')\varphi , \ (\text{at } r = R) . \end{cases} \tag{4.31}$$

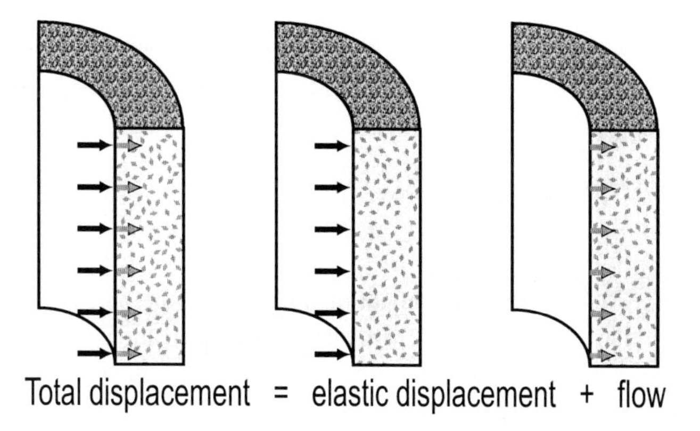

Figure 4.1. Under the excitation of a borehole Stoneley wave, the porous formation displacement contains two parts: elastic deformation and fluid flow. Consequently, the problem can be decomposed into two: one is the interaction with an equivalent elastic formation, the other is the flow exchange at the borehole interface.

The borehole wall conductance (u'/p') is the ratio of the displacement and pressure at the borehole boundary $r = R$. When the borehole wall is permeable, u' includes two contributions. The first is the elastic displacement of the wall, given by u_e'. The second is the fluid flow into pores that are open to the borehole wall, given by $\phi u_f'$, where ϕ is the porosity of the formation.

As illustrated in Figure 4.1, the present problem can be decomposed into two problems. The first problem is equivalent to that of a borehole with an equivalent elastic formation consisting of the porous skeleton and fluid. Only P and S waves in such a formation need to be considered. They are analogous to Biot's fast compressional and shear waves. For the second problem, one is mainly concerned with pore-fluid flow, which occurs in Biot's slow compressional wave. Splitting the problem of equation (4.31) into two, we have

$$\begin{cases} \varphi = \varphi_e + \varphi_f \\ \mathrm{v}^2 = \mathrm{v}_e^2 + \mathrm{v}_f^2 \end{cases} , \tag{4.32}$$

where φ_f and v_f are the perturbations to φ_e and v_e, respectively. The perturbations are the result of fluid flow at the borehole wall. The elastic potential φ_e satisfies the following boundary value problem

$$\begin{cases} \nabla_t^2 \varphi_e + v_e^2 \varphi_e = 0 , & (\text{with } v_e^2 = k_f^2 - k_e^2) \\ \partial_r \varphi_e = \rho_f \omega^2 (u_e' / p') \varphi_e , & (\text{at } r = R) \end{cases} . \tag{4.33}$$

The solution to the above elastic problem is known. The associated boundary condition leads to the borehole monopole-wave dispersion equation (equation (2.24), $n = 0$ case). Given the effective elastic moduli (e.g., equation (4.4)) (or equivalently, P- and S-wave velocities) and the density of the equivalent elastic formation, we can find the elastic Stoneley wavenumber k_e, as well as its radial counterpart v_e. To solve for v_f, we substitute equations (4.32) and (4.33) into (4.31) and obtain a boundary-value problem for φ_f:

$$\begin{cases}\nabla_t^2\varphi_f + v_e^2\varphi_f = -v_f^2\varphi \\ \partial_r\varphi_f = \rho_f\omega^2(u_e'/p')\varphi_f + \rho_f\omega^2(u_f'/p')\varphi, \text{ (at } r = R) \text{ .}\end{cases} \tag{4.34}$$

To this end, we can see that the effect of fluid flow at the borehole wall is to modify the boundary condition of the equivalent elastic formation, so as to change the fluid displacement potential from φ_e to φ and the wavenumber from k_e to k. Therefore, a boundary condition perturbation technique (Jackson, 1962; Morse and Feshbach, 1953) can be applied to incorporate the fluid-flow effects and to find the resulting changes in wave propagation characteristics.

The condition at the borehole boundary and its effect over the borehole area are related through the two-dimensional Green's theorem:

$$\iint_A \left(\varphi_e\nabla_t^2\varphi_f - \varphi_f\nabla_t^2\varphi_e\right) dA = \oint_S \left(\varphi_e\frac{\partial\varphi_f}{\partial r} - \varphi_f\frac{\partial\varphi_e}{\partial r}\right) dS \text{ ,} \tag{4.35}$$

where A is the borehole area and S is the borehole boundary at $r = R$.

Applying equation (4.35) to equations (4.33) and (4.34) and using their respective boundary conditions, we get

$$v_f^2 = -\rho_f\omega^2\phi\left(\frac{u_f'}{p'}\right)\frac{\oint_S \varphi\varphi_e dS}{\iint_A \varphi\varphi_e dA} \text{ .} \tag{4.36}$$

To evaluate v_f^2, we need to find the flow conductance u_f'/p' and the ratio of integrals in the above equation. As a first order perturbation, this ratio is approximately the ratio of bore perimeter to bore area (or, in the presence of a logging tool of radius a, the fluid annulus area):

$$\frac{\oint_S \varphi\varphi_e dS}{\iint_A \varphi\varphi_e dA} \sim \begin{cases}\dfrac{2}{R}, \text{ (without tool)} \\ \dfrac{2R}{R^2 - a^2}, \text{ (with tool) .}\end{cases} \tag{4.37}$$

To find u_f'/p', we make use of equations (4.10) and (4.3), as given previously. The first equation governs the dynamic pressure in the porous formation through the use of dynamic permeability in fluid diffusivity D (equation (4.11)). The second equation is a modification of Darcy's law, which states that the relative motion between pore fluid and solid is driven by the net effect of pore-fluid pressure gradient minus the acceleration of the solid. In the present problem, we consider the low-frequency regime of Biot's theory, since the effective frequency range of the Stoneley-wave propagation is usually below 5 kHz. At low frequencies, the pore-fluid flow is dominated by the slow-wave motion. Thus we can drop the second term in equation (4.3) and only consider the pore-fluid displacement. Furthermore, we use dynamic permeability instead of static permeability in this equation.

Under the excitation of a borehole propagation exp(ikz), the formation pore-fluid pressure has the form

$$p(r,z) = p(r)e^{ikz}, \; (r \geq R) \; . \tag{4.38}$$

Substitution of equation (4.38) into equation (4.10) results in a Bessel function for p,

$$\frac{d^2p}{dr^2} + \frac{1}{r}\frac{dp}{dr} + \left(\frac{i\omega}{D} - k^2\right)p = 0, \quad (r \geq R) \; , \tag{4.39}$$

for which the solution is

$$p(r) = p(R)\frac{K_0\left(r\sqrt{-i\omega / D + k^2}\right)}{K_0\left(R\sqrt{-i\omega / D + k^2}\right)} \; , \tag{4.40}$$

where $p(R)$ is the borehole pressure at the borehole wall. By differentiating the above equation with respect to r and using the modified Darcy's law given by equation (4.3) (note $fu'_f \sim w_r \sim \theta\,\partial_r p$; the term $\rho_f\omega^2 u_r \sim 0$ at low frequencies), the wall conductance is found to be

$$\frac{u'_f}{p'} = \frac{i\kappa(\omega)}{\omega\eta\phi}\sqrt{-i\omega / D + k^2}\;\frac{K_1\left(r\sqrt{-i\omega / D + k^2}\right)}{K_0\left(R\sqrt{-i\omega / D + k^2}\right)} . \tag{4.41}$$

A good approximation is replacing k in the above equation with k_e, since the difference between the two wavenumbers is an effect of second order. Using equations (4.36) and (4.37) and the following relation

$$\nu_f^2 = \nu^2 - \nu_e^2 = k_e^2 - k^2 ,$$

we get an expression for the Stoneley wavenumber for a porous formation

$$k = \sqrt{k_e^2 + \frac{2i\rho_{pf}\,\omega\,\kappa(\omega)R}{\eta(R^2 - a^2)}\sqrt{-i\omega / D + k_e^2}\;\frac{K_1\left(R\sqrt{-i\omega / D + k_e^2}\right)}{K_0\left(R\sqrt{-i\omega / D + k_e^2}\right)}} \; . \tag{4.42}$$

For a fluid-filled borehole without a tool, the tool radius a is set to zero (a=0). To include the tool effect in the Stoneley-wave problem, we can use an effective modulus model (see Chapter 5) to account for the tool's compliance, or a rigid tool model (Tang and Cheng, 1993b) if the tool is rigid compared to the borehole fluid. For the compliant or rigid tool, the above equation can be used to calculate the Stoneley wavenumber k by: 1) using a non-zero value for the tool radius a, and 2) calculating the elastic wavenumber k_e corresponding to the compliant (see Chapter 5) or rigid tool (e.g., Tang and Cheng, 1993b).

From the wavenumber k, the Stoneley-wave phase velocity and attenuation are calculated using equation (4.26). The results presented in equations (4.42) include both the effects of the formation elasticity and permeability on Stoneley waves. This is achieved by using the theory of dynamic permeability and the Stoneley wavenumber corresponding to an equivalent elastic formation. The dynamic permeability accounts for the frequency-dependent pore-fluid flow at the borehole wall, while the elastic wavenumber k_e accounts for the formation elasticity. The wavenumber k_e also allows us to model Stoneley-wave propagation in the presence of intrinsic attenuation due to anelasticity of the formation and borehole fluid (see equation (2.40) and Tang and Cheng, 1993b).

The advantage of the simplified theory, as compared to the full Biot-Rosenbaum theory, is that it reduces the complexity of root finding in the latter theory (equations (4.24 and 4.25)) to an analytical formula (equation (4.42)), greatly enhancing the speed of model calculation. In an inverse problem to extract formation permeability from the Stoneley-wave measurement, the decoupling of the flow effect from the elastic effect, as in the simplified model, is particularly advantageous. The elastic wavenumber k_e is calculated from measured elastic wave properties only once. The iterations in the inversion process only require the calculation of the flow effect, which is done efficiently using the analytical formula in equation (4.42). In contrast, in the Biot-Rosenbaum theory the flow and elastic effects are intimately coupled (see equations (4.24) and (4.25)). Each iteration of the inversion requires a search for the root of the dispersion equation for calculating the Stoneley wavenumber using both elastic and flow parameters.

Table 4.1. Porous formation and pore-fluid parameters used to calculate Figures 4.2 and 4.3. The parameters are defined in equations (4.1)–(4.6).

Formation	λ	μ	K_s	ρ_s	ϕ	τ	κ_0	K_{pf}	ρ_{pf}	η
	GPa	GPa	GPa	g cm^{-3}		Pa s	Darcy	GPa	g cm^{-3}	Pa s
fast	12.1	11.84	37.9	2.65	0.25	3	10	2.25	1	0.001
slow	7.54	3.22	37.9	2.65	0.25	3	.1, 1, 1.15	2.25	1	0.001

4.1.4 NUMERICAL COMPARISON BETWEEN THE EXACT AND SIMPLIFIED THEORIES AND A CORRECTION FOR FORMATION SOFTNESS

We now compare the results from the exact and simplified model theories. Figure 4.2 shows the Stoneley-wave velocity dispersion (a) and attenuation (b) curves for the two theories. To illustrate the effect of dynamic permeability, as given in equation

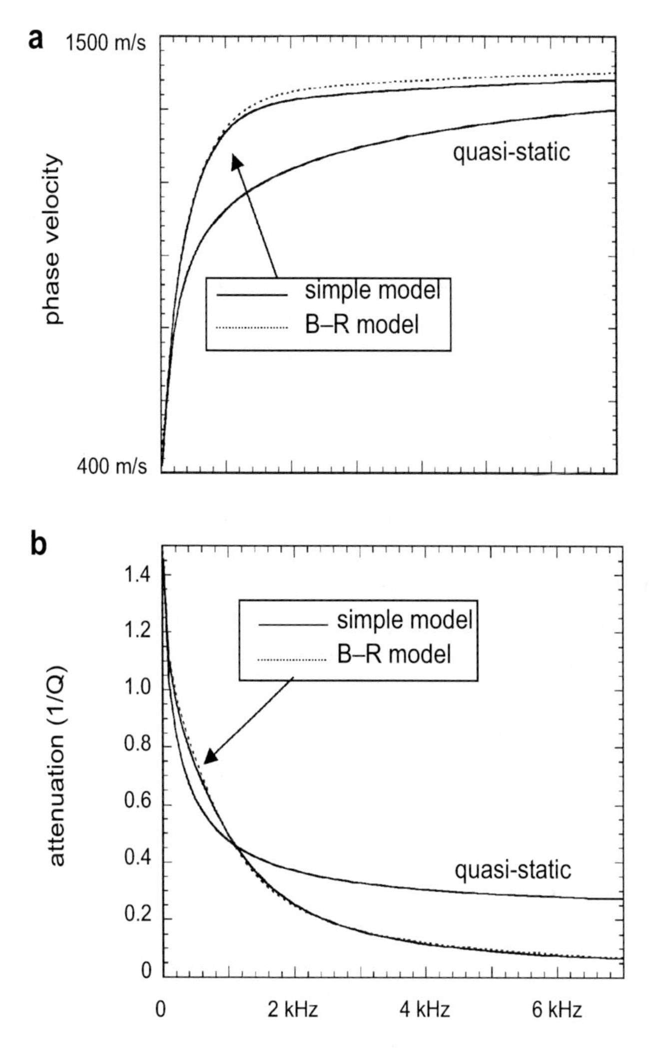

Figure 4.2. Comparison of the results of the *simplified* Biot-Rosenbaum model (solid curve) with the exact theory (dashed curve) and a test of the dynamic permeability effect for a highly permeable (fast) formation. Using dynamic permeability in the simple model, the calculated Stoneley phase velocity (a) and attenuation (b) agree with the exact theory quite well. The results using the static permeability (curves marked 'quasi-static') agree with the exact theory only in the low-frequency limit.

(4.5), we use a very high permeability value ($\kappa_0 = 10$ d, where d represents Darcy). The formation is a fast porous formation whose parameters are given in Table 4.1 (the borehole has a radius of 0.1 m and is filled with water). For comparison, a curve calculated by replacing $\kappa(\omega)$ with κ_0 is also shown (curves marked 'quasi-static'). The result from the simple theory fits that of the exact theory very well. The result obtained by using κ_0 agrees with the exact theory only at the low-frequency limit. It largely over-predicts the attenuation at higher frequencies. This difference can

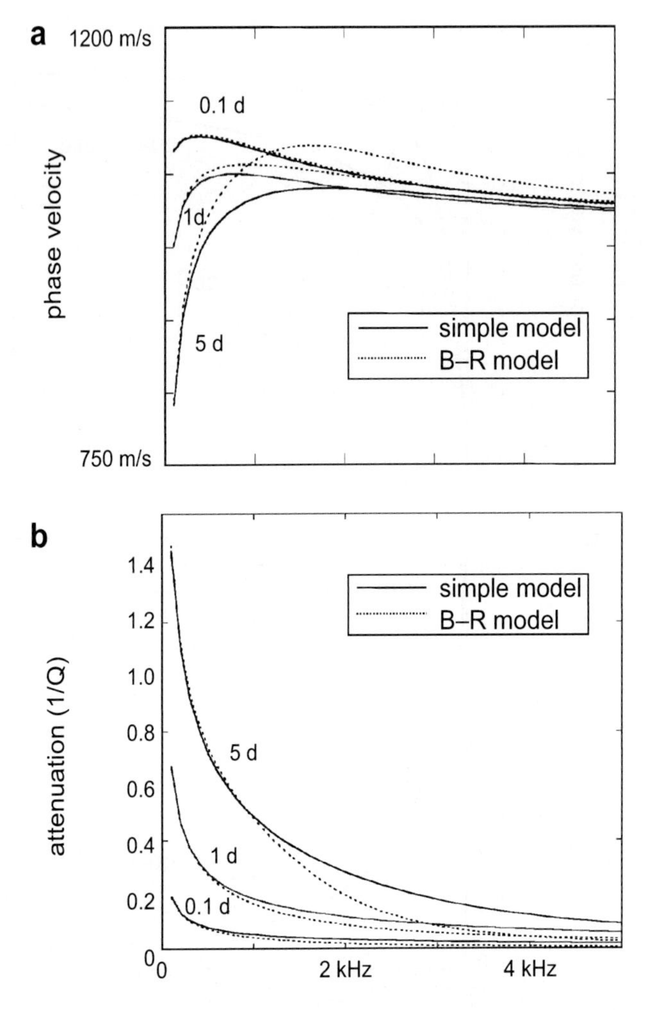

Figure 4.3. Comparison of the result of the simplified model (solid curves) with the result from the exact theory (dashed curves) for a soft formation case with three permeability values (0.1, 1, and 5 darcies, respectively). The simplified model underestimates the velocity (a) and overestimates the attenuation (b) at higher frequencies, because of the increased formation softness.

be explained by the frequency dependence of $\kappa(\omega)$. Since dynamic permeability decreases with frequency, the formation is less permeable under high-frequency excitations than it is under low-frequency excitations. For the Stoneley-wave dispersion curves (Figure 4.2a), the simplified dynamic model agrees with the exact model fairly well, with the simplified-model result being slightly lower than that of the exact theory. The velocity obtained using the static permeability κ_0 is significantly lower than that of the exact theory. This difference can be explained as follows: For the high permeability value used, the dynamic (or inertial) effect of pore fluid is increased, which tends to decouple the borehole wave propagation from the

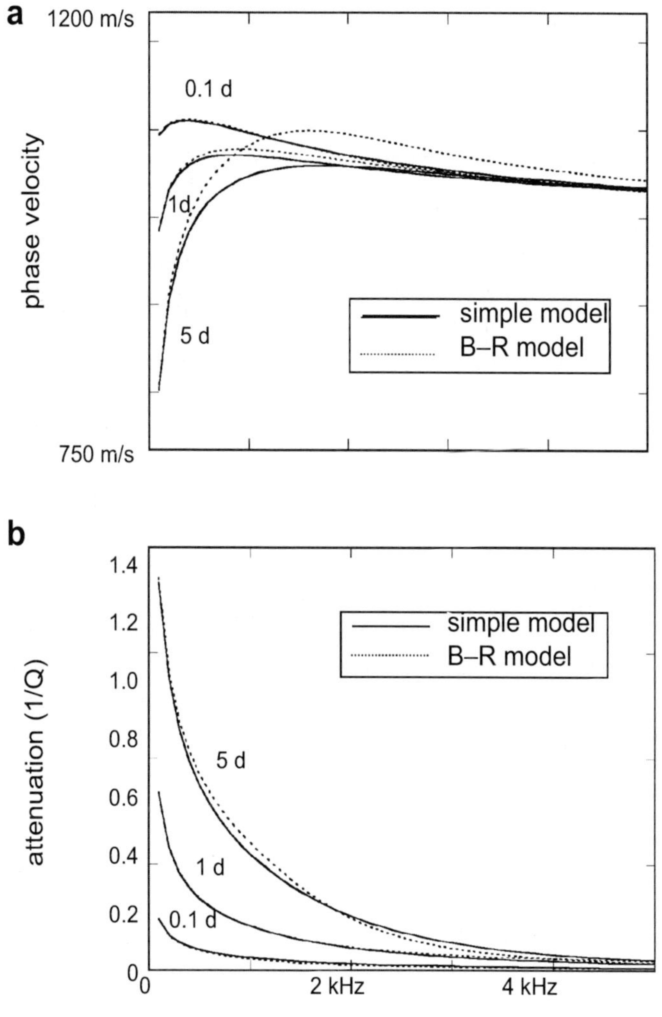

Figure 4.4. Correction for formation softness. After the correction, the attenuation curves from the simple model agree with those of the exact theory (b). The agreement for the phase velocity curves is also improved (a).

formation, resulting in the Stoneley-wave velocity approaching the borehole fluid velocity. This effect is accounted for by the dynamic permeability $\kappa(\omega)$ in both the exact and simplified theories. Using κ_0 in the simple theory assumes that the pore-fluid flow is still governed by viscous forces and therefore maintains significant borehole-formation coupling, resulting in a Stoneley-wave velocity that is lower than that of the dynamic models.

The simplified model over-predicts the permeability effects in soft formations, especially for the Stoneley-wave attenuation. Figure 4.3 shows the comparison between the simplified and exact theories for a soft or slow porous formation with three different flow properties (formation parameters are given in Table 4.1; borehole and fluid are the same as in Figure 4.2; permeability κ_0 is 0.1 d, 1 d, and 5 d, respectively; other parameters are kept the same for the three cases). As shown in Figure 4.3b, although the attenuation values predicted by both theories are identical at low frequencies and all decrease with frequency, the attenuation from the simplified model is significantly over-predicted at higher frequencies, as compared with the exact model. This discrepancy is caused by the increase of formation elastic compressibility. When the solid compressibility is closer to that of the pore fluid and the dynamic coupling between the two phases becomes strong, the coupling of the borehole propagation is affected more by formation elasticity than by permeability. It is therefore expected that the simplified model that accounts mainly for the permeability effect may become inadequate for soft formations.

The effect of formation softness can be corrected using a simple empirical approach by considering the borehole compliance. The borehole compliance is defined (Tang et al., 1991b) by a dimensionless quantity,

$$BC = f_e R \frac{I_1(f_e R)}{I_0(f_e R)} , \tag{4.43}$$

where $f_e = (k_e^2 - k_f^2)^{1/2}$ is the radial Stoneley wavenumber for the equivalent elastic formation. By numerical testing, we found a simple correction for the formation softness by dividing, in equation (4.42), the second term under the square root sign by a factor of $1 + BC^\gamma$, where $\gamma = V_S / V_f$ is the formation shear-to-borehole fluid velocity ratio. Equation (4.42) then becomes

$$k = \sqrt{k_e^2 + \frac{2i\rho_{pf}\omega\kappa(\omega)R}{\eta(R^2 - a^2)} \frac{\sqrt{-i\omega / D + k_e^2}}{1 + BC^{(V_s/\alpha_f)}} \frac{K_1\left(R\sqrt{-i\omega / D + k_e^2}\right)}{K_0\left(R\sqrt{-i\omega / D + k_e^2}\right)}} . \tag{4.44}$$

By this simple modification, the agreement between the simplified and the exact theories is improved significantly. Figure 4.4 shows the result. The calculation parameters are the same as those used in Figure 4.3. As shown for the Stoneley-wave velocity dispersion (a) and attenuation (b), for three permeability values ranging from medium (0.1 d), high (1 d), and very high (5 d), the simplified model results follow the complete theory quite well. The discrepancy between the attenuation curves from the two theories, as shown in Figure 4.3b, is largely reduced; the agreement in the velocity curves is also significantly improved. This indicates that the simple model has about the same sensitivity to formation permeability as the complete theory.

With the simple correction for borehole wall compliance, the simplified theory can be used to model Stoneley-wave propagation in permeable boreholes with both fast and slow formations. In particular, this simple and sufficiently accurate theory (compared to the full Biot theory) can be used to formulate inversion procedures to efficiently extract formation permeability from Stoneley-wave measurements.

4.2 Permeability Estimation from Borehole Stoneley Waves

Formation permeability is a key parameter for reservoir production and management, and estimation of permeability is an important topic in acoustic logging. Although the effect of formation permeability on borehole acoustic propagation has been studied for many decades, the validity of permeability measurement from borehole acoustic waves (i.e., Stoneley waves) was recognized only recently. Williams et al. (1984) first demonstrated the correlation of formation permeability with Stoneley-wave velocity and amplitude. Cheng et al. (1987) tried to fit the Williams et al. (1984) data with Biot-Rosenbaum theory, with limited success. The main obstacle was that the borehole Stoneley wave is affected by other effects that are unrelated to permeability. These effects, to mention a few, include formation elasticity, intrinsic attenuation, borehole size changes, mud cakes, and anisotropy. The challenge in estimating permeability from Stoneley waves is therefore the separation of the permeability effect from the non-permeability effects listed above. Methods that have been used include the wave phase or slowness method (Hornby, 1989), the amplitude attenuation method (Tang and Cheng, 1996), and most recently, a method based on both wave phase delay and amplitude attenuation measurements, as will be described below.

4.2.1 PERMEABILITY BASED ON STONELEY-WAVE SLOWNESS

During the propagation of a Stoneley wave in a permeable porous borehole, the Stoneley-wave slowness or velocity is affected by formation permeability. Formation permeability can therefore be estimated from the Stoneley-wave slowness measurement. Hornby (1989) noticed this effect from field data. He also found that the Stoneley-wave slowness is significantly affected by formation elasticity and borehole fluid properties. This effect is predicted by the theory described previously. To demonstrate borehole fluid and formation effects, we use the low-frequency approximation of the Stoneley-wave slowness S, obtained from equation (4.42) in the limit of $\omega \rightarrow 0$ (assuming no tool, $a = 0$):

$$S^2 = \left(\frac{\rho_f}{K_f} + \frac{\rho_f}{\mu} \right)_{elastic} + \left(\frac{2\rho_f i \kappa_0}{\eta \omega R} \sqrt{-i\omega / D} \frac{K_1\left(R\sqrt{-i\omega / D}\right)}{K_0\left(R\sqrt{-i\omega / D}\right)} \right)_{flow} . \tag{4.45}$$

It is clear that the Stoneley-wave slowness is controlled by borehole fluid modulus, formation shear modulus, and formation flow properties (permeability and porosity, viscosity, etc., which are also contained in D). To estimate formation permeability, an *elastic* Stoneley-wave slowness is first calculated using the measured formation shear slowness and an estimated borehole fluid slowness. Then the difference between the measured slowness and the calculated slowness is determined, and is used to estimate formation permeability.

The shortcoming of this approach is that the second term is generally small compared to the first term. Uncertainties in the borehole fluid and formation shear moduli, and other unknown effects, may be comparable or exceed the permeability effect. It is difficult to distinguish these non-permeability effects from the permeability effect using the slowness measurement alone.

4.2.2 PERMEABILITY BASED ON STONELEY-WAVE AMPLITUDE MEASUREMENT

Tang and Cheng (1996) described a method based on Stoneley-wave amplitude measurements. The Stoneley-wave amplitude spectrum in a permeable borehole is written as

$$A(\omega,z) = SR(\omega)E(\omega,z)\exp\left(-\frac{\omega d}{2QV_{ST}}\right), \tag{4.46}$$

where $SR(\omega)$ denotes the source and receiver response spectrum; d is the source-to-receiver distance; z is the depth location of the formation; and Q^{-1} and V_{ST} are, respectively, the Stoneley-wave attenuation and velocity, which can be theoretically calculated using equations (4.44) and (4.26). The Stoneley-wave excitation function (or borehole pressure response) $E(\omega, z)$ for the permeable borehole can be calculated from the excitation $E_e(\omega, z)$ of the equivalent elastic formation, where the latter is the excitation function of equation (2.26) evaluated for the formation elastic properties at depth z. The relation between $E(\omega, z)$ and $E_e(\omega, z)$ is given by

$$E(\omega,z) = \left(\frac{k_e}{k}\right)E_e(\omega,z) \ . \tag{4.47}$$

The above result is derived by considering the expression for Stoneley-wave pressure generated by an acoustic source (White, 1983). The pressure is

$$p = \rho_f V_{ST} v(t - z/V_{ST}),$$

where $v(t - z/V_{ST})$ is the axial borehole-fluid particle velocity. If we assume that the source at $z = 0$ generates the same particle velocity $v(t)$ in a borehole for both permeable and non-permeable (equivalent elastic formation) situations, then the respective pressure or excitation will be proportional to their respective Stoneley-wave phase velocity, leading to the above relation.

Further, by choosing a non-permeable reference depth z_0, (with $Q^{-1} = 0$), formation permeability at depth z can be estimated from minimizing the following objective function

$$\mathrm{Obj}[\kappa_0(z)] = \int_{\omega}\left\{A(\omega,z_0)\exp\left(-\frac{\omega d}{2Q(\kappa_0)V_{ST}(\kappa_0)}\right) - \left|\frac{k(z)E_e(\omega,z_0)}{k_e(z)E_e(\omega,z)}\right|A(\omega,z)\right\}^2 d\omega \ . \tag{4.48}$$

In this inversion procedure, the attenuation and velocity values are theoretically calculated as a function of the parameter κ_0; the value of κ_0 that minimizes the objective function is taken as the formation permeability at depth z. As shown in Tang and Cheng (1996), this method works well for highly permeable formations (permeability on the order of darcies). The change of Stoneley-wave amplitude due to changes in formation elastic property, especially in the shear modulus, is modeled well by the elastic excitation function $E_e(\omega, z)$. The shortcoming of this approach is that, for low to medium permeability values, the permeability-related amplitude change may be obscured by changes caused by other effects (e.g., intrinsic attenuation, scattering from borehole/formation boundaries, etc.). As in the slowness method, it is difficult to distinguish these non-permeability effects from the permeability effect using the amplitude measurement alone. It is therefore anticipated that a method using both phase and amplitude of the Stoneley wave will better indicate the permeability effect and allow a more reliable permeability estimation. Such a method is described in detail in the following sections.

4.2.3 PERMEABILITY ESTIMATION USING BOTH AMPLITUDE AND PHASE

The permeability estimation method consists of three major steps: wave separation, wave modeling, and permeability estimation/inversion. The main purpose of the first two steps is to separate effects that are unrelated to permeability from the permeability-related effect. The wave-separation step suppresses the effects caused by noise, borehole and formation scattering/reflection, etc. The wave-modeling step models the effects of formation elastic property and borehole variations on the wave propagation. The difference between the modeled and measured data reflects the effect of permeability. The difference is measured for both wave amplitude and phase. The correspondence between them will indicate the effect of permeability. The final step fits the theory to the difference data in order to estimate permeability.

4.2.3.1 Wave separation

During wave propagation from the transmitter to the receiver array on an acoustic tool, the Stoneley wave may be scattered or reflected at borehole reflectors, which may be caused by borehole diameter changes (washout), fractures, or bed boundaries. The wave data may also contain road noise due to tool movement during logging, and quantization noise, etc. These effects can be effectively suppressed using a wave separation procedure.

Wave separation separates the array Stoneley-wave data into direct, down-, and up-going Stoneley waves. The wave data array can be the actual receiver array on the acoustic tool, or a gather of the single receiver data over consecutive logging depths. The wave separation method is based on the moveout velocity (inverse of slowness) of wave events across the array.

Suppose there are p individual wave-mode events in the array consisting of N equally spaced waveforms (or traces) with a spacing d. To estimate the wave spectra of all p modes at the nth waveform location, one can connect the wave modes to the waveform data at other locations. To do so, one forward- or reverse-propagates these wave modes to other locations in the array using the respective slowness of the wave modes. For example, propagation of wave mode $h_r(\omega)$, where $r = 1, \ldots, p$, from location n to location m is mathematically expressed by $h_r(\omega)Z_r^{m-n}$, where $Z_r = \exp(i\omega s_r d)$ (s_r is the slowness of the rth wave mode), where the location index m can be smaller than, equal to, or greater than the index n. One then equates the sum of the propagated spectra to the measured wave spectra at each waveform location. This results in

$$n\text{th row} \rightarrow \begin{bmatrix} Z_1^{1-n} & \cdots & Z_p^{1-n} \\ \vdots & & \vdots \\ Z_1^0 & \cdots & Z_p^0 \\ \vdots & & \vdots \\ Z_1^{N-n} & \cdots & Z_p^{N-n} \end{bmatrix} \begin{bmatrix} h_1(\omega) \\ \vdots \\ h_p(\omega) \end{bmatrix} = \begin{bmatrix} W_1(\omega) \\ W_2(\omega) \\ \vdots \\ W_N(\omega) \end{bmatrix}, \tag{4.49}$$

or in matrix notation

$$\mathbf{Zh} = \mathbf{W}.$$

Because N, the number of wave traces in the array, is usually much greater than the number p of wave events, one can solve equation (4.49) using a least-squares method for each frequency. The least-squares solution of the wave-mode spectrum, in matrix notation, is given by

$$\mathbf{h} = \left(\tilde{\mathbf{Z}}\mathbf{Z}\right)^{-1} \tilde{\mathbf{Z}}\mathbf{W}, \tag{4.50}$$

where ~ denotes taking a complex conjugate. This gives the estimated wave spectrum h for each wave mode at the designated location n. Transforming the resulting p wave spectra into the time domain, one gets p individual waveforms for the p wave modes. The individual wave mode obtained using equation (4.50) is a stacked wave over the wave array. This can be understood from the operation $\tilde{\mathbf{Z}}\mathbf{W}$ in the least-squares solution, which means phase-shifting, with the wave-mode slowness s_r ($r = 1, \ldots, p$) as in matrix $\tilde{\mathbf{Z}}$, the data $\mathbf{W}$ of all receivers to the location n, and summing them.

The wave separation method can be applied to two configurations: 1) the receiver array on an acoustic tool (called Common Source Gather, CSG), and 2) a gather of single-receiver acoustic data over a number of consecutive logging depths (called Common Offset or receiver Gather, COG). Using these two configurations, two wave separation options can be devised, as explained below.

The first option is a two-step separation scheme, first in CSG, then in COG. The wave events in a CSG are shown in Figure 4.5a. This figure shows the synthetic Stoneley-wave seismograms with two borehole reflectors: one is below, the other above the acoustic logging tool (assuming the source transmitter is below the receiver array). The up-going waves contain the direct wave transmitted from source

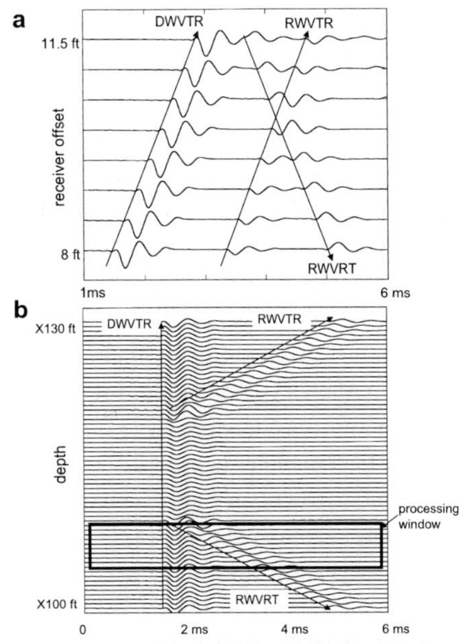

Figure 4.5. (a) Synthetic array acoustic waves with reflections from above the receiver array and below the source transmitter. The data show the direct wave along the transmitter-to-receiver direction (DWVTR), the reflected wave along the same direction (RWVTR, reflector below source), and the reflected wave from above the receiver array (propagation along the receiver-to-transmitter direction, RWVRT).
(b) The moveout of the three waves (DWVTR, RWVTR, and RWVRT) in the common-receiver-offset gather, and the processing window used to perform wave separation on these waves.

to receivers (DWVTR), and a reflected wave from the reflector below, traveling in the transmitter-to-receiver direction (RWVTR). The down-going wave is from the reflector above, propagating in the receiver-to-transmitter direction (RWVRT). The up-going waves (DWVTR and RWVTR) have the same moveout equal to $+s$, the Stoneley-wave slowness across the array. The down-going wave has a negative moveout, or an apparent slowness $-s$. Thus one can set $p = 2$, with $s_1 = +s$ and $s_2 = -s$ in equation (4.50) to perform a two-wave separation. The separated waves at the middle of the array (i.e., the location with $n = N/2$), which are already stacked over the N wave traces, are gathered over depth to form two individual COGs: 1) waves with $+s$ moveout in array (called "up COG"), and 2) waves with $-s$ moveout in the array (called "down COG"). In a COG, as shown in Figure 4.5b (this COG contains both up- and down-going waves), waves with fixed wave paths (e.g., source-to-receiver distance) have an almost zero moveout, while waves with varying paths (e.g., source-to-reflector and back to receiver) have a gradual moveout. Further, these latter waves transverse a depth interval twice; their apparent moveout is doubled as compared to the CSG case. Therefore, for the two individual COGs from the array wave separation, one sets $s_1 = 0$ for waves with fixed paths. For waves with varying paths, one sets an apparent slowness $s_2 = +2s$ for the up COG and $s_2 = -2s$ for the down COG, respectively. Applying again the two-wave separation to the two COGs separately, one gets the direct and up-going reflection waves from the up COG, and the down-going reflection wave from the down COG.

The second option is a one-step separation in COG for a single receiver on the tool. This option is designed for tools that do not have enough receivers to perform the array-wave separation (some existing tools have only two receivers). As shown in Figure 4.5b, this COG may contain three waves: a direct wave with an almost zero moveout, an up-going reflection wave, and a down-going reflection wave. Therefore, a three-wave separation scheme can be devised by setting $p = 3$, with $s_1 = 0$, $s_2 = +2s$ and $s_3 = -2s$ in equation (4.50). This yields the desired individual waves.

An application example is shown in Figure 4.6. Figure 4.6a shows the result of applying array-wave separation to a Stoneley-wave data set from a fractured and enlarged borehole segment (the borehole shape is illustrated in Figure 4.9). The left panel is the COG for the first receiver of an eight-receiver array. The data contain the direct wave, up-going, and down-going reflected waves. In the up-going (central panel) and down-going (right panel) wave data, the reflection from the lower and upper ends of the acoustic tool are also clearly identified. Application of the wave separation method to the up-going COG data removes the reflections due to borehole reflectors, resulting in the transmitted-wave data in Figure 4.6b (left panel). Application of wave separation to the down-going COG removes the tool-end reflection and yields the down-going reflection wave data shown in Figure 4.6b (right panel).

The wave separation processing has two major advantages for Stoneley-wave data processing. The first is that reflection or scattered waves are separated from the transmitted wave data. The second is that noise effects in the transmitted wave are suppressed because waves are stacked in both the CSG and COG during the processing. Further, by comparing the amplitude and time lag of the down-going reflected wave with those of the transmitted wave, one can obtain the reflectance (or reflectivity) of a borehole reflector and its depth location (see Tang, 1996b). This reflectance log is an important tool for characterizing borehole reflectors such as borehole changes, fractures, gas zones in formation (see Figure 1.11), and bed boundaries, etc.

4.2.3.2 Wave modeling by a propagator-matrix method

This section describes a fast method for modeling low-frequency Stoneley-wave propagation in an irregular borehole with varying formation elastic properties. The method allows the borehole and formation to vary arbitrarily along the borehole axial direction. Thus it is almost ideally suited to wave modeling using borehole-logging data (e.g., borehole caliper, compressional- and shear-wave slowness, and density, etc.), which record mainly axial borehole/formation variations. With this method, the wave amplitude and phase variation due to borehole and formation changes along the wave path can be simulated realistically.

In the modeling, the formation is discretized into a stack of layers. An irregular borehole penetrates these layers. An acoustic logging tool is centered in the bore-

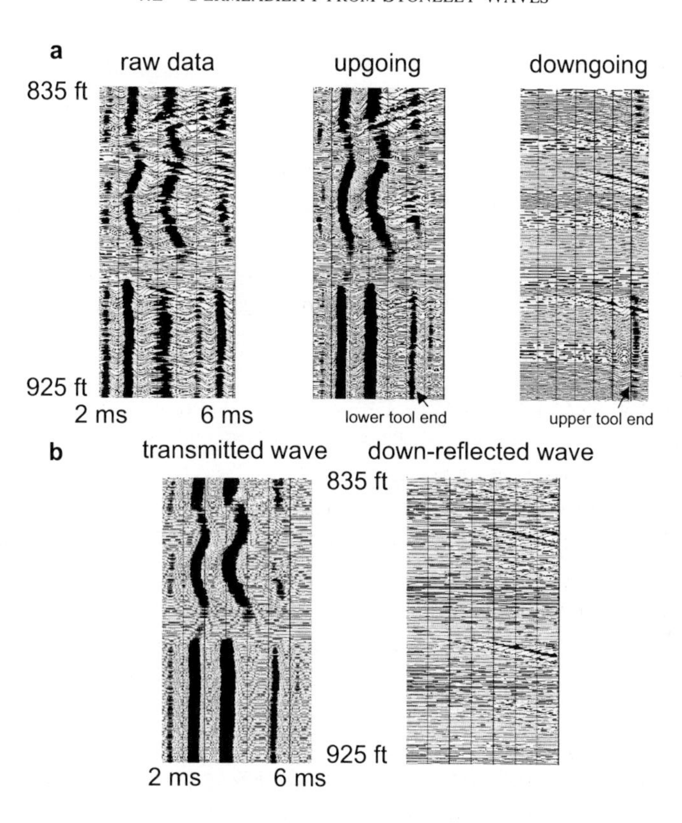

Figure 4.6. Example of wave separation. In (a), the array Stoneley wave data (receiver 1 data is shown in the left panel) is separated into up-going (central panel) and down-going (right panel) wavefields. On the common-receiver-offset displays, reflections from different origins (borehole/formation and tool ends, etc.) can be identified by their moveout. Applying the wave separation to the common receiver data yields the transmitted (left panel) and down-going reflected (right panel) wave data in (b).

hole. This model configuration is illustrated in Figure 4.7. Each layer in the model represents a depth interval whose thickness can be conveniently taken as the logging increment (typically 0.5 ft or 0.152 m), which is also the sampling interval of the log data. Therefore, the borehole diameter and the layer elastic properties are taken from the log measurements. The presence of the logging tool can be modeled using the method described in Chapter 5.

Stoneley-wave propagation is simulated using a propagator matrix method. The matrix elements describe the coupling of Stoneley waves across layers of different borehole radii and formation properties. Let $(b^+, b^-)^T$ be a vector containing up-going (represented by the "+" sign) and down-going (represented by the "–" sign) Stoneley-wave amplitude coefficients. The following equation describes the propagation of the coefficient vector at a borehole location z_1, through the discretized system to another location z_2:

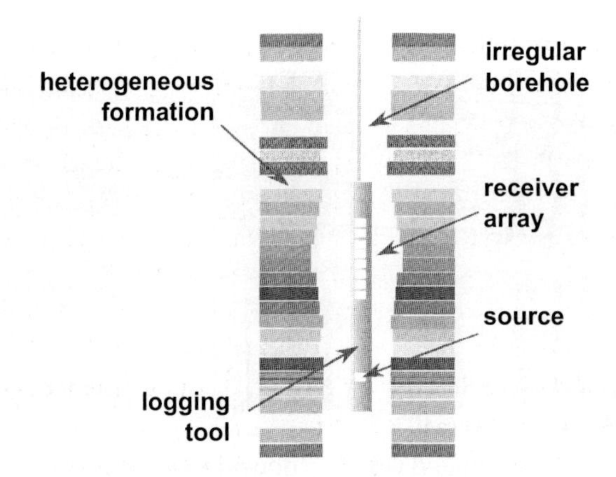

Figure 4.7. A representation of the model used for simulating Stoneley wave propagation in a heterogeneous formation with an irregular borehole. The formation is discretized into a stack of layers whose thickness is equal to the logging increment. For each layer, the borehole radius and formation properties are assigned from log data. Wave propagation from the source to receivers of a logging tool is simulated using a propagator matrix method.

$$\begin{pmatrix} b^+ \\ b^- \end{pmatrix}_{z_2} = \left(\prod_{l=1}^{L} G_l \right) \begin{pmatrix} b^+ \\ b^- \end{pmatrix}_{z_1} , \tag{4.51}$$

where L is the number of layers between z_1 and z_2. Tang (1996b) as well as Tezuka et al., (1997) derived an expression for the propagator matrix of the lth layer. It is given by

$$G_l = \begin{pmatrix} \frac{A_l}{A_{l-1}} \cos(k_e \Delta z) & -\frac{A_l}{A_{l-1}} \sin(k_e \Delta z) k_e \big/ \left(\rho_f \omega^2\right) \\ \left(\rho_f \omega^2 / k_e\right) \sin(k_e \Delta z) & \cos(k_e \Delta z) \end{pmatrix} . \tag{4.52}$$

In the propagator matrix, of special importance is the ratio of the cross-sectional borehole-fluid area of the lth layer to that of the (l–1)th layer, A_l/A_{l-1}; this ratio describes the effect of change in the borehole size on the wave propagation from the (l–1)th to lth layer. The result of equation (4.52) is derived using the conservation of fluid volume across a layer boundary (see, e.g., Tezuka et al., 1997). This is valid when the Stoneley-wave energy is mainly confined to the borehole fluid, as is the case for a hard (or fast) formation. When this condition holds, the effect of borehole-size change on the wave amplitude, and the formation elastic-property change on the wave phase, can be effectively and efficiently modeled using this equation. However, when the formation is soft (or slow), a significant portion of the Stoneley-wave energy is outside the borehole in the formation, making the above equation invalid. A general model is described by Gelinsky and Tang (1997). They derived expressions for the propagator matrix that is valid for an irregular borehole with an arbitrarily heterogeneous – fast or slow – formation (because the theory and the result are quite involved, the reader is referred to the Gelinsky and Tang (1997) for details).

We now describe the procedure for implementing the modeling method to the actual Stoneley-wave logging simulation. Let us refer again to the model in Figure 4.7, where the acoustic tool is embedded in the stack of the formation layers. In the modeling, the height of the stack is determined such that wave reflections from the upper and lower ends of the model lag behind the latest Stoneley-wave arrivals in the time window of the wave measurement. At one time during logging, the transmitter source occupies a layer called the source layer. We use $\mathbf{h}_{sr}$ to denote the product of the propagator matrices (i.e., matrix product in equation (4.51)) from the source layer's upper boundary to the receiver layer, and $\mathbf{h}$ to denote the product matrix from the same source boundary to the upper end of the stack. Similarly, we use $\mathbf{h}_b$ to denote the matrix product from the source layer's lower boundary to the lower end of the stack. The propagation of a Stoneley wave emitted from the source layer is completely described by $\mathbf{h}_{sr}$, $\mathbf{h}$, and $\mathbf{h}_b$. This can be understood from the following analysis.

In the source layer, the Stoneley-wave pressure field is related to the following 1D Green's-function determination problem:

$$\begin{cases} \dfrac{d^2p}{dz^2} + k_s^2 p = S(\omega)E_e(\omega, z_s)\delta(z - d_s) \\ \text{at } z = d_s, \quad p(z; z \le d_s) = p(z; z \ge d_s) \end{cases}, \tag{4.53}$$

where k_s is the source layer's Stoneley wavenumber and d_s is the distance from the transmitter to the layer's lower boundary; δ denotes Dirac's delta-function; $S(\omega)$ is the Fourier spectrum of the source and $E_e(\omega, z)$ the Stoneley-wave excitation function at the source layer (see equation (2.26)). The solution to the above problem is segmented into two parts at the source location: one is the pressure field above, and another is the pressure field below the source location. By connecting at $z = d_s$ the pressure below the source (related to $(b^+, b^-)^T{}_{lower}$) to the pressure above the source (related to $(b^+, b^-)^T{}_{upper}$), the wave-amplitude coefficient vectors at the upper and lower boundaries of the source layer are related to the source excitation by

$$\begin{pmatrix} b^+ \\ b^- \end{pmatrix}_{upper} - \begin{pmatrix} b^+ \\ b^- \end{pmatrix}_{lower} = \frac{S(\omega)E_e(\omega, z_s)}{ik_s} \begin{pmatrix} e^{-ik_s d_s} \\ -e^{ik_s d_s} \end{pmatrix}, \tag{4.54}$$

One now propagates the upper boundary vector in equation (4.54) to the upper end of the stack using $\mathbf{h}$, and the lower boundary vector to the lower end of the stack using $\mathbf{h}_b$. The boundary conditions specified at the two ends plus equation (4.54) suffice to determine these vectors (the boundary condition can be any type, e.g., $p = 0$, since the reflections generated at these ends are beyond the time window of interest at the receiver locations). After $(b^+, b^-)^T{}_{upper}$ is obtained, propagating the known coefficient vector from source to receiver using $\mathbf{h}_{sr}$ gives the wavefield at the receiver location.

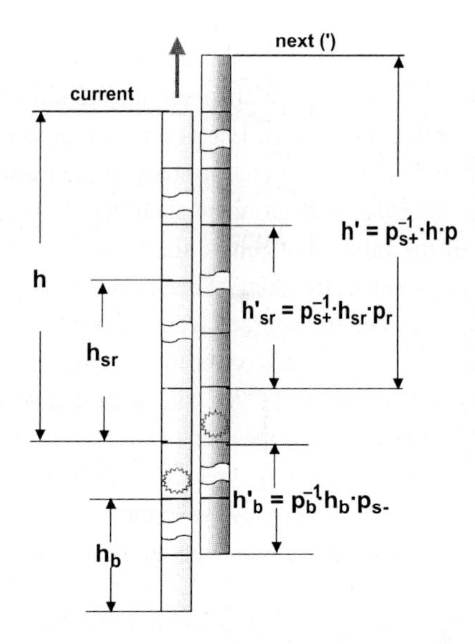

Figure 4.8. Model updating scheme for simulating the Stoneley wave propagation as the logging tool is moved trough the borehole. For the next logging position, the various propagator matrix products's are updated by convolving the propagator matrix of the top adjacent layer and deconvolving that of the lowermost layer. For this scheme, only one new propagator matrix (at the top of the model) needs to be calculated for each logging increment.

To simulate the logging process of an acoustic tool, the above procedure will be repeated for many consecutive source locations. This results in the following problem: Given $\mathbf{h}_{sr}$, $\mathbf{h}$, and $\mathbf{h}_b$, update these matrices to obtain the corresponding matrices $\mathbf{h}'_{sr}$, $\mathbf{h}'$, and $\mathbf{h}'_b$ when the source layer is incremented to the adjacent layer. To do this, we calculate the propagator matrix (equation (4.52)) adjacent to the top end of the current model stack (see Figure 4.8), and denote this matrix by $\mathbf{p}$. The new matrices are now updated by

$$\mathbf{h}' = \mathbf{p}_{s+}^{-1} \cdot \mathbf{h} \cdot \mathbf{p}; \;\; \mathbf{h}'_{sr} = \mathbf{p}_{s+}^{-1} \cdot \mathbf{h}_{sr} \cdot \mathbf{p}_r; \;\; \mathbf{h}'_b = \mathbf{p}_b^{-1} \cdot \mathbf{h}_b \cdot \mathbf{p}_{s-} \;, \tag{4.55}$$

where $\mathbf{p}_{s-}$, $\mathbf{p}_{s+}$, $\mathbf{p}_r$, and $\mathbf{p}_b$ are the propagator matrix at the source layer, the layer on top of the source layer, the receiver layer, and the lower-most layer of the stack, respectively. These matrices are already calculated at previous depths. The updating scheme is illustrated in Figure 4.8. Once $\mathbf{h}'_{sr}$, $\mathbf{h}'$, and $\mathbf{h}'_b$ are determined, wave propagation for the new source location is completely determined. This updating scheme, convolving a new layer at the front and deconvolving the last layer at the back, requires the least computational effort (only one matrix needs to be calculated with each increment). Thus it allows for the fast and accurate simulation of Stoneley waves in an actual borehole environment. The simulated wave spectral response can also be gathered into CSG and COG, which, after applying the wave separation method (i.e., equations (4.49) and (4.50)), will be split into direct, up- and down-

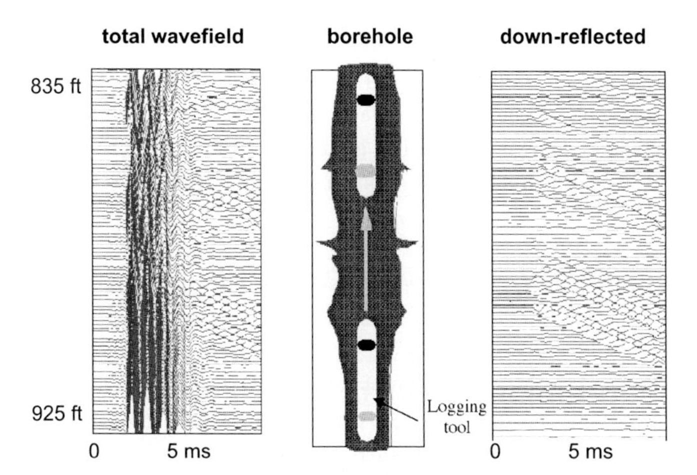

Figure 4.9. Synthetic common-receiver-offset Stoneley wave data across an irregular borehole segment. The borehole shape, as obtained from a caliper measurement, is depicted in the center panel. The total wavefield is shown in the left panel and the down-going reflected wavefield is shown in the right panel.

going waves. The simulated direct wave is compared with its measured counterpart in the permeability analysis.

Figure 4.9 shows an example of Stoneley-wave modeling using the above-described technique and procedure. The waveforms are displayed in a COG for a 90 ft borehole segment with severe irregularities. The borehole shape, as measured from caliper log data, is depicted in the central panel. A logging tool is moved up through the segment. The source is 8.5 ft below the receiver. The source wave is a decaying sine wave with a 1 kHz center frequency. The left panel shows the total wavefield, which includes the direct, up-going reflected, and down-going reflected waves. The synthetic Stoneley waves capture the effects of borehole diameter variation quite well. The significant amplitude variation in the direct wave and various up- and down-going reflected waves are realistically modeled, as compared with the real data in Figure 4.6a (left panel). The right panel of Figure 4.9 shows the down-going reflected wavefield. The origin of the down-going reflections corresponds to the locations of drastic borehole diameter changes.

4.2.3.3 Permeability indication

Indicating permeable formation intervals is sometimes very useful even without absolute permeability values. The permeability effect can be indicated through comparing the Stoneley-wave data from the wave separation processing and from the numerical simulation. By realistically simulating elastic wave propagation effects in an actual borehole environment, effects that are unrelated to permeability are taken into account. Then, the difference between the measured and simulated wave data will mainly reflect the permeability effects.

In the above wave simulation procedure, we need to specify a source signal or spectrum $S(\omega)$, as appears in equations (4.53) and (4.54). The wave source can be obtained from the actual Stoneley-wave data at a reference depth z_r. The permeability of the depth should be known or zero (i.e., an impermeable depth). Then the synthetic waveform log as a function of depth z can be simulated using

$$W_{syn}(f,z) = M(f,z)\left(\frac{W(f,z_r)}{M(f,z_r)}\right), \tag{4.56}$$

where M is the modeled direct Stoneley-wave spectral response (or transfer function) along the propagation path (the direct-wave response is obtained through wave separation), and W is the measured wave spectrum. The spectral ratio in equation (4.56) represents the deconvolution, at the reference depth z_r, of the wave-propagation transfer function from the measured wave. This yields the source spectrum $S(\omega)$ for the Stoneley-wave modeling. Then, convolving the source spectrum with the transfer function of any given depth z, as shown in equation (4.56), gives the synthetic wave at that depth.

For a given wave spectrum (synthetic or measured), the wave center (or centroid) frequency and the associated variance are calculated using

$$\begin{cases} f_c = \int f\, W(f)df \Big/ \int W(f)df \\ \sigma^2 = \int (f - f_c)^2\, W(f)df \Big/ \int W(f)df \end{cases}. \tag{4.57}$$

Furthermore, a center time (a robust measure of time of travel) is computed from the waveform of W using

$$T_c = \int t(W(t))^2 dt \Big/ \int (W(t))^2 dt\,. \tag{4.58}$$

The frequency shift and travel-time delay, between the measured wave and synthetic wave, are

$$\begin{cases} \Delta f_c = f_c^{syn} - f_c^{msd} \\ \Delta T_c = T_c^{msd} - T_c^{syn} \end{cases}. \tag{4.59}$$

The theory for Stoneley-wave propagation in a permeable borehole (e.g., equation (4.44) and Figure 4.2) shows that permeability has two direct effects on the Stoneley wave: 1) increase of wave attenuation and, 2) decrease of wave speed or velocity. The first effect is characterized by the frequency shift, and the second is by the travel-time delay, of the measured wave relative to the modeled wave. Therefore, these two quantities can be used to give a good indication of formation permeability. For instance, if the frequency shift corresponds to the travel-time delay for a depth

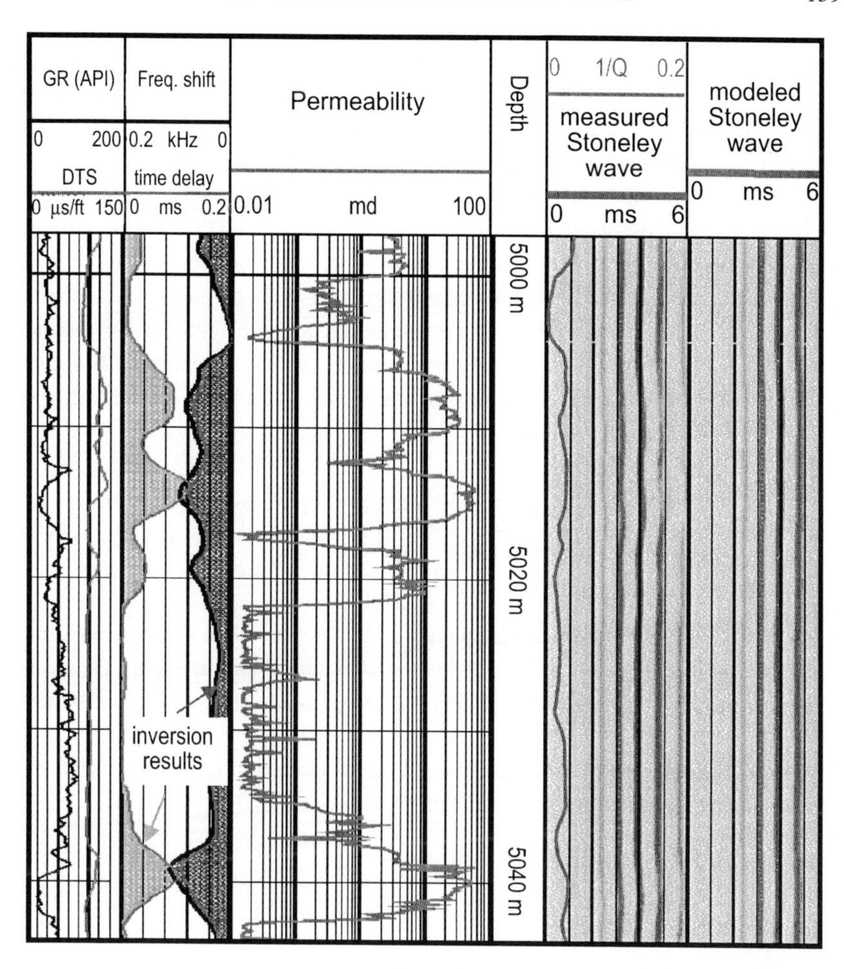

Figure 4.10. An example of permeability estimation from Stoneley-wave data. The Stoneley wave for the equivalent elastic formation is modeled (track 5) to compare with the measured data (track 4). Track 2 shows the time delay and frequency shift of the measured wave relative to the modeled wave; the correspondence between the delay and shift data indicates the existence of permeability. Theoretical fitting of the two data sets yields the permeability profile shown in track 3. Track 1 displays the gamma ray and shear slowness curves. The intrinsic wave attenuation curve is shown in track 4.

interval, one can almost be certain that these effects are related to formation permeability. The formation can thus be characterized as a permeable formation.

Figure 4.10 shows an example of permeability indication using Δf_c and ΔT_c log curves. Track 2 of the figure shows the frequency shift and travel-time delay (both are shaded) that are computed from the measured wave data in track 4 and the modeled wave data in track 5. The features on both curves correspond to (correlate with) each other quite well, indicating that they are related to formation permeability. The formation intervals sharing the correlation are thus characterized as a permeable formation.

4.2.3.4 Permeability estimation

The estimation of permeability uses the simplified theory (equation (4.44)). The Stoneley-wave attenuation due to borehole-fluid and formation intrinsic attenuation is also estimated, because it affects wave attenuation and travel-time in addition to permeability. Compared to permeability, intrinsic attenuation has a different frequency dependence and a different effect on wave attenuation and travel time. This allows it to be estimated simultaneously along with permeability. Including intrinsic attenuation in the inversion procedure allows us to account for this non-permeability effect. In a permeable borehole, the Stoneley wavenumber k is given by equation (4.44). The effect of intrinsic attenuation is included by the following transformation (equation (2.40))

$$k(\kappa_0, Q^{-1}) \rightarrow k(\kappa_0)\left(1+\frac{i}{2Q}\right)\left(1+\frac{1}{\pi Q}\ln\left(\frac{f}{f_0}\right)\right). \tag{4.60}$$

where f_0 is an arbitrary reference frequency which can be taken as the wave center frequency; Q is the quality factor; Q^{-1} is the attenuation that is mostly caused by borehole fluid and partially by formation (Cheng et al. 1982; Tang and Cheng, 1993b).

With the given theoretical model, the theoretical centroid (or center) frequency and associated variance are computed using

$$\begin{cases} f_c^{theo} = \int fPW^{syn}(f)\left|e^{ikd}\right|df \Big/ \int PW^{syn}(f)\left|e^{ikd}\right|df \\ \sigma_{theo}^2 = \int (f - f_c^{theo})^2 PW^{syn}(f)\left|e^{ikd}\right|df \Big/ \int PW^{syn}(f)\left|e^{ikd}\right|df \end{cases}, \tag{4.61}$$

where P is the excitation reduction due to permeability which can be calculated as $P = k_e/k$ (equation (4.47)); d is now the wave travel distance and $|e^{ikd}|$ is the amplitude decay caused by permeability and attenuation along the wave path. With equation (4.61), the theoretical frequency shift is defined as

$$\Delta f_c^{theo} = f_c^{syn} - f_c^{theo} . \tag{4.62}$$

The theoretical travel-time delay can be computed using the weighted spectral average slowness theorem, as given in equation (3.25).

$$\Delta T_c^{theo} = \int (kd/\omega - k_e d/\omega)(\omega W^{syn}(f))^2\, df \Big/ \int (\omega W^{syn}(f))^2\, df . \tag{4.63}$$

The above theoretical travel-time delay, computed using the frequency-domain formulation, should be a good approximation to its measured counterpart computed in the time domain, (as given by equations (4.58) and (4.59)), at least to the first order of the permeability-related effect. Therefore, equations (4.62) and (4.63) give the overall frequency shift and travel-time delay of the Stoneley wave in a permeable and attenuative borehole relative to an impermeable and non-attenuative borehole.

By comparing the measured frequency shift and travel-time delay given by equations (4.59) with the theoretical predictions given by equations (4.62) and (4.63), an inverse problem can be formulated which results in minimizing the following objective function

$$E(\kappa_0, Q^{-1}) = (\Delta f_c^{msd} - \Delta f_c^{theo})^2 / \sigma_{syn}^2 + 2\pi\sigma_{syn}^2 (\Delta T_c^{msd} - \Delta T_c^{theo})^2 + \alpha(\sigma_{syn}^2 - \sigma_{theo}^2), \quad (4.64)$$

where σ^2_{syn} and σ^2_{theo} are computed from equations (4.57) and (4.61), respectively. The first and second terms are scaled so that the frequency shift and travel-time delay have almost equal importance in the objective function. The third term is a penalty function. The coefficient α is very small so that this term has almost zero contribution in the inversion if κ_0 and Q^{-1} vary in the permissible range. If their values exceed the permissible range, the third term becomes large, driving the inversion parameters back to the permissible range.

The objective function comprises the summation of many frequency and time data points (see equations (4.57) and (4.58)), reliably recording the permeability and/or attenuation effects in the wave travel time and frequency data. In the inversion process any appropriate minimization algorithm can be used, for example, local algorithms like conjugate gradient, Levenberg-Marquardt (Press et al., 1989), or global algorithms like Simulated Annealing (Ingber, 1989; Chunduru et al., 1996). The local methods require starting values for the model parameters. They can be assigned using results from the previous depth. Global algorithms do not need starting values, but may require longer computational time in the inversion.

4.2.3.5 *Improving resolution*

In the above inversion formulation, the permeability effect is the average from the transmitter-to-center-receiver-array distance (typically about 10–13 ft), whereas the features in the processed Stoneley-wave data usually have a resolution of approximately the aperture (about 3–4 ft) of the processing window for CSG and COG, as used in the wave separation processing (see the processing window in Figure 4.5b). With the following scheme, the resolution of the permeability estimation can be enhanced to the resolution of the processing window aperture.

To improve the resolution, we divide the wave travel distance d (transmitter-to-center-of-receiver-array distance) into two parts:

$$d = d_0 + d' \quad (4.65)$$

where d' is a distance equal to the COG aperture, measured from the receiver toward transmitter; d_0 is the remaining distance linking to the transmitter. Assuming that the theoretical travel-time delay and frequency shift over d_0 are known from previous estimates, we subtract them from the measured data using

$$\begin{cases} \Delta f_c^{msd} & \rightarrow \Delta f_c^{msd} - \Delta f_c^{theo}(d_0) \\ \Delta T_c^{msd} & \rightarrow \Delta T_c^{msd} - \Delta T_c^{theo}(d_0) \end{cases} . \tag{4.66}$$

We then use d' to replace d in equations (4.61) for calculating the theoretical frequency shift and travel-time delay. Finally, the calculated quantities and the modified data (equations (4.62) and 4.63)) are used in the inversion formulation (equation (4.64)) to estimate the permeability over the reduced distance d'.

After the permeability over d' is estimated, the theoretical frequency shift and travel-time delay over d' can be calculated. These quantities are combined with those of the previous depths to give a new average of the quantities over the distance d_0 for estimating permeability in the next depth interval. This scheme is repeated for the entire depth of processing.

4.2.3.6 Pore-fluid parameter calibration

By performing an asymptotic analysis of the model theory given in equation (4.44), it can be shown that the permeability-induced travel-time delay and frequency shift are primarily controlled, besides other parameters (e.g., formation porosity and borehole and tool radii), by the following parameter combination

$$\frac{\kappa_0}{\eta K_{pf}^{1/2}} , \quad (K_{pf} = \rho_{pf} V_{pf}^2) , \tag{4.67}$$

where K_{pf} is the pore-fluid bulk modulus or incompressibility. This asymptotic analysis is valid for most formation permeability values and for the Stoneley-wave frequency range around 1 kHz. This parameter combination shows that formation pore-fluid parameters (i.e., viscosity, density, and acoustic speed) are required to obtain the absolute permeability value. These parameters, however, may be difficult to obtain if multiple fluids are present and the degree of fluid saturation is unknown. For this situation, we use a calibration method to determine these parameters.

The calibration consists of selecting a few (minimum of two) depths where permeability values have been estimated from other measurements (e.g., core, formation testing, Nuclear Magnetic Resonance (NMR), etc.). By choosing one of these depths as the reference depth in equation (4.56), synthetic wave data can be computed for the remaining depth(s). This process (i.e., reference selection and modeling) is rotated for all the depths. Then, by comparing the synthetic data with the measured data, we can estimate $\eta K_{pf}^{1/2}$ by minimizing the following misfit function

$$F(\eta K_{pf}^{1/2}) = \sum_{\substack{i,j=1 \\ i \neq j}}^{n} \left\{ (\Delta f_c^{msd} - \Delta f_c^{theo})^2 / \sigma_{syn}^2 + 2\pi\sigma_{syn}^2 (\Delta T_c^{msd} - \Delta T_c^{theo})^2 \right\} , \tag{4.68}$$

where n is the total number of depths chosen and the pair i, j means using depth i as the reference to compare with depth j (= 1, ..., n; $i \neq j$). In the minimization, the given permeability values at the calibration depths are used as a known parameter in the parameter combination given above. Only $\eta K_{pf}^{1/2}$ is varied as the fitting parameter. The value of this parameter that minimizes equation (4.68) is taken as the desired fluid-parameter value for the depth zone of interest and is used to estimate a continuous permeability curve for the entire zone.

4.2.4 EXAMPLE OF PERMEABILITY ESTIMATION

To demonstrate the validity of permeability estimation from Stoneley-wave logging data, the example in Figure 4.10 shows results from the various steps of the permeability estimation method. The synthetic transmitted Stoneley wave is shown in track 5 using a variable-density display (the shear-wave slowness curve used in the modeling is shown in track 1). The transmitted wave is obtained by applying the wave separation procedure to the synthetic wave data that contain both transmitted and reflected waves (see Figure 4.9 for an example). By comparing the measured transmitted-wave data (track 4) with the modeled data, permeability effects can be extracted.

Track 2 shows the excessive dispersion and attenuation of the measured data relative to the modeled data. The dispersion is measured as the travel-time delay, and the attenuation is measured as the frequency shift, between the measured and modeled waves. The correspondence between the delay (shaded curve from left to right) and shift (shaded curve from right to left) data gives a clear indication of permeability effects. The wave dispersion and attenuation data shown in track 2 are inverted to estimate permeability and attenuation. The goodness of this inversion can be measured by overlaying the inversion-fitted travel-time delay and frequency shift curves with their measured counterpart, as shown in track 2. The estimated intrinsic wave attenuation is shown in track 4. The estimated permeability profile is shown in track 3. The permeability profile correlates well with the sandstone intervals of this formation, as indicated by the gamma ray curve in track 1.

4.3 Joint Interpretation of Formation Permeability from Acoustic and NMR Log Data

With advances in wireline logging technology, continuous formation permeability profiles can now be measured from wireline logging. Acoustic logging and Nuclear Magnetic Resonance (NMR) logging are used to obtain two permeability profiles, referred to as Stoneley and NMR permeability, respectively. Stoneley permeability is obtained from Stoneley-wave travel-time and attenuation that are directly related to formation permeability, NMR permeability is derived from the NMR T_2-relaxation data that are related to pore-size distributions. We compare the Stoneley- and NMR-permeability estimation methods and results to further demonstrate the theory and application of the former method. Stoneley and NMR permeability estimates are based on fundamentally different measurements and physics. If the results of these techniques agree, the interpreter's confidence in the permeability profiles increases. On the other hand, if the results do not agree, the difference may be related to formation and pore-fluid properties. Specifically, the difference often indicates the presence of gas or fractures, or the presence of a stiff mud cake. In these situations, we can explain the difference using the respective physics and measurement principles.

4.3.1 Brief Summary of the Principles of Stoneley and NMR Permeability Measurement

From the theory described earlier, we see that the permeability-related Stoneley-wave attributes, e.g., travel-time delay and attenuation, are approximately controlled by the following parameter combination:

$$\left.\begin{array}{l}\text{Attenuation}\\ \text{Travel-time Delay}\end{array}\right\} \leftrightarrow \frac{\kappa}{\eta\sqrt{K_{pf}}}\,. \tag{4.69}$$

where κ is formation permeability, η is formation-fluid viscosity and K_{pf} is formation pore-fluid modulus or incompressibility. This relation indicates that, in addition to pore-fluid mobility κ/η, pore-fluid incompressibility K_{pf} also affects Stoneley-wave characteristics. However, the formation pore-fluid viscosity and incompressibility may be difficult to estimate given the unknown saturation states in the sensitive volume surrounding the borehole. For example, the sensitive volume may contain connate water, hydrocarbons, and invaded drilling fluid, etc. Therefore, in the Stoneley-wave permeability estimation, the pore-fluid parameters are treated as apparent or effective parameters for the different fluids saturating the sensitive volume. These parameters can be estimated or adjusted by calibrating the Stoneley-derived permeability with those from other measurements (e.g., formation testing, core, NMR, etc.). The calibration procedure has been described previously (see the discussions associated with equation 4.68).

Another important source of formation permeability is NMR-derived permeability. Undoubtedly, downhole NMR technology is the most significant advance in formation evaluation in the 1990s. However, the NMR tool does not directly measure permeability nor does NMR permeability follow from Darcy's law. The NMR signal amplitude is proportional to the hydrogen proton index associated with oil, gas, and water in the formation. Modern NMR tools utilize a Carr-Purcel-Meiboom-Gill pulse sequence to collect a time series of echoes. The rate of decay of the echo amplitudes is a measure of the pore space's surface-to-volume ratio (Kenyon et al., 1986). Small pores have a large surface-to-volume ratio, while large pores have a small surface-to-volume ratio. Typically, pores in a reservoir vary in size; hence the pore-size distribution can be derived by decomposing the NMR echo signal into a number of decaying exponentials and computing the distribution of their relaxation time T_2 (or rate of decay). The pore-size distribution is the key to determining irreducible water saturation, moveable fluid saturation, and permeability.

NMR permeability can be estimated using the Coates equation that is based on the observation that rocks with high irreducible water saturation generally have low permeability. The Coates equation uses the ratio of the moveable-to-bound fluid saturation derived from the T_2 distributions:

$$\kappa[\text{md}] = \left(\frac{\phi}{C}\right)^m \left(\frac{BVM}{BVI}\right)^n . \tag{4.70}$$

Here, κ is permeability in millidarcy, ϕ is the porosity in percent, and BVM and BVI are the bulk-volume moveable and bulk-volume irreducible fluids, respectively. The constant C and exponents m and n can be adjusted or calibrated to match other permeability estimates, e.g., formation testing, core, and Stoneley-wave, etc.

We have briefly summarized the principles for the Stoneley-wave and NMR permeability measurements. We have also demonstrated that both methods have parameters that can be adjusted or calibrated to match each other, or match with other permeability estimates. When the calibrated Stoneley and NMR permeability profiles correspond to each other both in magnitude and variability, the two profiles are deemed as in good agreement. If they differ in magnitude or in character, the difference can be analyzed to find its cause, which may also provide valuable information for formation evaluation.

4.3.2 INTERPRETATION WITH FIELD EXAMPLES

4.3.2.1 Good correspondence between Stoneley- and NMR-permeabilities

The Stoneley-wave permeability method has been applied to many acoustic data sets from various reservoirs around the world. The Stoneley-derived permeability profiles were compared with NMR permeability profiles whenever possible. In general, when both the Stoneley and NMR data are of good quality, there is good correspondence between the two permeability profiles, especially for sandstone formations.

Figure 4.11 shows an example of comparing Stoneley-derived permeability with NMR permeability for a sandstone formation, as indicated by the gamma-ray curve in track 1 (in sand/shale formations, the sandstone intervals are often characterized by low gamma-ray counts). Track 4 shows the measured Stoneley-wave data from the wave separation processing. Track 5 shows the modeled Stoneley-wave data. The modeling accounts only for the formation elasticity and borehole changes, while the measured data contain both permeability and elasticity/borehole-change effects. Comparison of the measured and modeled wave data yields the centroid frequency shift and travel-time delay of the measured wave relative to the modeled wave (the delay of the modeled wave relative to the measured wave in the shale forma-

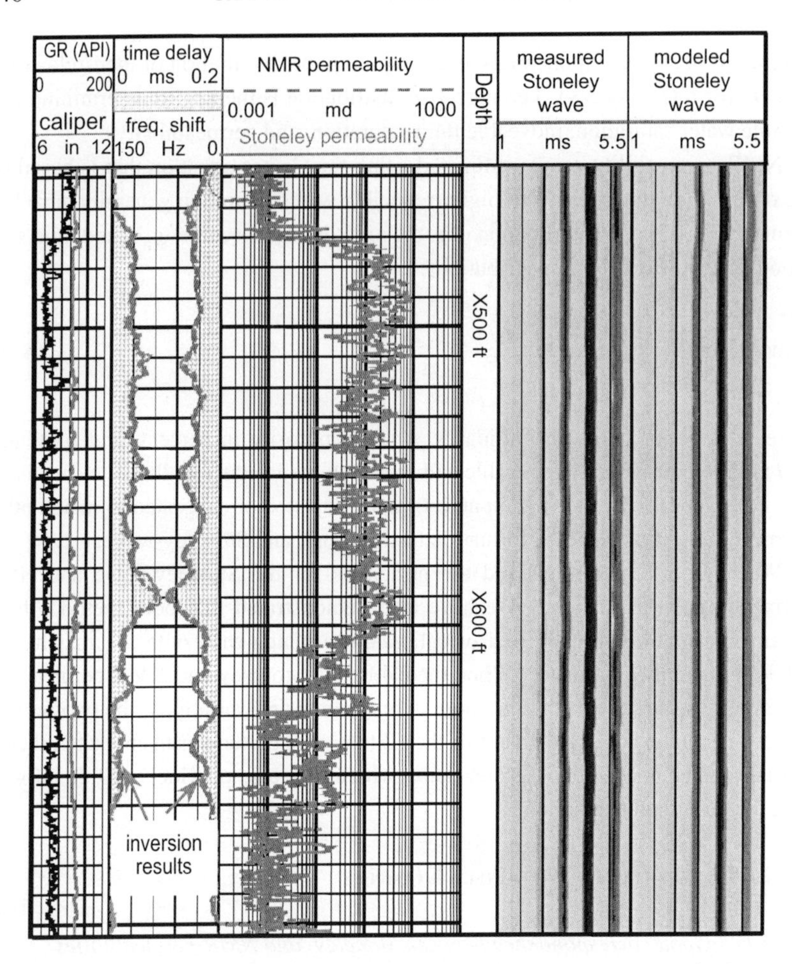

Figure 4.11. Comparison of permeability profiles (track 3) from Stoneley (solid curve) and NMR (dashed curve) measurements for a sand/shale formation.

tion above X457 ft is an anisotropy effect, as will be discussed in Chapter 5). The frequency shift gives a robust measure of wave attenuation. Track 2 shows the frequency shift and travel-time delay. The correspondence between the frequency shift and travel-time delay gives a good indication of permeability effects, as is expected by the theoretical relation given in (4.69). Therefore, the correspondence between these two curves is used as the quality control (QC) for Stoneley-wave permeability estimation. Finally, applying the inversion method to the data in track 2 gives the Stoneley permeability profile shown in track 3 (solid curve). The NMR permeability is also shown in the same track (dashed curve). The Stoneley and NMR permeability profiles agree remarkably well. Since the two permeability profiles are based on fundamentally different physical concepts and derived from different measurements, the agreement gives confidence that the derived permeability profiles are correct.

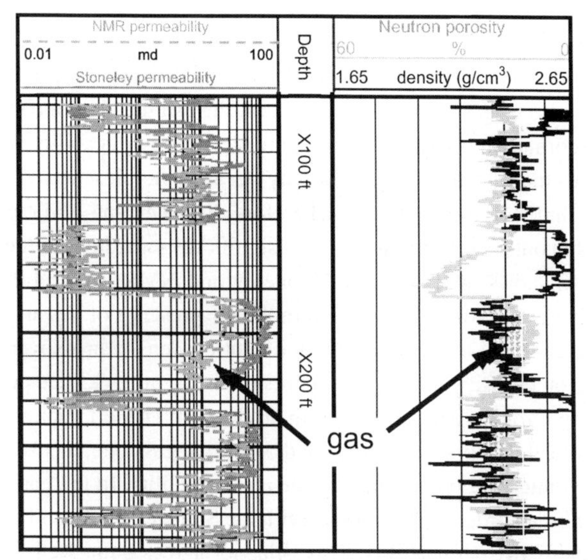

Figure 4.12. Example of Stoneley and NMR permeability difference in the presence of formation gas saturation. The Stoneley method tends to over-estimate – while the NMR method tends to underestimate – the formation permeability, resulting in the separation of the two permeability profiles. The right track shows gas identification using the classical neutron-density crossover method.

4.3.2.2 *Effects of gas saturation on Stoneley and NMR permeability profiles*

One application of jointly interpreting Stoneley and NMR permeability measurements is for formation gas saturation detection. In the presence of gas saturation, the Stoneley-wave attenuation and travel-time delay will be enhanced due to an increase of pore-fluid mobility and compressibility (see relation (4.69)), while the NMR signal may be reduced due to the decrease of hydrogen index of gas. Thus, the Stoneley and NMR permeabilities will reflect the changes of fluid saturation due to the presence of gas. In the Stoneley permeability estimation, the permeability profile can be calibrated to the NMR permeability for a depth zone using the calibration method associated with equation (4.68). The calibrated parameters are used to estimate permeability from other depths. If, however, the pore-fluid system is different from that of the calibrated depth zone, the estimated Stoneley permeability deviates from the true formation permeability. The theoretical relation (4.69) shows that the estimated permeability κ_{est} is related, approximately, to the true formation permeability κ_{true} via the following equation:

$$\kappa_{est} = \kappa_{true} \frac{\left(\eta\sqrt{K_{pf}}\right)_{cali}}{\left(\eta\sqrt{K_{pf}}\right)_{true}}, \tag{4.71}$$

where the subscripts *cali* and *true* refer to the calibrated fluid parameter value and true fluid parameter value for the depth interval of estimation, respectively. Suppose the calibrated value is for oil and the depth interval is fully or partially saturated with gas. Then the denominator value in equation (4.71) decreases because the viscosity and bulk modulus of gas are small. As a result, the estimated permeability is higher than the true formation permeability. In contrast, NMR permeability in the gas zone underestimates the true permeability simply because the presence of gas decreases the hydrogen index of the pore fluid. Therefore, Stoneley and NMR permeabilities, when calibrated outside a gas zone, exhibit significant separation in the gas zone.

Figure 4.12 shows an example of Stoneley and NMR permeability difference caused by the presence of gas. In this example, the presence of gas around the depth of X200 ft is identified by the conventional neutron-density crossover plot, as seen from the neutron porosity and density curves on the right track. The presence of gas around this depth zone is also verified from production. The left track shows Stoneley and NMR permeability profiles. Above and below the gas zone, the two permeability profiles agree quite well both in magnitude and variability. In the gas zone, the two profiles differ by an order of magnitude, with the Stoneley permeability consistently higher than the NMR permeability. This behavior agrees quite well with the above analysis. This example demonstrates that joint interpretation of Stoneley and NMR permeability can be useful for detecting gas saturation.

4.3.2.3 Effects of fractures

Another useful application of comparing Stoneley and NMR permeability profiles is the characterization of fractured reservoirs. Because the Stoneley wave is related to fluid flow at the borehole interface, regardless of whether the fluid flow occurs in a porous formation or in fractures, the Stoneley wave can sense both matrix and fracture permeability. This is clearly demonstrated by Tang and Cheng (1993a), who showed that a permeable fracture zone can be alternatively modeled using a permeable porous zone in the Stoneley-wave measurements. Conversely, the NMR measurement cannot adequately measure fracture permeability. The NMR-measured porosity in fractured reservoirs can be subdivided into matrix porosity and fracture porosity. Generally, fracture porosity is small (< 0.5 %), and the NMR signal predominantly reflects the surface-to-volume ratio of the matrix pore system. Hence the NMR method measures mainly the matrix permeability and largely underestimates fracture permeability. Thus, by comparing Stoneley and NMR permeability profiles for a fractured reservoir, it is possible to separate permeability contributions arising from the rock matrix and the fractures.

Figure 4.13 shows an example of Stoneley and NMR permeability profiles across a fractured depth interval. The presence of fractures is identified from two sources of measurement. The first is the fracture image from a circumferential acoustic-imaging tool, as shown in the right image of the figure. The second are the Stoneley-wave reflection events at the identified fracture locations (Track 3, displayed using a variable-density image). The reflection of Stoneley waves caused by fractures is also an important means of characterizing permeable borehole fractures (Hornby et al.,

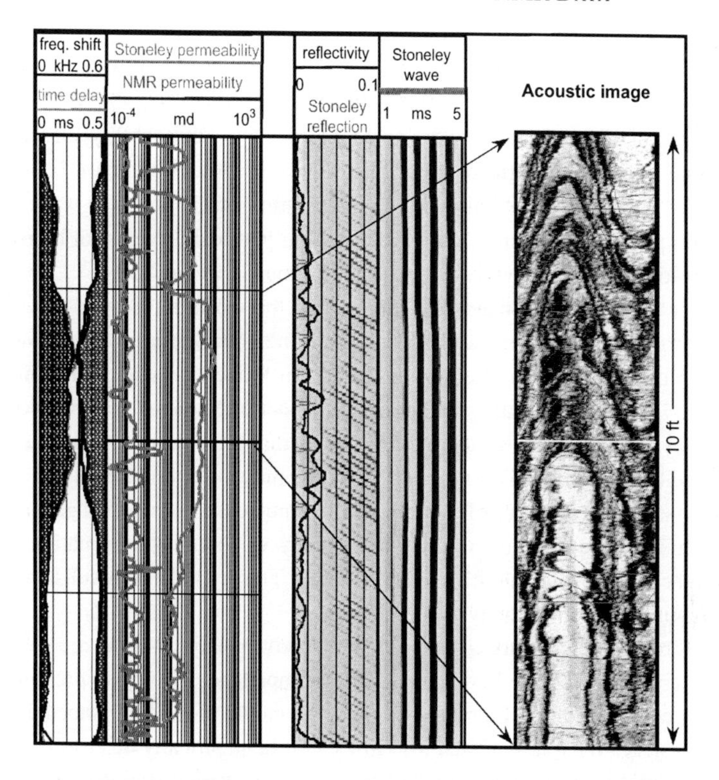

Figure 4.13. Stoneley and NMR permeability profiles (track 2) in the presence of borehole fractures. The fractures are identified by Stoneley wave reflection (track 3) and borehole acoustic image (right panel). Note the significant response in the Stoneley wave results (tracks 1 through 4) and the lack of response in the NMR profile (track 2).

1989; Tang and Cheng, 1993a) and is commonly used for fracture analysis. Track 3 shows the (down-going) Stoneley-wave reflection image and the reflectivity (or reflection coefficient) profiles, obtained from the wave separation processing. The raw reflectivity profile (solid curves) is obtained by comparing the amplitude of the reflected wave (track 3) with that of the transmitted wave in track 4, while the processed reflectivity (spikes) is determined by the location of the zero time lag between the two waves. In this depth zone, there are many reflection events corresponding to the fracture locations. The Stoneley- and NMR-derived permeability profiles are shown in track 2. As anticipated from the above discussion, the NMR permeability (left curve) shows almost no response to the fractures, the measured permeability corresponding mainly to the rock matrix. However, the Stoneley attributes (frequency shift and travel-time delay in track 1 and wave image in track 4) and the derived permeability (right curve in track 2) show an excellent response to the fractures, with each peak of the permeability profile corresponding to the fracture location quite well. This example shows the value of Stoneley-wave measurement in fracture characterization.

4.3.2.4 *Effects of mud cake*

The effects of borehole mud cakes on Stoneley-wave permeability measurements have been a major concern for more than three decades. A common belief is that the presence of mud cakes will block the interaction of the Stoneley wave with the formation pore-fluid system, such that formation permeability cannot be measured by means of the Stoneley wave or any type of acoustic waves.

A large number of acoustic data sets around the world have been processed using the Stoneley wave permeability estimation method. Surprisingly, in only a few cases can one observe the effect of mud cake on Stoneley-wave permeability measurements. The unexpectedly small mud cake effect on Stoneley waves, as observed from many field data sets, leads to a hypothesis that the effect of mud cake may be controlled by its rigidity. Two types of mud cake may exist: one is soft and the other is hard or stiff. The presence of a soft or flexible mud cake does not, in general, block the communication between a borehole Stoneley wave and the formation, while a hard mud cake may prevent this communication. This hypothesis is substantiated by the following theoretical analysis.

Let us recall the pressure communication condition expressed in equation (4.22), which is controlled by the borehole-wall flow impedance T and the pressure difference between borehole and formation $p_f - p$. The nature of the flow impedance T has been a mystery: What is T related to? How can it be practically determined? A theoretical model illustrated in Figure 4.14 helps answer these questions. In this model, the pore space is modeled as a tube of radius a and the mud cake is modeled as an elastic half space with a shear rigidity or modulus μ_c and Poisson's ratio ν_c. The half space approximation for modeling the mud cake is justified because the mud cake thickness (on the order of millimeters to centimeters) is much greater than the pore size (or the order of several to tens of micrometers). According to the theory of elasticity, a pressure load (i.e., the pressure difference $p_f - p$) on a circular disk of radius a (the pore radius) at the surface of an elastic half space (the mud cake) will produce an average displacement w given by

$$w = \frac{4a(1-\nu_c)(p_f - p)}{3\pi\mu_c} . \tag{4.72}$$

A mud cake is usually very soft compared to common solids (rocks and metals, etc.). It is thus a good approximation to set $\nu_c = 0.5$, the value for a fluid. The displacement in equation (4.72) is for a single pore. Multiplying this displacement with porosity ϕ gives the volumetric fluid-flow displacement in equation (4.22). Comparing the fluid-flow displacement with that of equation (4.22) and noting that w in equation (4.72) is the relative displacement between pore-fluid and solid at the pore opening, one immediately obtains an expression for T, as

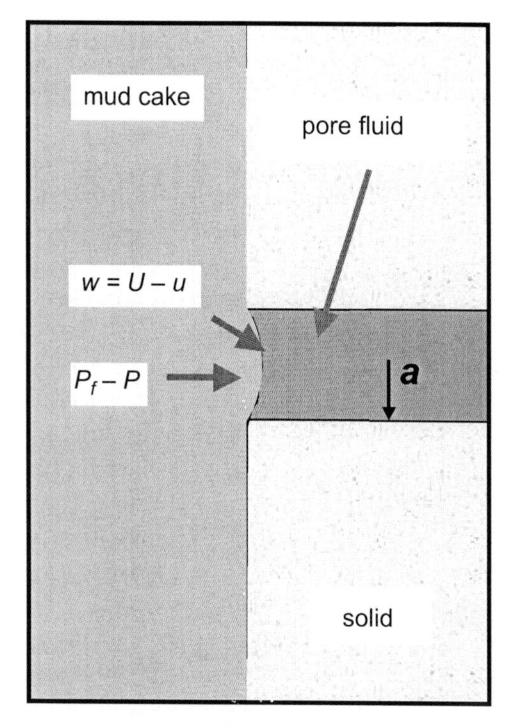

Figure 4.14. Modeling the effects of mudcake on the permeability measurement using borehole acoustic waves. The pressure difference between borehole and pore fluid causes the mud cake to deform at the pore opening. The relative displacement between fluid and solid is controlled by the mud-cake's shear rigidity and by the pore size.

$$T = \gamma \frac{\mu_c}{a} , \tag{4.73}$$

where the proportionality constant γ is a pore-shape factor. It is $3\pi/2$ if the pore is a circular tube and may take other values for different pore shapes. Equation (4.73) suggests that the borehole interface flow impedance T, in the presence of a mud cake, is proportional to the mud-cake shear rigidity and inversely proportional to the pore size. With this result, one can interpret Stoneley-wave permeability measurements for the mud-cake effect.

The laboratory work of Tang and Martin (1994) showed that for a fresh, thin, and soft mud cake, the measured μ_c is virtually zero, resulting in the vanishing of T. In this situation the borehole pressure is equal to the pore-fluid pressure at the borehole interface, imposing no impedance on the hydraulic exchange at the interface. This suggests that a soft mud cake will not affect Stoneley-wave permeability measurements. This may explain why, in general, we can reliably measure permeability from Stoneley waves. However, the same work of Tang and Martin (1994) showed that a dense, heavy (i.e., stiff) mud cake exhibits a measurable shear rigidity, which, according to equation (4.73), results in a finite value for T. The value of T is significant enough to prohibit the hydraulic communication between borehole and formation.

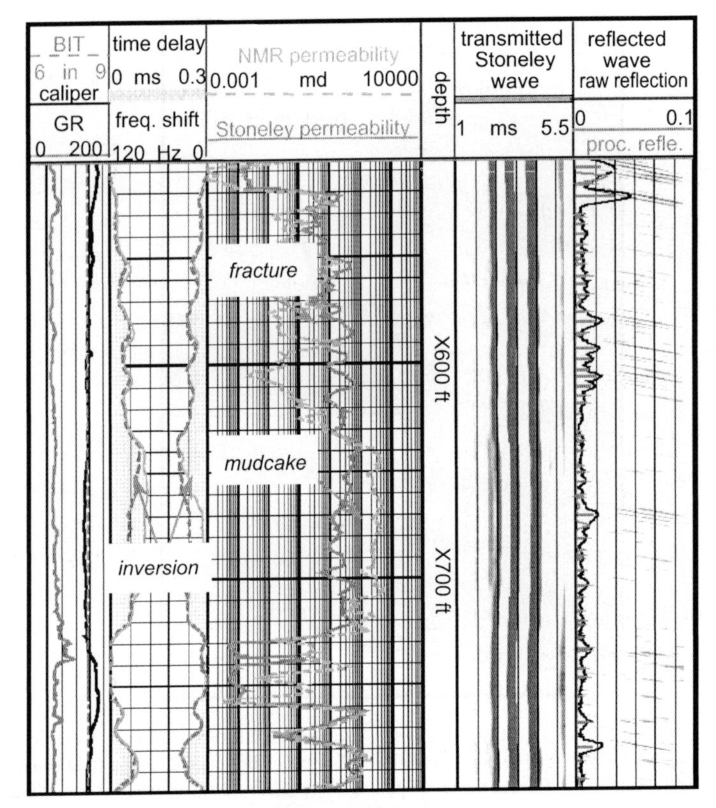

Figure 4.15. Stoneley and NMR permeability profiles (track 3) in the presence of borehole fractures (upper portion) and a stiff mudcake (lower portion).

Under these circumstances, comparing Stoneley- and NMR-derived permeability profiles will not only identify the presence of mud cake, but also correctly indicate the formation rock permeability from the latter measurement.

Figure 4.15 shows an example of the Stoneley-wave measurement in the presence of a stiff mud cake and fractures. Track 3 shows Stoneley (solid) and NMR (dashed) permeability profiles. The two profiles agree very well at the top and bottom portions of the figure, but differ significantly in the middle part. From X570 ft–X630 ft, the Stoneley permeability is much higher than the NMR permeability, while from X640 ft–X710 ft, the NMR permeability is much higher than the Stoneley permeability. The causes of these differences are discussed below.

The high Stoneley permeability and low NMR permeability in the X570 ft to X630 ft interval present another example of Stoneley and NMR response in fracture formations, in addition to the one shown in Figure 4.13. The presence of fractures can be substantiated by analyzing the Stoneley-wave reflection data (Hornby et al., 1989; Tang and Cheng, 1993a). Track 5 shows the (down-going) Stoneley-wave reflection image and the reflectivity log curves. In the depth interval of X570 ft–X630 ft, there are many reflection events. These events can be distinguished from other events (e.g., events at the top and bottom of the figure) that may be caused by borehole and lithology changes, as seen from the caliper and gamma-ray data in track 1. Further, the two quality-control curves in track 2, the travel-time delay and frequency shift, show excellent correspondence and overlap with inversion-fitted curves, indicating a good permeability response. We can conclude that the X570 ft–X630 ft zone is a permeable zone and the observed permeability difference is caused by fractures.

The Stoneley and NMR permeability difference between X640 ft and X710 ft is caused by a stiff mud cake layer at the borehole interface. The first evidence is the caliper in this interval: it is slightly smaller than the drilling bit size (see track 1), a phenomenon frequently associated with the presence of stiff mud cakes. Furthermore, in this interval Stoneley travel-time delay is high but the frequency shift is low, showing no correlation, as would be expected for a permeable interval, e.g., the intervals above and below. The inversion results cannot fit either curve; the model theory under-predicts the travel-time delay and over-predicts the frequency shift. The behavior of the Stoneley-wave attributes is also consistent with the mud-cake hypothesis. For a stiff mud-cake layer at the borehole, the Stoneley wave is delayed. Because the mud-cake seals the borehole interface, the hydraulic communication between borehole and formation is decreased. This causes an increase of travel-time delay and decrease of centroid-frequency shift (or attenuation), as we see from the data in track 2. As a result, the permeability estimated from the Stoneley wave is below the actual value. This is demonstrated by comparison with the NMR permeability. The NMR method measures the pore-size distribution in the formation, instead of fluid-flow. Therefore, permeability derived from the NMR measurement is reliable with or without mud cake. The permeability obtained from NMR shows quite high values compared to other depths (after drilling a well, mud-cake buildup is often found in high-permeability intervals because permeability facilitates mud filtration into the formation). This example demonstrates that, although the effect of mud cake can be detected by the abnormal behavior of the Stoneley-wave attributes, a stiff/hard mud cake results in the underestimation of formation permeability from the Stoneley wave. In this situation, the NMR method provides a reliable indication of permeability.

4.3.3 REMARKS ON PERMEABILITY MEASUREMENT USING STONELEY WAVES

In this chapter, we described the theory and method for estimating formation permeability from borehole Stoneley waves. The theoretical foundation for this measurement is the hydraulic exchange between the borehole and the formation pore-fluid system. Because the slow compressional wave, as in Biot's theory, is primarily borne in the pore fluid, this wave and its characteristics control the hydraulic exchange (this is why the simplified theory, which focuses mainly on the slow wave, captures the effects of a porous borehole on Stoneley-wave propagation quite well). Based on slow-wave characteristics, we can discuss the merit and limitation of the permeability measurement using Stoneley waves.

It is remarkable that recent advances made it possible to derive a continuous permeability profile from borehole Stoneley-wave log data. The theory and measurement results have evidently established the strong correlation between formation permeability and borehole Stoneley waves. Because acoustic logging is fast and relatively inexpensive to perform, this provides a useful and economical method for estimating formation fluid transport properties.

The Stoneley-wave method, however, may have some drawbacks because it depends on the slow-wave characteristics in the formation. This is caused by the fact that the formation slow wave, when excited by a low-frequency Stoneley wave in the borehole, is a diffusive wave with strong attenuation. The slow wave exists only in a small region close to the borehole wall (Zhao et al., 1993, Zhao 1994). In other words, the wave phenomenon resembles a "skin-depth" effect, having a very limited depth of penetration into the formation (the skin-depth phenomenon is common for waves with a diffusive nature, e.g., thermal waves). Thus the Stoneley-wave measurement is strongly affected by the conditions of the borehole wall. The effect of mud cakes is an example of borehole-wall effects. As discussed earlier in this chapter, a mud cake with finite shear rigidity can seriously impair the sensitivity of the Stoneley wave to formation permeability, resulting in invalid permeability estimates (fortunately, in many situations mud cakes appear to have virtually no shear rigidity). Drilling-induced effects can also affect Stoneley-wave permeability measurement. Drilling may induce fissures or micro-cracks in the formation, which form a damaged zone with high permeability around the borehole. Zhao et al. (1993) and Zhao (1994) has modeled this situation. The result shows that the Stoneley-wave attenuation and dispersion are largely controlled by the high-permeability zone, even when the zone is only 1–2 inches thick. This result is confirmed by an unpublished report of a laboratory experiment performed at the Earth Resources Laboratory, M.I.T. In this scale-model acoustic-logging experiment, near-borehole micro-cracks, resulting from the thermal effect in drilling a plexiglass model, strongly attenuate the Stoneley waves measured in the borehole model, although the formation (plexiglass) has no permeability. The above discussions suggest that borehole conditions should be taken into account when interpreting the Stoneley-wave-derived permeability profile.

It is worthwhile to discuss the effect of anisotropy on Stoneley-wave permeability measurement. Schmitt (1989) has modeled this effect using a poroelastic transversely isotropic (TI) model. In this model, not only the elastic constants, but also the permeability, are modeled as transversely isotropic, with the TI-symmetry axis along the borehole (assuming a vertical borehole). The result shows that Stoneley-wave attenuation and dispersion are mainly controlled by horizontal permeability, showing almost no sensitivity to vertical permeability. This result is quite understandable in terms of slow-wave characteristics. The slow wave, when excited by a borehole Stoneley wave, is dominated by the radial motion rather than the vertical motion (see equation (4.40) and (4.41)). Thus the wave is controlled by the permeability in the radial (horizontal) direction.

Finally, we mention an interesting observation that may lead to further research.

The observation is that the Stoneley-wave propagation model based on Biot's theory appears to underestimate the permeability effect, especially in low-permeability formations. According to the theory, in order to produce a measurable effect on Stoneley waves, the permeability of the formation (water- or oil-saturated) must be no less than several millidarcies. However, it is commonly observed that the permeability effect can still be measured from Stoneley waves when the formation permeability is well below 1 md (e.g., Qobi et al., 2001). There are three possible explanations for this observation. The first is that it is possible that the fluid (oil or water) saturating the formation may contain some gas. As a result, the pore-fluid system may have an apparent modulus/viscosity that is much smaller than that of pure water or oil. According to equation (4.69), the reduced fluid modulus/viscosity can produce significant Stoneley-wave attenuation and dispersion even with a small permeability value. The second explanation is that the (low-permeability) formation may sometimes be damaged by drilling. As discussed above, the drilling damage may create a high-permeability zone around the borehole and significantly affect the Stoneley-wave propagation. A third possibility is the existence of some non-Biot mechanism(s) that contribute to the hydraulic interaction between borehole and formation. This is perhaps an interesting topic of research in this important area. For example, Prof. Kexie Wang and his students (personal communication) at Jilin University, China, find that adding a pore-fluid squirting mechanism in the Biot theory (Dvorkin and Nur, 1992) can explain the above observation.

CHAPTER 5: ACOUSTIC LOGGING IN ANISOTROPIC FORMATIONS: THEORY, METHOD, AND APPLICATIONS

5.1 Anisotropy in a Borehole Environment

Many rocks in earth formations exhibit anisotropic characteristics. A common situation is the anisotropy existing in many sedimentary rocks, such as shales. Fortunately, in many if not most cases, anisotropy is modeled by its simplest form: transverse isotropy (TI). This anisotropy has a symmetry axis such that along any direction transverse to this axis one sees the same material property (velocity/slowness). Between the symmetry axis direction and the direction perpendicular to it, one sees a material property difference. The TI-type of anisotropy is frequently encountered in borehole acoustic logging. For a borehole penetrating an earth formation, two TI cases are commonly encountered. In one case the axis of symmetry of the TI formation coincides with the borehole axis. This case is referred to as vertical TI or VTI (one can think of a vertical borehole penetrating horizontally-layered formations). The existence of VTI in earth formation rocks is closely related to the depositional process of sediments over geological times. Another TI case is the situation where its axis of symmetry is perpendicular to the borehole axis. This anisotropy is often called azimuthal anisotropy because one now sees different material properties along different azimuthal directions from the borehole. The azimuthal anisotropy for a vertical borehole is also called horizontal TI or HTI, which is appropriate since the TI symmetry axis is horizontal. The HTI around a borehole can be caused by fractures parallel with the borehole. Fracture-induced HTI is commonly encountered in deep formations where the stress field is dominated by the vertical overburden. Since fractures tend to initiate along the maximum stress direction (the vertical direction), the resulting fracture system tends to align with the vertical direction. An unbalanced formation stress field can also cause azimuthal anisotropy around a borehole (vertical or deviated). Many rocks in earth formations have porosity and micro-cracks, such as sandstones, making their elastic properties sensitive to the ambient stress field. For a borehole penetrating such rocks, if the two principal stresses transverse to the borehole are not equal, the elastic property of the surrounding rock shows an azimuthal variation around the borehole, resulting in an apparent azimuthal anisotropy. Moreover, the azimuthal anisotropy can also be seen in a deviated or horizontal borehole that is inclined or parallel to the beds of vertical (TI) formation layers.

Acoustic logging in VTI formations was studied in Chapter 2 for monopole and dipole cases, and by Ellefsen (1990) for higher-order modes (e.g., screw waves). It has been shown that acoustic logging can only determine vertical wave-propagation velocity for most types of wave modes used, such as the P- and S-waves in monopole logging, the flexural wave from dipole logging, and the screw wave in quadrupole logging, etc. The vertical propagation wave velocity is $V_P^2 = c_{33}/\rho$ for the P wave, and $V_S^2 = c_{44}/\rho$ for the S wave (including flexural and screw waves). This is because most borehole waves (monopole and dipole, etc.) involve wave motion or vibration along or transverse to the borehole. The only exception is the Stoneley wave. At low frequencies, the Stoneley wave, or tube wave, involves radial displacements that distort the circumference of the borehole. This circumferential distortion involves the shear modulus c_{66}. Later in this chapter, we will show that, by determining vertical propagation S-wave velocity V_{Sv} from acoustic logging, V_{Sh} can be estimated from the measured Stoneley-wave data to determine the shear-wave VTI property.

The recent development of cross-dipole acoustic-logging technology has made it possible to measure shear-wave azimuthal anisotropy in a downhole environment. Because of this development, the shear-wave HTI has become an important topic. Ellefsen et al. (1991) and Sinha et al. (1994) performed theoretical analyses of the HTI formation surrounding a borehole. Cheng and Cheng (1995) modeled the synthetic acoustic waveforms in a HTI formation using a finite-difference technique. This chapter will deal primarily with the shear-wave HTI, or the shear wave azimuthal anisotropy, in the borehole environment. We will describe the basic results of the analyses that form the basis of shear-wave HTI determination from cross-dipole acoustic waveform data.

Cross-dipole anisotropy measurements have now been routinely made in common well logging practice. Several important applications have been identified for the cross-dipole measurement. One important application is fracture detection and analyses in open and cased boreholes. The cased-hole application is particularly important because it provides an effective means for evaluating hydraulic fracture stimulation, and finding fracture orientation and extent. Another important application is formation stress analyses to identify stress-induced anisotropy and estimate stress orientation and magnitude. In deviated boreholes, cross-dipole measurements can also measure the effect of VTI projected along the borehole. In this chapter, we will describe the theoretical foundations for fracture and stress measurements using cross-dipole logging. Field data examples will be shown to demonstrate the applications.

5.2 Analysis of Cross-Dipole Acoustic Waveform Data for Shear-Wave Anisotropy Determination

5.2.1 Basic Theory for Flexural Waves in Azimuthally Anisotropic Formations

Let us consider dipole acoustic logging in a circular borehole with an azimuthally anisotropic formation. For a borehole configuration, the azimuthually anisotropic formation is considered to be transversely isotropic (TI) with the axis of symmetry perpendicular to the borehole axis, i.e., the anisotropy is HTI. If the TI symmetry axis is at an arbitrary angle with the borehole, an apparent azimuthal anisotropy may still be measured but the interpretation in terms of TI parameters is complicated (Ellefsen et al., 1991; Sinha et al., 1994). Figure 5.1 shows a diagram illustrating the four-component dipole logging in such a formation. The tool is assumed to be perfectly centered in the borehole. The dipole logging is a directional measurement with the direction of source-transmitter vibration making an angle θ with the fast shear-wave azimuth. When θ is non-zero, the borehole flexural-wave motion induced by the source splits into fast and slow flexural waves which propagate along the borehole axial direction. During logging in such a formation, the fast and slow flexural waves are received by in-line and cross-line receiver arrays aligned in the borehole axial direction. For the in-line receivers, the maximum receiving sensitivity is in the source vibration direction, while for the cross-line receivers, this sensitivity direction is perpendicular to the source direction. When θ is 0° or 90°, only fast or slow flexural waves are excited. The waves generated in these situations are called principal flexural waves, denoted by FP and SP, respectively.

Four-component dipole data are commonly acquired during logging using a cross-dipole tool. A cross-dipole tool consists primarily of two sets of transmitter-receiver arrays, each pointing to the X- and Y-axes of the Cartesian coordinates, as shown in Figure 5.1. The four-component wave data are two in-line components (XX and YY), and two cross-line components (XY and YX), where the first letter refers to source orientation, and the second letter refers to receiver orientation. We now describe the relationship between the four-component data (XX, XY, YX, and YY) and the principal waves. Assume that both X- and Y-transmitters emit a source wave signal $S(t)$. Take the X-source transmitter as an example. As shown in Figure 5.2, the transmitter emits a shear wave in the X-direction. Upon entering the formation, this wave will split into fast and slow shear waves that are polarized in fast and slow principal direction, respectively. Fast and slow waves due to the X-source are obtained by projecting the source shear wave, which polarizes in the X-direction, onto the fast and slow directions, as

$$\begin{cases} F_x = S(t)\cos\theta \\ S_x = S(t)\sin\theta \end{cases} . \tag{5.1}$$

The fast and slow waves, after propagating a distance z from the source to the

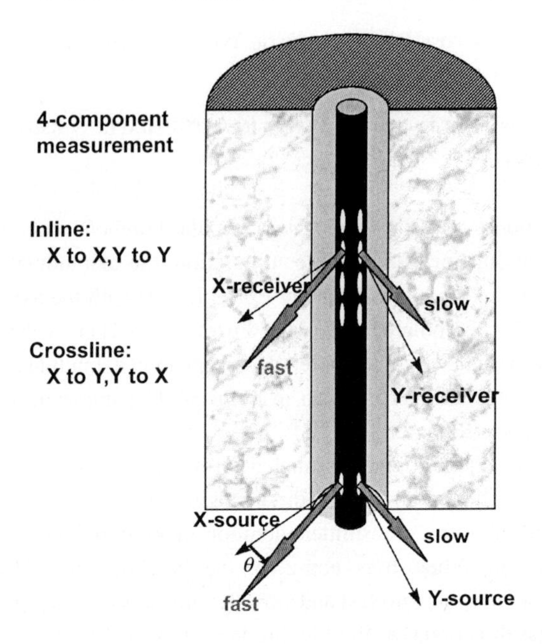

Figure 5.1. Diagram illustrating four-component (4C) cross-dipole logging in an azimuthually anisotropic (HTI) formation. The 4C data XX, XY, YX, and YY are acquired to determine the fast shear polarization direction θ and the anisotropy magnitude (difference between fast and slow velocities).

receiver location with their respective slowness s_1 and s_2, are received by the X- and Y-transmitters. The XX and XY wave data are obtained by respectively projecting the fast and slow waves in equation (5.1) back to the X- and Y-axis, as

$$\begin{cases} XX(t) = \left(S(t - s_1 z)\cos\theta\right)\cos\theta + \left(S(t - s_2 z)\sin\theta\right)\sin\theta \\ XY(t) = -\left(S(t - s_1 z)\cos\theta\right)\sin\theta + \left(S(t - s_2 z)\sin\theta\right)\cos\theta \end{cases}. \tag{5.2}$$

The expressions for YY and YX data can be obtained by replacing θ with $\theta + 90°$ in the above expression. Obviously, the fast and slow principal waves are $FP(t) = S(t - s_1 z)$ and $SP(t) = S(t - s_2 z)$, respectively (for simplicity, we ignore the borehole response difference in the fast and slow directions). We can therefore construct the fast and slow principal waves from the four-component data using

$$\begin{cases} FP(t) = \cos^2\theta\, XX(t) + \sin\theta\cos\theta\left(XY(t) + YX(t)\right) + \sin^2\theta\, YY(t) \\ SP(t) = \sin^2\theta\, XX(t) - \sin\theta\cos\theta\left(XY(t) + YX(t)\right) + \cos^2\theta\, YY(t) \end{cases}. \tag{5.3}$$

It is clear from the above derivations that the analysis of the four-component data is very similar to the analysis of a 2×2 stress or strain system. The data can therefore be expressed in a matrix form, which, after rotating by an appropriate angle θ, can be diagonalized, the two diagonal terms being the fast and slow principal waves, respectively.

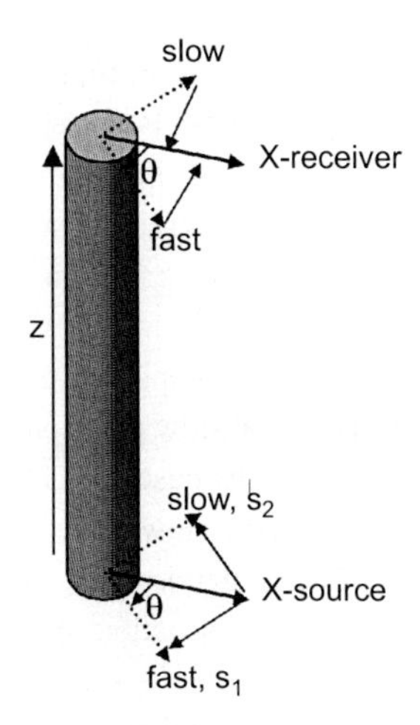

Figure 5.2. Borehole shear-wave measurement in a HTI formation. The dipole source in the X-direction emits a shear wave. Upon entering the formation, the wave splits into fast and slow shear waves. After propagating a distance z along the borehole with respective slowness s_1 and s_2, the fast and slow waves project back to the X- (Y-) receiver direction to yield the XX-(XY-) component wave data.

$$\begin{pmatrix} XX & XY \\ YX & YY \end{pmatrix} \xrightarrow{\theta} \begin{pmatrix} FP & 0 \\ 0 & SP \end{pmatrix} . \tag{5.4}$$

The angle θ between the X-direction and fast shear azimuth (see Figure 5.2) is the angle required to diagonalize the 2×2 matrix, which is obtained using the following expression:

$$\sin 2\theta (XX(t) - YY(t)) - \cos 2\theta (XY(t) + YX(t)) = 0 . \tag{5.5}$$

In conventional four-component data processing, equation (5.5) is first solved to determine the unknown orientation angle θ (e.g., Alford, 1986; Mueller et al., 1994). This angle is then substituted into equations (5.3) to combine the four-component data into the principal time series $FP(t)$ and $SP(t)$. There is an ambiguity in the determination of fast shear azimuth θ. This is because both θ and $\theta + 90°$ are a solution of this equation. Thus, the determined angle is either the fast or slow shear-wave azimuth. To overcome this ambiguity, principal wave series are processed across the receiver array to obtain the velocities. If the velocity associated with $FP(t)$ is smaller than that associated with $SP(t)$, then the identities of $FP(t)$ and $SP(t)$ are interchanged, and the fast shear azimuth is $\theta + 90°$ instead of θ. However, because the anisotropy can be a small quantity to measure, the velocity (slowness) difference between $FP(t)$ and $SP(t)$ can be smaller than the sum of the velocity (slowness) errors incurred in the processing of the principal wave arrays. These errors are commonly caused by dispersion effects of flexural waves, poor data quality, etc. In these situations, it is difficult to determine the shear-wave anisotropy, and resolve the ambiguity of the fast shear azimuth. A solution to this problem is presented by formulating an array waveform inversion procedure, which will be described in the next section. The essence of the waveform inversion is matching the fast and slow waveforms to estimate the anisotropy parameters. This match is based on the similarity of the fast and slow principal waves, as described below.

Theoretical studies show that the two principal waves have the same polarity and similar waveforms. An example of fast and slow principal flexural waves is shown in Figure 5.5a (solid curves; this figure will be described in more detail later.), the two waves having a similar wave shape with the slow wave lagging behind the fast wave. The similarity of the principal waves and their time lag are used to formulate a waveform inversion procedure for estimating formation anisotropy. In some cases, the similarity of the fast and slow flexural waves may degrade due to large anisotropy, differential attenuation, and/or flexural wave dispersion. This will introduce amplitude mismatch between the fast and slow waves. However, the phase mismatch (or time lag) between the waves, compared to the amplitude mismatch, is the first-order effect that is captured in the waveform inversion procedure. The degradation in similarity can also be reduced by matching the fast and slow waves toward the wave onset.

5.2.2 WAVEFORM INVERSION ANALYSIS

The array waveform inversion formulation below is used for simultaneous determination of anisotropy and fast shear azimuth. Initially, two auxiliary principal time series are generated by differentiating equations (5.3) with respect to θ. These auxiliary waves will be minimized to locate the azimuth of the fast or slow wave. Mathematically, the fast and slow principal waves represent a state of maximum or minimum attained by the flexural wave at certain azimuth θ. Therefore, by finding the zero of (or minimizing) the wave's derivative with respect to θ, this azimuth can be located. The auxiliary waves are given by

$$\begin{cases} FP^{\theta}(t) = -\sin 2\theta\,(XX(t) - YY(t)) + \cos 2\theta\,(XY(t) + YX(t)) \\ SP^{\theta}(t) = \;\;\sin 2\theta\,(XX(t) - YY(t)) - \cos 2\theta\,(XY(t) + YX(t)) \end{cases} . \quad (5.6)$$

The significance of equations (5.6) will be discussed later in conjunction with the inversion analysis. Next, we make use of the similarity between fast and slow principal waves in equations (5.3) and equations (5.6). For the moment, we assume that the formation spanned by the transmitter-to-receiver-array distance is homogeneous and the slowness values of the fast and slow shear waves are s_1 and s_2, respectively. Assuming that the fast and slow waves are similar in shape, the slow wave, after being properly advanced in time, should substantially match the waveform of the fast wave. For example, at receiver m, the expression for the waveform matching is

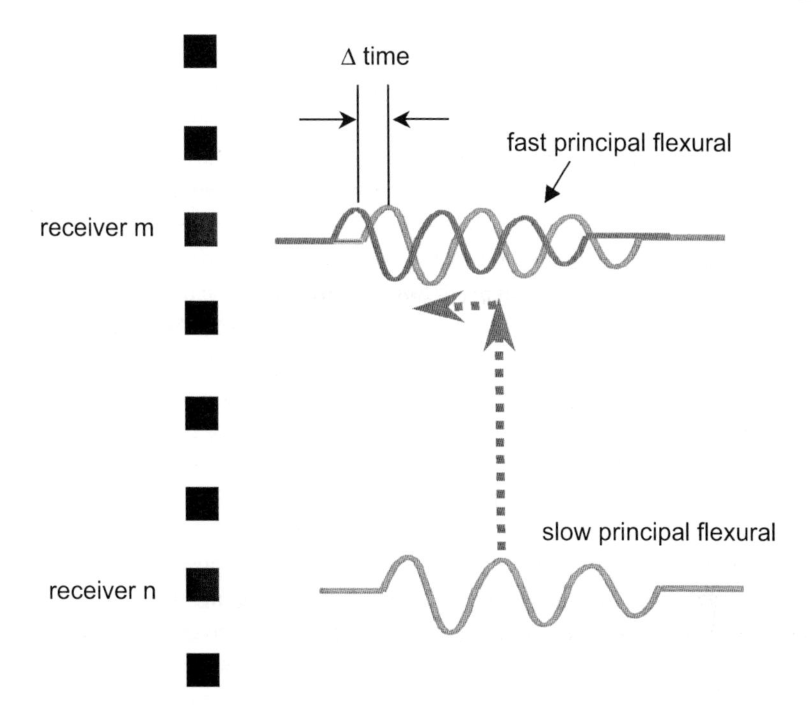

Figure 5.3. Matching fast and slow principal shear waves across a receiver array to determine anisotropy. The slow wave of receiver n is first propagated to receiver m, then advanced by a time shift (controlled by anisotropy) to match with the fast wave there. For all receiver combinations ($1 \le n,\ m \le N$), there are total 64 waveform pairs for an eight ($N = 8$) receiver array.

$$SP_m(t+\delta s\, z_m) \approx FP_m(t) \quad ; \quad SP_m^{\theta}(t+\delta s\, z_m) \approx FP_m^{\theta}(t)\ , \tag{5.7}$$

where $\delta s = s_2 - s_1$; subscript m denotes the receiver index, and z_m is the distance between the source and the mth receiver in the array. This waveform-matching configuration is illustrated in Figure 5.3 at the receiver m location. Now, by means of wave propagation, the wave data of the entire array are used in the waveform matching. The principal wave data at any receiver, say, receiver n, can be propagated to receiver m to perform the waveform matching, as is also illustrated in Figure 5.3. For example, propagating the slow principal wave $SP_n(t)$ at receiver n to receiver m using slowness s_2, and then advancing it in time by $\delta s z_m$, the resulting waveform should match that of the fast wave $FP_m(t)$, as

$$SP_n(t-s_2(m-n)\delta z+\delta s\, z_m) \approx FP_m(t)\ ;\ SP_n^{\theta}(t-s_2(m-n)\delta z+\delta s\, z_m) \approx FP_m^{\theta}(t)\ , \tag{5.8}$$

where δz is receiver spacing; the receiver index m can be smaller than, equal to, or greater than the receiver index n. We propagate $SP_n^{\theta}(t)$ by also using s_2, because this wave, being generated from $SP_n(t)$ by operating on θ only, should have the same propagation velocity (or slowness) as $SP_n(t)$. Treating the auxiliary waves (controlled by θ) in the same way as treating the principal waves (controlled by anisotropy) ensures that they have equal importance in the inversion formulation, so that both azimuth and anisotropy are properly determined. The wave propagation procedure connects waveforms at different receivers using wave slowness across the array. Thus, by matching the principal waves across the entire array, an inversion procedure is formulated to estimate wave slowness s_2, anisotropy δs, and angle θ. Later in this chapter, we describe a numerical result that further reduces the inversion parameters to δs and θ only. The objective function for the inversion is constructed as

$$\begin{aligned} E(\delta s, \theta, s_2) = & \sum_{m,\,n=1}^{N} \int_T \left(SP_n(t - s_2(m-n)\delta z + \delta s\, z_m) - FP_m(t)\right)^2 dt \\ & + \sum_{m,\,n=1}^{N} \int_T \left(SP_n^{\theta}(t - s_2(m-n)\,\delta z + \delta s\, z_m) - FP_m^{\theta}(t)\right)^2 dt \quad . \end{aligned} \tag{5.9}$$

The integration is over the time window T, in which the waveforms are matched. The summation in equation (5.9) includes all possible combinations of fast and slow principal wave traces in the array consisting of N receivers. For example, for a typical eight-receiver array ($N = 8$), the fast waves of any receiver are compared with the slow waves at all eight receivers by means of propagation and time shifting. Thus, for all eight receivers, the summation in equation (5.9) includes $8 \times 8 = 64$ waveform pairs. This data combination, or stacking, is an effective way to obtain reliable anisotropy estimates. Because the anisotropy is usually a small quantity to measure, it is easily obscured by the noise in the wave data. By stacking an enormous amount of data using equation (5.9), any incoherent signal, such as noise, will be largely suppressed or eliminated. This allows the (small) anisotropy effect to be unambiguously and reliably estimated. This approach of using all data combinations in a receiver array is also utilized in the waveform inversion analysis for estimating slowness across the array (see Chapter 3). In the formulation of equation (5.9), the direct estimation of anisotropy significantly reduces the anisotropy error compared to the method using Alford (1986) rotation, because we now deal with only one uncertainty instead of two, as incurred in the fast and slow wave velocity analysis procedure of the conventional method. Besides, this one uncertainty should be reduced by the data-stacking scheme of equation (5.9).

We now discuss the structure of the objective function and its role in the inversion. The first term in equation (5.9) is mostly sensitive to s_2 and δs, while it is moderately sensitive to θ. When proper values of these parameters are reached, this term will be minimized. The second term, however, is primarily controlled by θ. Note that the functional forms of $SP^{\theta}(t)$ and $FP^{\theta}(t)$ are the same as in equation (5.5). When the value of θ corresponds to the fast or slow shear azimuth, both $SP^{\theta}(t)$ and $FP^{\theta}(t)$ tend to vanish and the second term is minimized. Although minimizing either $SP^{\theta}(t)$ or $FP^{\theta}(t)$ can yield the desired θ value, minimizing the residue between $SP^{\theta}(t)$ and $FP^{\theta}(t)$ has two additional advantages besides finding the correct θ value:

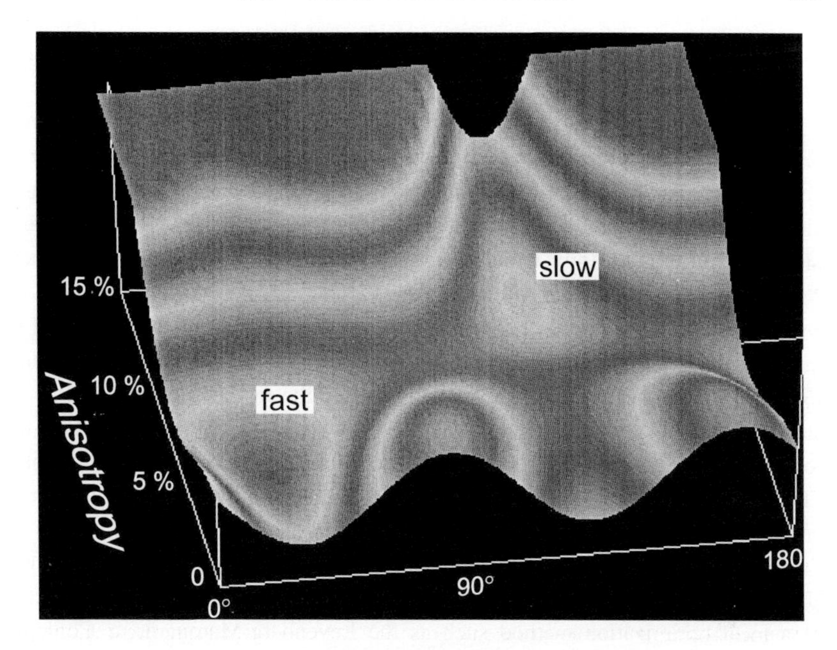

Figure 5.4. Example of objective function for cross-dipole data inversion. Vertical axis is rms wave mismatch residue (arbitrary scale). This function is minimized to find the global minimum at the fast shear polarization azimuth. Note that a local minimum also exists at the slow shear azimuth.

1. Before the convergence of the inversion process, $SP^{\theta}(t)$ and $FP^{\theta}(t)$ are sensitive to s_2 and δs. This helps drive the parameters towards correct values.
2. Minimizing the wave residue, instead of the waveform of $SP^{\theta}(t)$ or $FP^{\theta}(t)$, ensures that the first and second terms in equation (5.9) have the same order of magnitude, so they have equal importance in the inversion process.

Figure 5.4 shows a plot of the error surface given by equation (5.9) for a field data set. The error surface is plotted against anisotropy (defined by $2\delta s/(s_1+s_2)$) and angle θ by setting s_2 to an appropriate value. The advantage of simultaneous determination of θ and anisotropy can now be explained. As discussed earlier, whether $FP(t)$ represents a fast or slow principal wave is ambiguous, because $FP(t)$ can become $SP(t)$ if we replace θ with $\theta + 90°$, or if $\delta s = s_2 - s_1$ is negative. With the formulation given in equation (5.9), this ambiguity can be avoided or reduced. In the inversion process, we find $FP(t)$ as the true fast wave by solving the problem in the domain of $\delta s \geq 0$. For example, let us assume that the inversion finds a θ value that corresponds to the slow shear azimuth (e.g., the local minimum marked **Slow** in Figure 5.4). For this θ value, the objective function will not be minimized in the $\delta s \geq 0$ domain since its true minimum is in the negative δs domain (see also Figure 5.4; at the slow shear azimuth, a minimum is forming toward the negative δs direction). The inversion then proceeds to adjust the value of θ and calculate the principal waves until both terms of equation (5.9) are minimized for $\delta s \geq 0$, which corresponds to the true minimum marked **Fast** in Figure 5.4. To this end the value of θ corresponds to the fast shear azimuth, which is about 90° away from the previous θ value.

The minimum at the slow azimuth is only a local minimum. At the slow azimuth, $SP(t)$ becomes the actual fast wave and $FP(t)$ becomes the slow wave, as can be verified by replacing θ with $\theta + 90°$ in equations (5.3). However, because the inversion using equation (5.9) is confined to the $\delta s \geq 0$ domain, the inversion process always advances $SP(t)$ (now the fast wave) to match with $FP(t)$ (now the slow wave). A partial match can be obtained when the second wave peak or trough of the fast wave matches the first peak or trough of the slow wave. Since the first cycle of the fast wave is skipped, this partial match produces only a local minimum, as seen in Figure 5.4. Only when the θ value corresponds to the fast azimuth do the fast and slow waves attain the best match in the chosen window T. This shows that the true (or global) minimum of the objective function defines the solution of the problem. Therefore, by finding the global minimum, we simultaneously determine anisotropy and the fast shear azimuth.

The inversion process uses a global minimization method described in Ingber (1989) and Chunduru et al. (1996). The use of a global optimization method, instead of a local optimization method such as the Levenberg-Marquardt or Conjugate Gradient algorithm (Press et al., 1989), is justified by the nature of the problem defined by equation (5.9). This problem is strongly nonlinear and multiple local minima exist. As shown in Figure 5.4, for the parameter θ alone, there exist two minima corresponding to fast and shear wave azimuths, respectively. For $\delta s \geq 0$, there may exist other local minima at larger δs values. For an inversion problem with multiple minima, a local minimization method may get trapped in a local minimum, failing to find the true solution (the global minimum) of the problem. A global method, on the other hand, will find the global minimum of the objective function. Although global optimization methods are generally computationally more intensive than local methods, efficient algorithms exist to speed up the global minimization procedure. A Very Fast Simulated Annealing algorithm (Ingber, 1989; Chunduru et al., 1996) is used to find the global solution of equation (5.9).

5.2.3 INCORPORATING IN-LINE SLOWNESS

The number of inversion parameters can be reduced from three (s, δs, and θ) to two (δs, and θ) if the slowness value from an in-line component (XX or YY) is obtained. An in-line wave array measures the fast (slow) shear slowness when it points to – or is polarized in – the fast (slow) direction. What slowness value will an in-line wave measure when it points at an angle θ from the fast direction? The answer to this question will help us understand the uncertainties in conventional dipole (i.e., in-line only) logging in an azimuthally anisotropic formation and will provide a useful application in the present cross-dipole anisotropy analysis.

Numerical modeling is used to study the problem. We first compute the synthetic in-line flexural waves for an anisotropic formation (fast and slow shear velocities are

1190 m/s and 1090 m/s, respectively) using a finite difference method formulated for a HTI formation (Cheng and Cheng, 1995). The synthetic in-line waves are computed with θ varying from fast direction ($\theta = 0°$) to slow direction ($\theta = 90°$) at 10° increments. The waves are shown in Figure 5.5a, with the fast and slow waves denoted by solid curves and the in-line waves by dashed curves. There are a total of eleven array waveform sets (a data set is also computed for $\theta = 45°$). These wave data sets are processed to find their associated velocity across the array using the waveform inversion method (see Chapter 3). The obtained velocity as a function of θ is shown in Figure 5.5b (solid circles). Note the computed fast and slow flexural-wave velocities are a little lower than the respective shear velocities because of dispersion effects; but the relative difference, i.e., anisotropy, is almost the same as computed from the shear velocities. The computed velocity shows a regular relation with θ. It has the value of the fast velocity at $\theta = 0°$, then changes dramatically between 30° and 60°, and finally reaches the slow velocity at $\theta = 90°$. This behavior is well characterized with a simple empirical relation:

$$v(\theta) = v_1 + \frac{v_2 - v_1}{2}\left(1 - \tanh(8(\theta/90° - 0.5))\right) ,$$

(subscripts 1 and 2 refer to fast and slow velocity, respectively), as shown by the solid curve in Figure 5.5b. This relation holds for modeling results done for formations with different values of shear velocity and anisotropy. With this relation, we now understand the velocity an in-line wave will measure when it points at an angle from the principle direction.

Let us return to the problem defined in equation (5.9). Assume we have an in-line slowness value $s(\theta)$, measured at an unknown azimuth θ. With the result from the above numerical study, we can express the slow shear-wave slowness using the following relation:

$$s_2 = s(\theta) + \frac{\delta s}{2}\left(1 - \tanh(8(\theta/90° - 0.5))\right) , \tag{5.10}$$

where θ is in degrees. Substituting this expression into equation (5.9), we see that the inversion parameters reduce to δs and θ only, which significantly improves the speed of the inversion. Besides reducing the size of the inversion problem, the in-line slowness is integrated from source to receiver to calculate an approximate shear-wave travel-time. This travel time is used to define the start time of the waveform window T used in equation (5.9). In this way, the start time of the window T for each subsequent depth is updated in the data processing.

5.2.4 Improving Resolution

In the previous analysis, we assume that the formation from source to receiver array is homogeneous. That is, the splitting of the fast and slow waves occurs at the source; thereafter, the two waves propagate with respective uniform slowness values of s_1

and s_2 toward the receiver array. By this assumption, the anisotropy parameter δs determined from inversion is an average value from source-to-center-array distance (about 10 ft–13 ft (3 m–4 m)). However, the parameters θ and $s(\theta)$ are determined from the receiver array having an aperture of typically 3.5 ft (1 m). It is therefore desirable to determine the anisotropy with the same resolution as θ and $s(\theta)$ or s_2, as this will help resolve thin anisotropic intervals in the formation.

We now describe how the resolution of anisotropy can be improved within our inversion formulation. The essence of this formulation is to utilize the time difference between fast and slow waves across the receiver array to determine the anisotropy. At any receiver, say, receiver m, this time difference δt can be decomposed into two parts as

$$\delta t = \delta s z_m = \delta t_0 + \delta s'(z_m - z_1) \ , \tag{5.11}$$

where $\delta s'$ is the slowness difference between slow and fast shear waves within the array, z_1 is the source-to-first-receiver distance, and δt_0 is the initial time difference between the fast and slow waves at the first receiver. The difficulty is that δt_0 is an unknown parameter. Inverting this parameter from data is one possible solution but would add computation time. However, if we know the average value of δs from source to first receiver, we can compute δt_0 as the product of this average δs and z_1. This leads to a processing scheme described below.

At the starting depth, we invert δs for the entire source-to-receiver spacing. The result is designated as $\overline{\delta s}$. This gives an estimate of $\delta t_0 = \overline{\delta s}\, z_1$ for equation (5.11). Making the substitution

$$\delta s\, z_m \rightarrow \overline{\delta s} z_1 + \delta s'(z_m - z_1)$$

in equation (5.9), we can invert $\delta s'$ for the interval spanned by the receiver array for the next depth. At each depth after inversion, the estimated $\delta s'$ is combined with previous $\delta s'$ estimates to produce a new estimate of $\overline{\delta s}$. for the following depth. Processing data for the entire depth segment of interest in this manner, we can obtain anisotropy to the same resolution as slowness and angle, which is about 3.5 ft (1.1 m).

5.2.5 PROCESSING EXAMPLES

In this section, we present the results of anisotropy analysis using the inversion technique formulated in equation (5.9). We demonstrate the improved resolution of anisotropy using equation (5.11). We also compare the results from the present analysis with those from the conventional analysis. Finally, we demonstrate the basic quality indicators of the present technique using a field example.

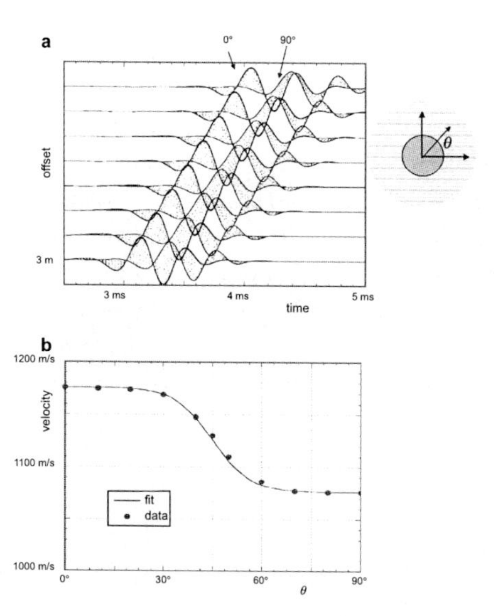

Figure 5.5. (a) Synthetic in-line array flexural waves by pointing the dipole at various azimuths between fast ($\theta = 0°$) and slow ($\theta = 90°$) directions (see the right diagram).
(b) In-line flexural wave velocities (dots) from processing the data in (a). The solid curve is the theoretical fit using a hyperbolic tangent function

5.2.5.1 Comparing inversion and conventional techniques

Figure 5.6 shows the four-component cross-dipole data (XX, XY, YY and YX), across a very slow diatomite formation segment. Only data from receiver 1 of an eight-receiver array are displayed. The presence of anisotropy is evident by the significant cross-component wave amplitude. The cross-line wave amplitude varies with depth as a result of tool rotation and change of formation properties.

It is both interesting and instructive to compare the anisotropy analysis results of the present technique with those of the conventional technique. Figure 5.7 compares the two techniques using the cross-dipole data shown in Figure 5.6. The left track shows the gamma-ray curve and the tool's X-dipole azimuth curve. The latter curve clearly shows the spinning of the tool during data acquisition. The middle track shows the fast shear direction (Fastdir), or azimuth, and the right track displays the associated anisotropy determined for the array aperture resolution. In both tracks the thick-solid curves are from the inversion technique, while the thin-dotted curves are from the conventional technique.

There is some degree of agreement between the two results in intervals where the anisotropy value is relatively high. In these intervals the two fast shear azimuths overlap or track each other quite well. However, when the anisotropy value is relatively low the conventional analysis exhibits a marked instability in the fast shear azimuth determination. As described earlier, conventional analysis finds the fast azimuth as the azimuth of the principal wave whose estimated slowness is smaller. However, angle flip between fast and slow directions occurs when the velocity error

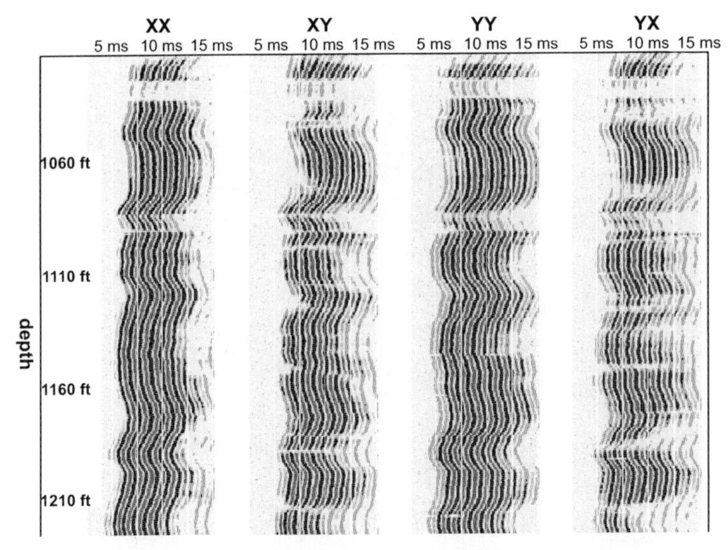

Figure 5.6. Field example of 4C cross-dipole data in an anisotropic formation, plotted using a variable-density display (only data from receiver 1 of an eight-receiver array are displayed). Note the significant cross-component wave amplitude.

is comparable or exceeds the velocity difference between fast and slow shear waves. This situation is most likely to occur when the anisotropy value is low. When this happens, the conventional technique can determine neither the fast azimuth nor the anisotropy value correctly. In contrast to conventional analysis, the inversion technique provides a more stable fast azimuth curve. There are only two angle flips over the entire depth interval. The determined azimuth shows a well-defined direction despite the rotation of the tool (see tool azimuth in left track). This example demonstrates that the inversion technique provides a more robust and reliable estimate of formation shear-wave anisotropy and azimuth.

5.2.5.2 Interpreting Anisotropy With Quality Indicators

In the following example, we discuss some quality indicators associated with the inversion analysis technique. These indicators are processing window overlay with fast and slow shear wave data, inversion residue error differences, fast and slow wave comparison, and anisotropy azimuth versus tool azimuth.

Figure 5.8 shows an example of processed anisotropy results and quality indicators. The formation consists of shale (upper portion) and sand/shale (lower portion) sequences, as can be seen from the gamma-ray curve in track 1. This well is nearly vertical (well deviation < 5°) and circular (no borehole breakout and washout were reported for this depth section). This figure is a good example of interpreting shear-wave azimuthal anisotropy. The lower portion of this figure corresponds to significant estimated anisotropy values, while the upper portion has an almost zero anisotropy value. The processed anisotropy is shown in track 5 as shaded curves, for

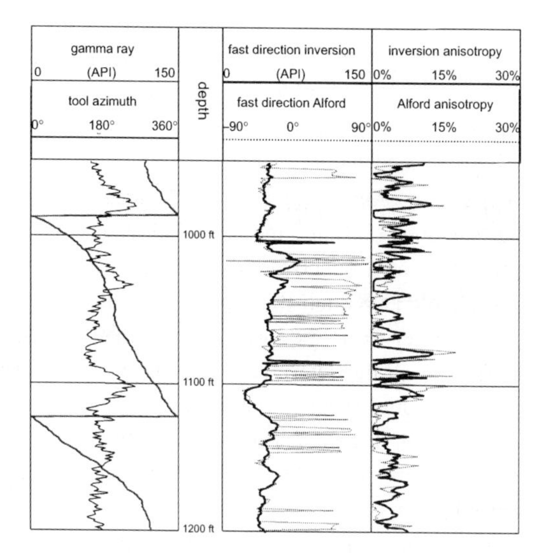

Figure 5.7. A comparison of the fast shear azimuth (track 2) and anisotropy (track 3) for both the Alford-rotation and inversion techniques. The inversion results are thick (black) curves, and the rotation results are thin-dotted (red) curves. Track 1 shows the gamma-ray and tool azimuth curves.

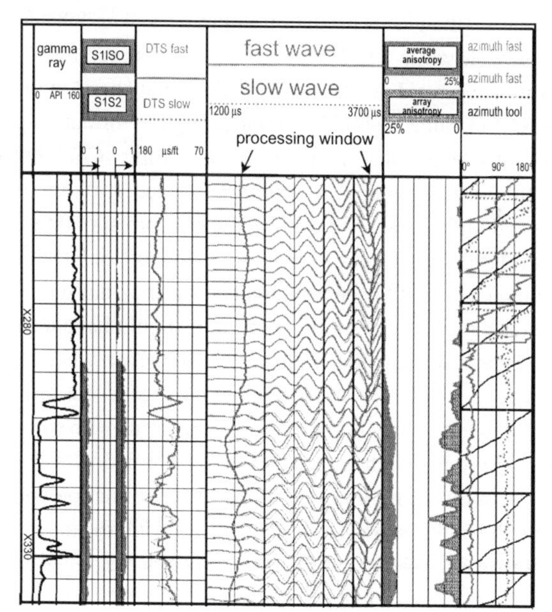

Figure 5.8. Anisotropy analysis results along with associated quality indicators. Track 1 plots the gamma-ray curve. Track 2 shows the quality indicators for anisotropy (S1ISO) and fast shear azimuth (S1S2). Track 3 plots the fast and slow slowness curves. Track 4 shows the fast and slow flexural waves along with the processing window. Track 5 shows the average anisotropy from transmitter to receiver (left to right) and the anisotropy over the array aperture (right to left). Track 6 plots the fast and slow shear azimuth curves and the tool azimuth curve.

the transmitter-to-receiver (ANIS_AVE) and array aperture (ANIS_ARRAY) resolutions, respectively. The fast and slow shear slowness curves, denoted by DTS_FAST and DTS_SLOW, respectively, are shown in track 3. Track 4 shows the fast and slow shear waves and the processing window given by two (start and end) time curves. The fast and slow waves are computed by stacking the array data to the center of the receiver array using equations (5.3) and the inversion results. These two waves are well contained in the processing window, indicating that the computed shear slowness from the in-line dipole (XX) and the input wave data are appropriate. Shear-wave splitting is clearly observed in the lower sand formation. Splitting disappears in the upper shale formation. If the determined anisotropy is true, the splitting of the fast and slow waves should start at the initial portion, or onset, of the waveform, as shown in track 4. Conversely, if the splitting occurs in the later portion of the waveform, the anisotropy is probably caused by such effects as dispersion, mode interference, and noise contamination.

The minima of the objective function (5.9) or the waveform misfit residues also serve as a good quality indicator. These indicators are denoted by S1ISO and S1S2, as shown by the two shaded curves in track 2. We use the relative residue error difference instead of the errors to reduce the effects of wave amplitude variation with depth so that it is more indicative of anisotropy. S1ISO is the relative difference between the fast-wave residue (the global minimum in Figure 5.4) and the data-fitting residue of an isotropic model, while S1S2 is the relative difference between the fast-wave residue and slow-wave residue (the local minimum at the slow azimuth in Figure 5.4). When the S1ISO value is significant, modeling the formation as anisotropic fits the data significantly better than modeling it as isotropic. Conversely, if the S1ISO value is small, the isotropic model is appropriate since it fits the data equally well. Indeed, the low and high values of S1ISO in track 2 correspond to the low- and high-anisotropy intervals very well. Let us now discuss the role of S1S2. The high value of the S1S2 curve indicates that the determined fast shear azimuth is well distinguished from the slow shear because the global minimum should have a significantly lower residue than the local minimum (see Figure 5.4). However, when S1S2 is small, it is difficult to determine which one of the minima belongs to the fast shear wave. The behavior of S1S2 is seen clearly in Figure 5.8. In intervals with significant anisotropy values, the S1S2 curve has high values and the fast and slow shear azimuths (track 6) are quite stable, showing no flipping. In the upper formation with an almost zero anisotropy value, the S1S2 value is low and the two azimuth curves become unstable. The formation anisotropy and azimuth are well-characterized by the quality indicators computed from the data-fitting residues.

Comparison of the determined anisotropy azimuth with tool azimuth may sometimes be a good indication of anisotropy. As shown in track 6, the drastic cycling of the tool azimuth curve shows that the tool was spinning during data acquisition. However, the fast shear azimuth shows a well-defined direction in the anisotropic formation despite the drastic rotation of the tool. In contrast to this, the fast (or slow) azimuth in the upper low anisotropy formation becomes undefined, showing a tendency of spinning with the tool.

The above example demonstrates the application of the quality indicators associated with the inversion analysis. These indicators provide a practical and useful tool for controlling the quality of log-derived azimuthal shear-wave anisotropy estimates. The analysis results, together with the quality indicators, verify the effectiveness and validity of the inversion technique.

5.3 Application of Cross-Dipole Anisotropy Measurement to Fracture Analyses in Open and Cased Holes

A significant application of cross-dipole measurement is in the analysis of formation fractures in a borehole environment. Characterization of fractures in earth formations is an important task in hydrocarbon exploration because fractures provide conduits for reservoir fluid flow. Detecting fractures behind casing is also important for enhancing reservoir recovery. Formation fractures parallel to or inclined with a borehole create azimuthal shear-wave anisotropy around the borehole. As described later, a shear wave polarized in the fracture-strike direction sees a higher modulus and travels faster than a shear wave polarized in the direction normal to the fracture surface, giving rise to anisotropy for the two polarization directions. The amount of anisotropy gives a measure of fracture intensity, and the fast shear-wave polarization azimuth gives the fracture strike direction. This is the theoretical basis for fracture measurements using cross-dipole logging.

In the following, we first verify the theoretical foundation for the cross-dipole measurement of fractures by modeling dipole acoustic waveforms in a fractured formation. We then compare the theoretical results with field measurements to demonstrate the applicability of the theory to open and cased boreholes. The problem of orthogonal fractures is also investigated as it sometime occurs in the borehole logging measurement.

5.3.1 Dipole Logging in Fractured Formation: Theory and Modeling

Fracture-induced seismic anisotropy can be described by Schoenberg and Sayers' (1995) theory. The presence of fractures changes the property of the elastic tensor c_{ijkl} of the medium. For a fracture system aligned in the x-direction and a borehole along the z-direction (see Figure 5.9a), the resulting anisotropy corresponds to a transversely isotropic (HTI) medium. The axis of symmetry of this HTI medium is perpendicular to the borehole axis in the y-direction. The most important elastic constants of the TI medium relevant to the dipole logging are

$$\begin{cases} c_{44} = \mu - \mu\delta \\ c_{55} = \mu \\ c_{66} = \mu - \mu\delta \end{cases}, \tag{5.12}$$

where we have assumed that the background medium is isotropic with Lamé constant λ and shear modulus μ. The parameter δ is a measure of the degree of fracturing or fracture intensity, as given by

$$\delta = \mu Z_T / (1 + \mu Z_T) \ . \tag{5.13}$$

The parameter Z_T is the fracture tangential compliance given by (Schoenberg and Sayers, 1995)

$$\sigma_T Z_T = \frac{1}{V} \sum_n \iint_{S_n} [u] \, dS \ . \tag{5.14}$$

As Figure 5.9a shows, σ_T is the tangential (shear) stress and $[u]$ is the transverse displacement discontinuity (dislocation) across the fracture surface S; the integration is first over the fracture surface S of each individual fracture, then averaged in a volume V containing n fractures. Equation (5.14) shows that the average of the displacement discontinuity is linearly proportional to the shear stress traction acting on the fracture planes. The proportionality constant is the fracture compliance Z_T. It is then understood that the fracture compliance is a measure of the degree of fracturing in terms of $[u]$, the fracture surface dislocation under the shear stress σ_T, and n, the number of fractures in the volume of fracturing. This compliance, combined with equation (5.13), gives the dimensionless parameter δ which is used to characterize the fracture intensity, or the degree of fracturing. Other elements of the elastic tensor, such as c_{11}, c_{12}, etc., are also affected by fracturing, but are insignificant in this dipole shear-wave logging configuration.

Equations (5.12)–(5.14) provide a simple physical explanation for the interaction of dipole-shear waves with formation fractures. As is also shown in Figure 5.9a, for a shear wave that propagates in the z-direction and is polarized in the fracture strike (x) direction, the shear stiffness or modulus is that of the background medium, as given by c_{55} of equations (5.12). However, if the shear wave, still propagating in the z-direction, is polarized in the (y) direction normal to the fracture surface, tending to shear across the fracture surface, the wave encounters a weaker modulus caused by fracturing, as described by c_{44} of equation (5.12) (the modulus c_{66} corresponds to a shear wave that is polarized in the y-direction and propagates in the x-direction; $c_{66} = c_{44}$ because of the TI symmetry). Thus, a velocity difference or anisotropy exists between the two polarization directions (x and y), with the anisotropy proportional to fracture intensity. Conversely, by determining anisotropy from shear-wave logging, we obtain a measure of fracture intensity from the anisotropy magnitude, and the fracture strike direction as the fast shear-wave polarization azimuth.

In earth formations, fractures often occur as orthogonal fracture sets or joints (see Figure 5.9b). An example are conjugate shear fractures. Mathematically, a system containing two orthogonal fracture sets is treated as the superposition of two TI-systems (Schoenberg and Sayers, 1995). Further, for the borehole configuration shown

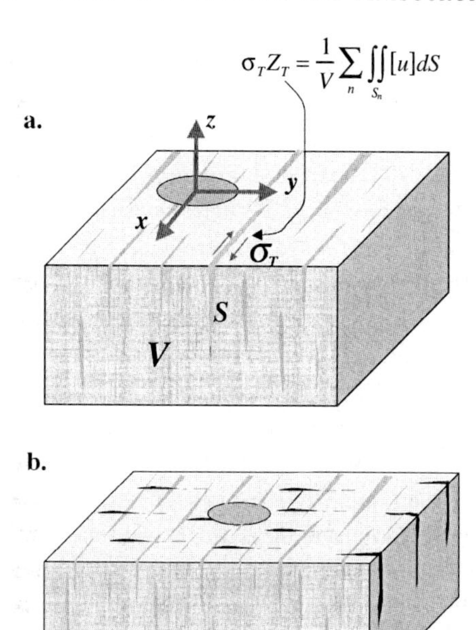

Figure 5.9. (a) A diagramatic view of a fractured formation penetrated by a borehole. The average of the (shear) displacement discontinuity across the parallel fractures in volume *V* is linearly proportional to the shear stress traction acting on the fracture planes; the proportionality coefficient is the fracture tangential compliance. The fracture compliance gives rise to shear-wave anisotropy, such that shear waves polarized along and perpendicular to the fracture planes will see different velocities.
(b) A scenario with two orthogonal fracture sets.

in Figure 5.9b, the two fracture sets correspond to two HTI-systems whose axes of symmetry are aligned in the *x*- and *y*-directions, respectively. The linear superposition of the two HTI-systems using equations (5.12) results in a new system whose shear-wave moduli are given by

$$\begin{cases} c_{44} = \mu - \mu\delta_1 \\ c_{55} = \mu - \mu\delta_2 \\ c_{66} = \mu - \mu(\delta_1 + \delta_2) \end{cases}, \tag{5.15}$$

where δ_1 and δ_2 are fracture intensities of fracture sets aligned in *x*- and *y*-directions, respectively. Equations (5.15) show that orthogonal fracture sets reduce the shear modulus regardless of the shear-wave polarization direction. However, the anisotropy, as governed by the difference between c_{44} and c_{55}, is also reduced. In particular, the anisotropy vanishes if the two fracture sets have the same fracture intensity $\delta_1 = \delta_2$.

Applying equations (5.12) and (5.15) to model a fractured formation surrounding a fluid-filled borehole (open or cased), we can simulate the dipole-shear wave propagation in the borehole using a finite difference method (Cheng et al., 1995). In the simulation, we use $\rho = 2600$ kg/m^3, $v_P = 3600$ m/s, and $v_S = 1400$ m/s for the density and for the compressional and shear velocities of the background medium, respectively. The fracture intensity is $\delta = 0.2$. The borehole has a diameter of 20 cm and is filled with water.

The open-hole case is discussed first. Figure 5.10a shows the waveform modeling results for an eight-receiver array 3.35 m away from the source. The waveform center frequency is 2 kHz. Shear-wave anisotropy is clearly seen from the simulated waveforms. Indeed, the dipole-shear waves polarized in the x-direction (solid curve) travel faster than the waves polarized in the y-direction (dashed curve). The relative difference between their respective velocities is almost exactly 10%, as specified by the model parameters c_{44} and c_{55}.

The effects of casing and cement can now be evaluated. A 1 cm thick steel and a 2.54 cm thick cement layer (casing and cement elastic properties are given in Table 2.1) are placed inside the borehole. The modeled waveforms are shown in Figure 5.10b. Although the waveforms and the wave arrival time are changed compared to the open-hole case in Figure 5.10a, the waves pointing in the x-direction (solid curve) still travel faster than the waves pointing in the y-direction (dashed curve). Moreover, the anisotropy measured from the fast and slow waves is 10 %, almost the same as the open-hole case. This means that the propagation of dipole waves in a cased borehole is largely controlled by the formation shear-wave properties. In other words, we can still reliably determine formation shear-wave anisotropy through casing.

Finally, we model dipole logging in a formation with orthogonal fracture sets. We assume that the fracture intensities for both sets of fractures are the same, i.e., $\delta_1 = \delta_2 = 0.1$. Other formation and borehole parameters are the same as the open-hole case of Figure 5.10a. Figure 5.10c shows that anisotropy vanishes because the two shear waves polarized in the fracture alignment directions travel at the same speed. Waveforms for the two polarizations (solid and dashed curves) overlay with each other (the last trace shows only the solid curve, indicating there are two sets of waveforms). However, the wave velocity or arrival time is significantly delayed compared to the unfractured case (the solid waveforms in Figure 5.10a travel with the speed of the background medium, see equations (5.12)), indicating a reduced shear modulus. The overall results of this simulation are correctly described by equations (5.15).

5.3.2 FIELD FRACTURE MEASUREMENT EXAMPLES

5.3.2.1 *Open-hole fracture analysis example*

An open-hole fracture analysis study was conducted in a well in West Texas, USA. The subject formation is a tight mixed carbonate from the Clear Fork Limestone sequence in the central platform of the Permian Basin. The final analysis is displayed in Figure 5.11. In this section we present the anisotropy from the cross-dipole data and dip analysis result along with an acoustic borehole image. Throughout the entire section, the acoustic image reveals open, natural fractures at 5693 ft, 5698 ft, 5728 ft, 5734 ft, and from 5734 ft–5,760 ft. The fractures are interpreted as conjugate shear

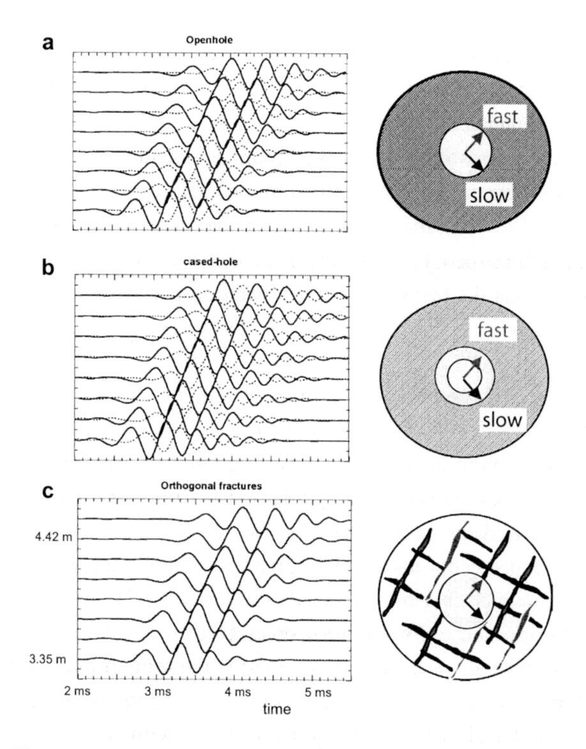

Figure 5.10. Theoretical modeling of dipole logging in fractured formation.
(a) Open hole: note the splitting of fast (solid) and slow (dashed) shear waves.
(b) Cased-hole: note the same splitting as in (a).
(c) Orthogonal fractures of equal intensity: note the vanishing of wave splitting.

fractures, dipping about 70°–80° with an average strike of N30°E (see fracture dip analysis, track 3). A second set of open, almost horizontal load-release fractures, seen at 5704 ft, 5721 ft, and 5736 ft and some stylolites (5731 ft and 5756 ft) are associated with the conjugate shear fractures. This classical fracture system indicates that the maximum stress direction is vertical. The strike of the shear fractures is the direction of the maximum horizontal stress axis, N30°E. The stress direction is consistent with the breakout direction as seen in nearby zones.

From the acoustic image analysis, it may be predicted that the maximum anisotropy interval is from 5740 ft–5753 ft with the shallower intervals exhibiting lower values of anisotropy associated with the conjugate shear fractures. The direction of anisotropy from the dip analysis would be N30°E.

This interpretation confirms the anisotropy analysis from the cross-dipole acoustic data. In track 1, there are anisotropy profiles with two resolutions, each scaled to a maximum of 50 %. Over the lower dominant fracture interval, the anisotropy averages 15 %–20 %, which is typical for open fractures of this character. The less dominant fractures above, which are just recognizable, average 6 %–10 % anisotropy. The visual comparison of the fracture image with the measured anisotropy confirms, at least qualitatively, our theory-based intuition that a fracture-induced anisotropy is proportional to fracture intensity. The fast-shear polarization angle is compared with the fracture strike direction. Acoustic processing for a fast shear angle (the direction along the strike of a fracture) is shown in track 2 which, shows that it is relatively stable and consistent over the fracture zones. Over the entire interval, the computed angle compares well with the image processing result, in the vicinity of N30°E. Overall, we see that the two analyses are in good agreement. This open-hole example confirms the theory that fracture-induced anisotropy is governed by fracture intensity, and that the fast shear-wave polarization azimuth is along the fracture strike direction.

5.3.2.2 An open versus cased hole example with orthogonal fractures

In this section, we compare the fracture analysis results before and after casing the borehole. A potential limitation with orthogonal fractures will also be indicated. A field study was carried out in Michigan, USA, to evaluate the feasibility of cased-hole fracture measurements using cross-dipole logging. The subject formation is the Antrim shale, which is a highly impermeable rock. Gas production is through the fracture networks present in the formation. Detecting fracture systems in cased boreholes will help locate potential gas-producing depth intervals.

For comparison purposes, open-hole logging was first run to locate fractures. The open-hole data included cross-dipole acoustic logging, which was used to obtain anisotropy for fracture characterization, and acoustic imaging, which was used to identify fractures at the borehole wall. The well was then cased and cemented. Cross-dipole logging was repeated and the data were processed to compare with the open-hole results.

The open-hole, cross-dipole analysis detects three high anisotropy intervals around X030 ft, X080 ft, and X130 ft in the depth interval of interest. The anisotropy represents the average value over the wave travel distance from transmitter to receiver array. The anisotropy is plotted in track 2 (left shaded curve; scale range is from 0–30 %, left to right) of Figure 5.12. Acoustic imaging results confirm that these high-anisotropy intervals are fracture locations. The acoustic image for the X080 ft interval is shown by the lower-right image of Figure 5.12. This image and the image processing results (right of the cross-dipole results) show that the fractures have medium to high dip angles and a dominant strike direction (see the rose diagram). Across the high-anisotropy fracture intervals, we can clearly observe the shear-wave splitting phenomenon (track 1). The fast (dark or solid curve) and slow (light or dashed curve) dipole-shear waves, as obtained from cross-dipole data processing, show significant separation at the fracture intervals, in good agreement with the theoretical modeling result (Figure 5.10a).

Cased-hole, cross-dipole analysis results are shown in track 2 (anisotropy, shaded curve from right to left) and track 3 (fast and slow shear waves). The cased-hole anisotropy result is in close agreement with the open-hole result. The same three high-anisotropy intervals with similar anisotropy to their respective open-hole values

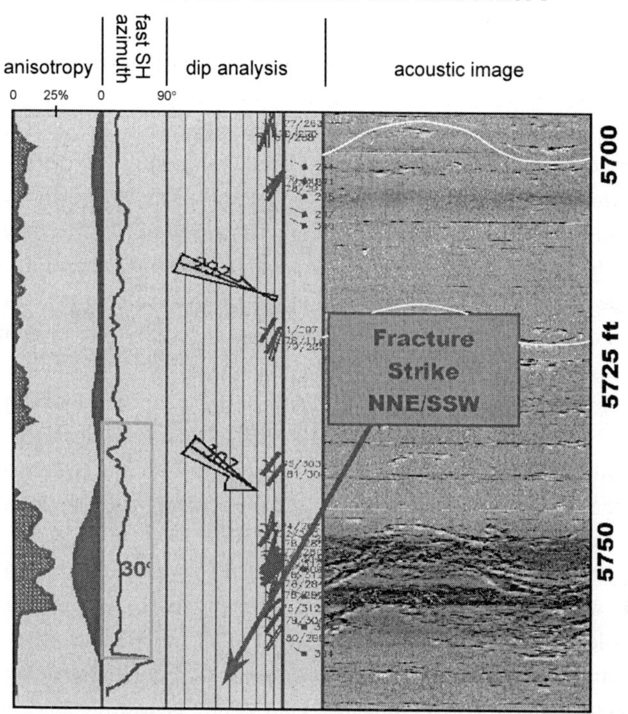

Figure 5.11. Cross-dipole analysis result over a fractured carbonate interval, along with dip analysis and acoustic imaging results. Comparison of the results shows that the determined fast shear polarization azimuth is in the fracture strike direction and that the magnitude of anisotropy is proportional to fracture intensity.

are clearly identified. Although the waveforms (track 3) show significant distortion compared to the open-hole case, presumably caused by imperfect cement bonding, the splitting of fast and slow waves is still quite discernable, indicating the measurable effect of anisotropy through casing (some aspects associated with cased-hole dipole logging will be discussed in the next section). The cased-hole result confirms the theoretical modeling result in Figure 5.10b, demonstrating the feasibility of shear-wave anisotropy measurement through casing.

A conflict is seen when interpreting the imaging and cross-dipole results for the X050 ft interval. The acoustic image shows this interval to have been intensely fractured (top image of Figure 5.12). However, unlike other fracture intervals, both open- and cased-hole anisotropy curves (track 2) exhibit minimal anisotropy for this interval. A clue was found from inspecting the image and the associated processing results. The image shows the existence of multiple conjugate fractures. As the rose diagram to the left of the image indicates, the azimuths of the fractures fall into two dominant directions that are orthogonal to each other and of equal intensity. The reduction of anisotropy due to orthogonal fractures is well-described by the theory, as in equations (5.15), and numerical modeling, in Figure 5.10c. This example points out a potential limitation in the measurement of fractures using cross-dipole logging.

5.3.2.3 *Evaluating hydraulic fracture stimulation behind casing*

Evaluating hydraulic fracture stimulation in cased boreholes presents an effective and economical application for the cross-dipole technology. Hydraulic fracturing is often performed in the final stage of well completion. Pressuring perforations through casing can effectively create hydraulic fractures, as illustrated by a diagram in Figure 1.16. However, evaluating the stimulation result presents a formidable task. The presence of casing makes it difficult to detect and evaluate the vertical extent and azimuth of the stimulated fractures. This section demonstrates a solution to this problem using the cross-dipole acoustic-logging technology.

In open-hole situations, fracture-induced anisotropy can be effectively measured using cross-dipole acoustic logging. Application of this technology to cased holes has been hindered by two factors. The first is the concern of the effect of casing and cement on the cross-dipole measurement, and the second is the lack of an effective device to measure the tool's orientation inside casing.

The result of numerical modeling (Figure 5.10b) demonstrates the feasibility of shear-wave anisotropy measurement through casing. The results show that a cross-dipole tool centered in the borehole can measure shear-wave anisotropy through casing and cement, provided that they are well bonded with the formation. In the cased-hole dipole logging practice, poor casing-cement bonding and tube (or Stoneley) wave contamination are a common problem that degrades the measurement results. Even a slightly off-centered dipole tool tends to generate tube waves in a cased hole to interfere with the dipole waves, especially when the tube- and dipole-wave velocities are close. Therefore, care should be taken to minimize the tube wave generation during logging.

The tool azimuth measurement in cased holes is now made with a gyrocompass device. Most open-hole logging tools use a magnetic device to measure the tool azimuth in a borehole. This device does not operate inside casing because the metal casing obscures the earth's magnetic field. A gyroscope maintains its orientation relative to the earth's north, while the compass frame moves or rotates with the tool. Using this device, we can record the amount of tool rotation or azimuth at each logging depth of a cased borehole. With these foundations, we can apply cross-dipole technology to cased-hole problems.

A well in southeastern New Mexico was chosen to evaluate cross-dipole technology for cased-hole applications. This well was drilled into a carbonate formation at a depth of approximately 6800 ft. Petrophysical analysis was done on this particular well to illustrate the geological formations inherent in the area. This example is located in the northwest edge of the Central Basin platform and is bordered to the west by the Delaware Basin, to the east by the Midland Basin and near the northwestern shelf and the Capitain Reef Trend. These basins are part of what makes up the Permian Basin.

The goal of this test was to determine the fracture trends in a water-flooded environment. Open-hole logging in the well was first performed to locate the zones of

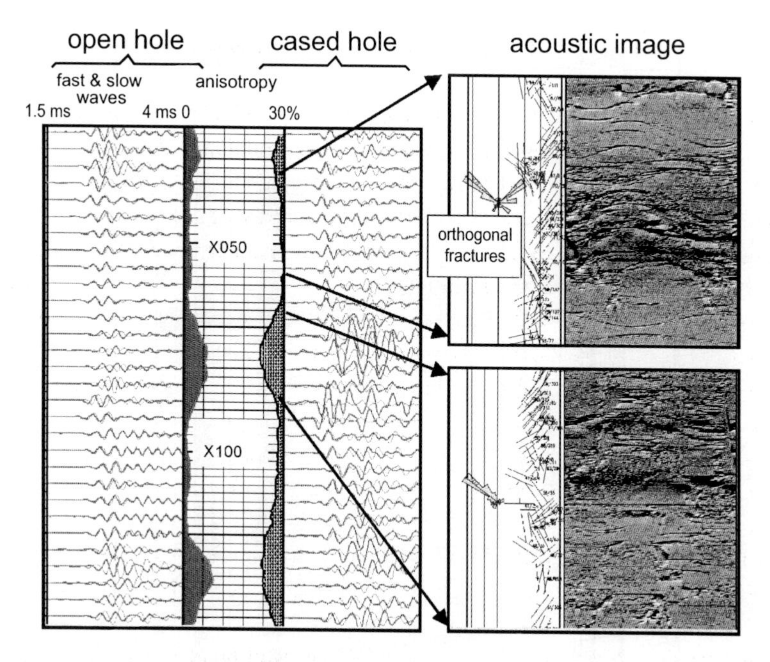

Figure 5.12. Field results from cross-dipole and acoustic imaging processing. The open- and cased-hole results give similar anisotropy values for fracture intervals. Note the anisotropy almost vanishes across the X050 ft interval because of orthogonal fractures.

interest and to determine the amount of anisotropy that existed prior to casing. After the borehole was cased, the same logging measurements were repeated for comparison purposes. Then the intervals of interest were completed. A completed interval was hydraulically fractured and tagged with three radioactive isotopes: Scandium (Sc-46), Iridium (Ir-192), and Antimony (Sb-124). The radioactive isotopes can be detected with a gamma-ray device in a cased hole so that the fracture extent along the borehole can be traced. A post-stimulation logging was repeated. The results of this logging, together with those of the open-hole and pre-stimulation cased-hole logging runs, were analyzed to determine the extent and azimuth of the stimulated fractures.

Fracture stimulation creates dramatic anisotropy effects behind casing. A large amount of anisotropy is measured throughout the depth interval. Figure 5.13 shows post-stimulation anisotropy analysis results. Immediately above the lower stimulated interval (6650 ft–6770 ft), the anisotropy exceeds 10 %. The quality indicators S1ISO and S1S2 (track 2) have quite significant values. The fast and slow shear-wave splitting is clearly visible (track 4). More interesting, the overall fast-shear azimuth shows a well-defined trend along almost the east direction, despite the rapid rotation of the tool azimuth (track 6). All these results indicate a well-defined and reliably measured azimuthal anisotropy behind casing and cement.

The injection of the radioactive isotopes from the lower interval (6650 ft–6770 ft) resulted in drastically high gamma-ray count values (note the gamma-ray curve scale in track 1, Figure 5.13, is 0–1200 API). However, these high values only occur in

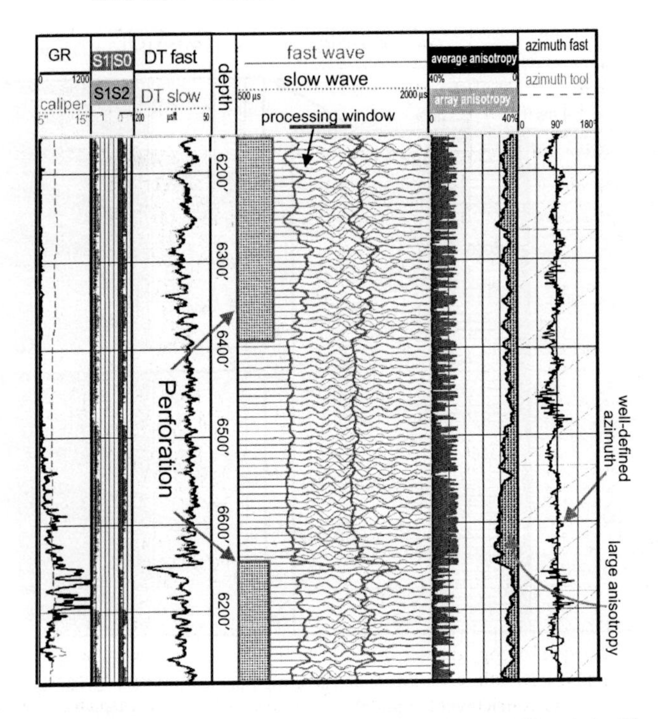

Figure 5.13. Post fracture-stimulation measurement results in a cased borehole. The stimulation creates substantial anisotropy that is reliably determined by the cross-dipole analysis. Note that the anisotropy is accompanied by shear-wave splitting in track 4 and a well-defined azimuth in track 6.

the lower portion of Figure 5.13. The gamma-ray curve in Figure 5.13 starts to drop below its open-hole values (given in track 1 of Figure 5.14) at about 6475 ft, suggesting that the radioactive tracers did not migrate beyond this depth and that the lower fracture is not well-connected with the upper one. It is also interesting to note that in the vicinity of 6475 ft, the anisotropy curves (track 5) show a minimum and the associated azimuth (track 6) shows some instability pertaining to minimal fracturing. In the following, the cross-dipole and radioactive tracer measurement results will be compared and interpreted to find the fracture trends and connectivity.

A gamma-ray spectral analysis was performed to identify the concentrations of the injected radioactive tracers along the borehole. The results, together with those of the cross-dipole analysis, are presented in Figure 5.14 (depth between 6470 ft–6775 ft).

We now explain the presentations for the analysis results in Figure 5.14. Track 1 displays the open-hole gamma-ray and caliper curves. Track 2 shows the neutron porosity and density log curves. The tracer analysis results in track 4 are presented in a cylindrically symmetric fashion, by flipping the shaded concentration curves of the isotopes – Ir-92: medium gray; Sb-124: dark gray; Sc-46: light gray – at both sides of the borehole. The cross-dipole results are shown using an image presentation called "anisotropy map" (track 5). The anisotropy map combines the derived average anisotropy and its azimuth to make an image display in the azimuthal range

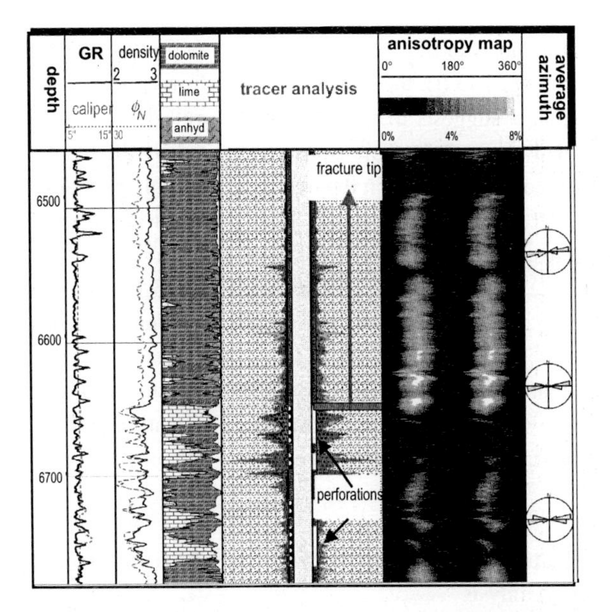

Figure 5.14. Comparison of radioactive tracer (track 4) and cross-dipole (tracks 5 and 6, presented using anisotropy map and rose diagrams) analysis results for a fracture stimulation treatment. Both results show the upward fracture extension along the borehole and a possible restriction (or fracture tip) around 6475 ft. The map and the rose diagrams give the fracture (also the maximum stress) orientation.

of 0°–360°. The brightest color of the map corresponds to the anisotropy value and its azimuth. With the anisotropy map, an analyst can quickly assess depth intervals of interest by looking at the brightness, direction, and continuity of the features on the map. The map also facilitates comparison with other borehole image logs (e.g., acoustic and resistivity). Rose diagrams (track 6) are plotted along with the image map to accurately indicate the anisotropy azimuth over each labeled depth interval of 100 ft.

An important result of the post-stimulation cross-dipole measurement is the determination of the fracture orientation and the in-situ stress orientation. The overall anisotropy azimuth, as seen from the rose diagrams, is in N80°E direction. Perforation/fracture tests (Behrmann and Elbel, 1991) show that fractures initiated from perforations, regardless of their initial direction and perforation orientation, orient toward the maximum in-situ stress direction within one borehole diameter. Because a dipole wave can penetrate through casing and cement (Figure 5.10b) and reach deep into the formation (several borehole diameters), the anisotropy azimuth (i.e., fracture azimuth) seen in track 5 should correspond to the maximum stress direction. This direction is indeed in agreement with the regional stress orientation of the field.

Another important result is the mapping of fracture extent along the borehole. Since the shear-wave anisotropy in a fractured formation gives a measure of fracture intensity, the anisotropy magnitude and continuation can be used to evaluate the stimulated fractures. In the interval of 6490 ft–6650 ft (except for a disruption of fracture growth at 6545 ft), the anisotropy map shows a high-valued, almost continuous feature. This suggests that the fracture is well developed in this dolomite-dominated formation (track 3). This is also supported by the tracer analysis result, which shows high concentrations in the treatment interval and significant concentrations (especially Ir-92 and Sc-46) across the high anisotropy interval. The fracture may have been restricted around 6475 ft, for the anisotropy map there shows much reduced magnitude (with poor indicator values, Figure 5.13) and distorted azimuth. The lithology map in track 3 indicates that there are several limestone streaks in this interval. This may have acted as a restriction to the fracture growth. As seen from the anisotropy map for 6650 ft–6770 ft, even in the treatment interval the fracture corresponds to lower anisotropy in the limestone formations. The anisotropy map, coupled with the tracer analysis, gives an effective mapping of fracture trends along the borehole.

5.4 Application of Cross-Dipole Anisotropy Measurement to Formation Stress Analysis

The horizontal stress field and its orientation are an important aspect of formation evaluation and a governing factor in the optimization of the development and drainage of a reservoir. This is especially true if the formation is hydraulically stimulated or is naturally fractured. In this case the stress orientation governs the directional aspects of permeability and hence production. The standard methods for determining formation stress include the evaluation of borehole breakout from image logs, packer and micro-fracture testing, and anealstic stress relaxation from whole core, etc. Here we demonstrate the use of cross-dipole acoustics in assessing the horizontal stress field.

5.4.1 Stress Determination From Acoustic Logging: A Theoretical Foundation

The foundation for measuring formation stress from cross-dipole logging is the effect of stress-induced shear-wave anisotropy. Theoretical modeling of stress-related elastic wave propagation provides a tool for analyzing this effect. In this aspect, two theoretical approaches have been used in the modeling. The first approach (Sinha and Kostek, 1995; Winkler et al., 1998) uses a nonlinear elastic wave theory, in which the elastic property of a stressed solid depends on the applied stress (external, or static, stress and wave-induced, or dynamic, stress) and the resulting stress-strain relationship is nonlinear. In the second approach (Tang et al., 1999b), the problem is decomposed into two parts. The first part is concerned only with the change of elastic properties caused by the external stress, which can be effectively modeled by a stress-velocity relation. The second part is the wave propagation through the pre-stressed solid. Because the wave-induced stress and strain are infinitesimally small compared to the static stress and strain, the linear stress-strain relationship can be used to model the wave propagation, for which the elastic properties involved are those of the pre-stressed solid.

The second approach is simpler and more straightforward than the first one. Nevertheless, both theories give similar results in terms of modeling the stress-induced anisotropy and its effect on borehole acoustic wave propagation. For this reason, we will describe the second approach that is based on the stress-velocity relation.

5.4.1.1 *Stress-Velocity Relation*

The stress-induced shear-wave anisotropy effect shows that a shear wave polarized in the maximum stress direction will travel faster than a shear wave polarized perpendicular to this direction (Rai and Hanson, 1988). This phenomenon is well-described by a stress-velocity relationship. Consider a biaxial loading situation where two orthogonal stresses σ_x and σ_y are acting in the x- and y-directions, respectively (see Figure 1.14b). The propagation direction of a shear wave is in the z-direction. The propagation velocity of a shear wave polarized in the x- or y-directions is given, respectively, by the following stress-velocity relation (Mao, 1987)

$$\begin{cases} v_x^2 = v_{0x}^2 + S_{\parallel}\sigma_x + S_{\perp}\sigma_y \\ v_y^2 = v_{0y}^2 + S_{\parallel}\sigma_y + S_{\perp}\sigma_x \end{cases}, \tag{5.16}$$

where $S_{\parallel}$ is the stress-velocity coupling coefficient for a stress that is parallel to the shear-wave polarization direction, and $S_{\perp}$ is the cross-coupling coefficient for a stress perpendicular to the polarization direction; v_{0x} and v_{0y} are the unstressed shear-wave velocity for the x- and y-polarization directions, respectively. Intrinsic anisotropy exists if v_{0x} and v_{0y} are not equal. The stress-coupling coefficients are analogous to the "third-order elastic constants" used in the theory of non-linear elasticity (see Sinha and Kostek, 1995). Note that the coupling coefficients are defined in terms of "velocity squared versus stress", while the third-order elastic constants are defined in terms of "density times velocity squared versus stress". Equations (5.16) provide a simple linear relation to describe the stress-induced shear-wave velocity changes. It is very important to demonstrate that this relation is applicable to real rocks before one applies it to the formation stress evaluation problem. We will use laboratory data to test the relation. We will also provide a method for determining the coupling coefficients in equations (5.16) from the laboratory data.

In the laboratory, the coupling coefficients $S_{\parallel}$ and $S_{\perp}$ of a rock sample can be determined from a shear-wave velocity measurement under uniaxial loading. In the setup shown in Figure 5.15, a uniaxial stress σ_x is applied to the sample in the x-direction. Shear-wave transmitter and receiver transducers facing the z-direction are mounted at the opposite sides of the sample. The shear-wave velocity is measured at two polarization directions: one in the x-direction and the other in the y-direction

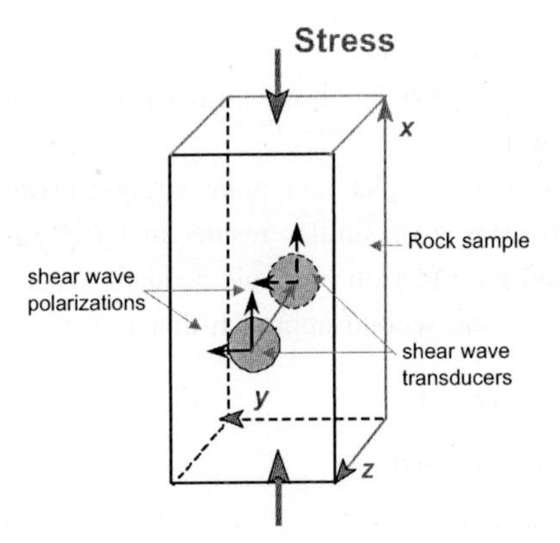

Figure 5.15. Laboratory configuration for measuring stress-induced shear-wave anisotropy. A uniaxial stress is applied in the *x*-direction. Shear-wave velocity is measured along the *z*-direction for the two polarization directions *x* and *y*.

that is perpendicular to the applied stress. For this configuration, equations (5.16) reduce to

$$\begin{cases} v_x^2 - v_{0x}^2 = S_{\parallel}\sigma_x \\ v_y^2 - v_{0y}^2 = S_{\perp}\sigma_x \end{cases} . \tag{5.17}$$

If the shear velocities v_x and v_y are measured for a range of stresses σ_x, the coefficients $S_{\parallel}$ and $S_{\perp}$ can be determined by linearly fitting the data of $v_x^2 - v_{0x}^2$ versus σ_x, and the data of $v_y^2 - v_{0y}^2$ versus σ_y, respectively. We use the published laboratory data of Rai and Hanson (1988) to demonstrate the method.

Figure 5.16 shows the measured squared shear-velocity difference versus uniaxial stress for three different sedimentary rocks. The first is a sandstone rock of 2.5 % porosity (solid circles: parallel with stress direction, open circles: perpendicular to stress direction). With increasing stress, there is clearly a linear relationship between the squared velocities and stress, as described by equations (5.17). Linearly fitting the data versus stress and evaluating the slope of the fitted lines give $S_{\parallel}$ = 40 414 m²/s²/MPa and $S_{\perp}$ = 18 663 m²/s²/MPa for this sandstone rock. One notices that $S_{\parallel}$ is at least twice as great as $S_{\perp}$, meaning that stress produces a much greater velocity change parallel to the wave polarization direction than perpendicular to it. The same conclusion can also be drawn from the laboratory results of Nur and Simmons (1969), although their measurement was made with a different configuration.

The second data set is for a sandstone of 18.6 % porosity (solid squares: parallel with stress direction, open squares: perpendicular to stress direction. Using the same method, we get give $S_{\parallel}$ = 89 213 m²/s²/MPa and $S_{\perp}$ = 31 867 m²/s²/MPa. These values are much larger than their counterparts associated with a sandstone of 2.5 % porosity, indicating that stress produces a greater velocity change on rocks with more imperfections (pores, cracks, etc.) than on rocks with fewer imperfections. Again, $S_{\parallel}$ is much greater than $S_{\perp}$. The fit to the parallel data shows some deviation from the

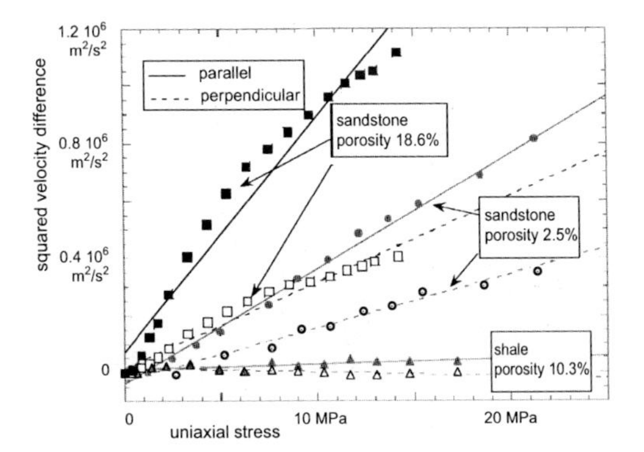

Figure 5.16. Test of the stress-velocity relation using the laboratory data of Rai and Hanson (1988). The parallel and cross-coupling coefficients are determined by linearly fitting the data versus stress. Solid symbols: parallel, open symbols: cross coupling.

linear theory. Nevertheless, the theory fits the data reasonably well as a first order approximation.

The third data set is for a shale sample with 10.3 % porosity (solid triangles: parallel with stress direction, open triangles: perpendicular to stress direction). This sample exhibits strong shear-wave transverse anisotropy between z- and x- (or y-) directions (Rai and Hanson, 1988). Surprisingly, there is very little stress-dependent anisotropy between the x- and y-directions as well as stress-induced velocity changes, as shown in Figure 5.16. The linear fitting gives $S_{\parallel} = 1850$ m^2/s^2/MPa and $S_{\perp} = -1600$ m^2/s^2/MPa. Compared to the previous two examples, these two stress-velocity coupling coefficients are quite small. This result suggests that the stress-induced azimuthal anisotropy in shales is insignificant. As we will show later, this result is very useful in interpreting cross-dipole anisotropy logs in sand-shale formations.

The evaluation of the linear stress-velocity relationship using laboratory data not only validates the theory but also provides a method for determining the coupling coefficients. It also shows that the stress-velocity coupling effect is greater in rocks with high porosity and/or crack density than in rocks with low porosity/crack density. In particular, we have seen that this effect is insignificant in shale. Having established the validity of the stress-velocity relation (equations (5.16)), we apply it to investigate the stress-induced velocity variation around a borehole, as will be discussed in the following section.

5.4.1.2 Application to stress-induced shear-wave velocity variation around a borehole

In this section we develop a simple and effective theory for stress-induced shear-wave velocity changes around a borehole. Consider a fluid-filled borehole configuration, with the borehole axis in the vertical direction (Figure 1.14b). There are three stresses acting in and around the borehole: p, σ_x, and σ_y, where p is the fluid pressure inside the borehole, and σ_x and σ_y are the two principal stresses in the x- and y- directions, respectively. In terms of p, σ_x, and σ_y, the corresponding cylindrical stress elements for the formation, i.e., the radial stress σ_r, the tangential stress σ_θ, and the shear stress $\sigma_{r\theta}$ can be expressed as (Jaeger and Cook, 1977)

$$\begin{cases} \sigma_r = \dfrac{\sigma_x + \sigma_y}{2}\left(1 - \dfrac{R^2}{r^2}\right) + \dfrac{pR^2}{r^2} + \dfrac{\sigma_x - \sigma_y}{2}\left(1 + 3\dfrac{R^4}{r^4} - 4\dfrac{R^2}{r^2}\right)\cos 2\theta \\ \sigma_\theta = \dfrac{\sigma_x + \sigma_y}{2}\left(1 + \dfrac{R^2}{r^2}\right) - \dfrac{pR^2}{r^2} - \dfrac{\sigma_x - \sigma_y}{2}\left(1 + 3\dfrac{R^4}{r^4}\right)\cos 2\theta \\ \sigma_{r\theta} = -\dfrac{\sigma_x - \sigma_y}{2}\left(1 - 3\dfrac{R^4}{r^4} + 2\dfrac{R^2}{r^2}\right)\sin 2\theta \end{cases}, \quad (5.18)$$

where θ is the angle from the x-direction, r is the radial distance, and R is the borehole radius. To illustrate the variation of the various stress components around a borehole, we calculate equations (5.18) for the maximum and minimum stress values of 40 MPa and 30 MPa, respectively, and a borehole pressure of 30 MPa. The calculated radial, tangential, and shear stress distributions are shown in Figure 5.17 as contour plots. A common characteristic of the stress distributions is the stress concentrations in the near-borehole region, where the stresses vary drastically. By contrast, in the region far from the borehole the radial and tangential stresses become homogeneous and the near-borehole shear stress anomaly (or concentration) vanishes. The stress concentrations, when coupled with the stress-velocity relation, create substantial shear-velocity variations around the borehole, which can be detected with borehole acoustic logging measurements. With the given stress-velocity relation (equations (5.16)) and the stress field (equations (5.18)), we can describe the stress-induced velocity variations around the borehole.

We now apply the stress-velocity relation (equations (5.16)) to the borehole environment. To apply the stress-velocity relation to the maximum (x) and minimum (y) stress directions, we need to rotate the stress elements of the cylindrical coordinates to the x- and y-directions of the Cartesian coordinates using the following equations:

$$\begin{aligned} \sigma_{xx}(r,\theta) &= \sigma_r(r,\theta)\cos^2\theta - \sigma_{r\theta}(r,\theta)\sin 2\theta + \sigma_\theta(r,\theta)\sin^2\theta \\ \sigma_{yy}(r,\theta) &= \sigma_r(r,\theta)\sin^2\theta + \sigma_{r\theta}(r,\theta)\sin 2\theta + \sigma_\theta(r,\theta)\cos^2\theta \end{aligned}. \quad (5.19)$$

Using the stress-velocity relation and assuming that the rock is isotropic in the unstressed state (i.e.,$v_{0x} = v_{0y} = v_0$), the velocities of a shear wave polarized in the x- and y-directions are, respectively, given by

$$\begin{aligned} v_x^2(r,\theta) &= v_0^2 + S_\parallel \sigma_{xx}(r,\theta) + S_\perp \sigma_{yy}(r,\theta) \\ v_y^2(r,\theta) &= v_0^2 + S_\parallel \sigma_{yy}(r,\theta) + S_\perp \sigma_{xx}(r,\theta) \end{aligned}. \quad (5.20)$$

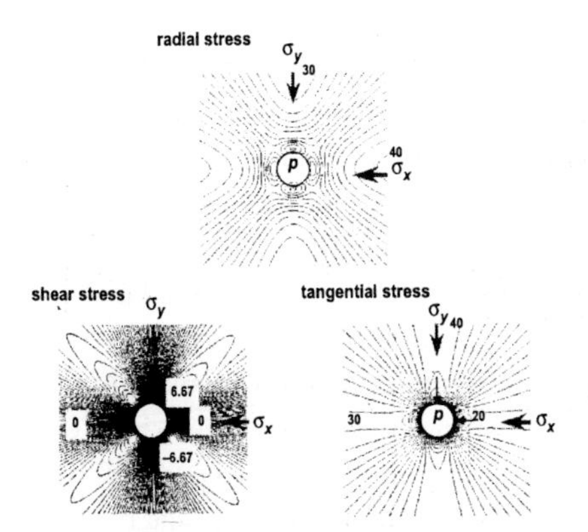

Figure 5.17. Radial, shear, and tangential stress distributions around a borehole, displayed using contour plots. The maximum and minimum stresses far from the borehole and the borehole pressure are 40 MPa, 30 MP, and 30 MPa, respectively. The numbered labels on the plots are in MPa.

Equations (5.20) indicate that shear waves polarized in orthogonal directions, e.g., x and y, have two different propagation velocities, i.e., there is anisotropy. We model this anisotropy using the transverse isotropy (TI) model. The use of the TI model is based on the rock physics of a cracked porous rock, as described in Walsh (1965). For a porous rock containing randomly oriented cracks or elongated pores (e.g., sandstones), the application of an oriented stress will close/suppress the cracks normal to the stress. The stress preserves existing, or even initiates new, cracks along the stress direction. Such a solid with aligned cracks, as studied by Hudson (1981), is effectively a TI medium, whose axis of symmetry is normal to the crack alignment direction, or the applied stress direction. This justifies the use of the TI model for a stressed formation surrounding a borehole.

From the above description of the TI model related to the horizontal stress, it is clear that this is a horizontal TI (HTI) scenario. The axis of symmetry of this HTI medium is perpendicular to the borehole axis in the y-direction. An HTI medium has five elastic moduli: c_{11}, c_{12}, c_{13}, c_{44}, and c_{55}. The last two, i.e., c_{44} and c_{55}, are of primary importance for the shear-wave propagation (e.g., White, 1983). According to equations (5.20), these two moduli are given by:

$$\begin{cases} c_{55} = \rho\, v_x^2(r,\theta) \\ c_{44} = \rho\, v_y^2(r,\theta) \end{cases}, \tag{5.21}$$

where ρ is formation density. The remaining moduli are approximately calculated as $c_{12} = c_{13} = c_{11} - 2c_{55}$, $c_{66} = c_{44}$ (because y is the symmetry axis), and $c_{11} = c_{33} = \rho v_P^{\,2}$, where v_P is the compressional wave velocity which can be set to a proper value. The reason that we can make the approximation is that dipole-shear logging is largely insensitive to these parameters (Ellefsen, 1990). When using equations (5.21) to model the azimuthal anisotropy around a borehole, the borehole axis is taken as the z-direction that is perpendicular to the xy-plane.

Equations (5.21) define a medium that is both anisotropic and inhomogeneous, because the velocities given by equations (5.20) change with r and θ. To illustrate the stress-induced shear velocity variation around the borehole, we numerically evaluate formation stresses (equations (5.18)), and the resulting anisotropic velocity distributions (equations (5.20)). Figure 5.18 shows the shear velocity distributions around the borehole, where we used $S_{\parallel}$ = 89 213 m^2 s^{-2} MPa^{-1} and $S_{\perp}$ = 31 867 m^2 s^{-2} MPa^{-1}, as determined from the sandstone of 18.6 % porosity (Figure 5.16). As can be expected, v_x and v_y change significantly with azimuth. An interesting feature in the v_x and v_y variations is that along the x-direction the shear velocity is lower near the borehole and becomes higher about 1–2 radii away from the borehole, while for the y-direction the velocity is higher near the borehole and becomes lower away from the borehole. This feature can be seen clearly in Figure 5.20a, which will be discussed in more detail in the next section. This velocity variation characteristic is caused by the stress concentrations in the near borehole region. The characteristic of the shear-velocity distributions has significant implications for acoustic measurements in a borehole, as will be described below.

5.4.1.3 *Characterizing and estimating effects of formation stress on acoustic measurements*

The stress-induced velocity variation around the borehole produces measurable effects on monopole and dipole/cross-dipole acoustic logging waveforms. Conversely, these measurable effects allow us to estimate stress-field parameters. In the following, we discuss the influence of velocity distributions (equations (5.20)) on dipole and monopole acoustic waves. We demonstrate that a cross-dipole anisotropy measurement can determine the orientation of the principal stresses and their difference, and that a monopole shear-wave measurement or dipole dispersion curve comparison can distinguish the stress-induced anisotropy from other sources of anisotropy.

We now consider stress-induced shear-wave anisotropy measured by a cross-dipole acoustic tool. This tool consists of two directional dipole transmitter and receiver systems that are pointed 90° apart (see Figure 5.1). The dipole tool, being a low-frequency sounding device, can penetrate deep into the formation. According to equations (5.18) and (5.20) (see Figures 5.17 and 5.18), the r-dependent terms rapidly diminish at 2–3 borehole radii. This shows that a low-frequency dipole polatized in x- and y-directions measures, respectively, two velocities given by

$$\begin{cases} V_x^2 \approx v_x^2\,(\infty,\ 0^o) = v_0^2 + S_{\parallel}\sigma_x + S_{\perp}\sigma_y \\ V_y^2 \approx v_y^2\,(\infty, 90^o) = v_0^2 + S_{\parallel}\sigma_y + S_{\perp}\sigma_x \end{cases}, \tag{5.22}$$

where $v_x^{\,2}$ and $v_y^{\,2}$ are given in equations (5.20). Considering the fact that $\sigma_x > \sigma_y$ and $S_{\parallel} > S_{\perp}$ (see Figure 5.16), we immediately conclude that .$V_x > V_y$.

This indicates that away from the borehole the low-frequency dipole tool mea-

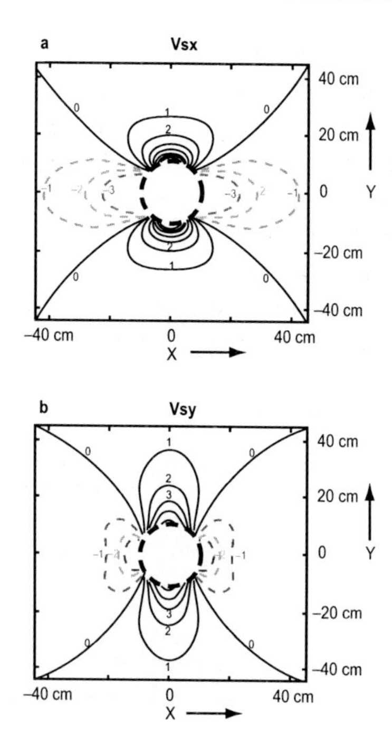

Figure 5.18. Stress-induced shear-wave velocity variation around the borehole, displayed using contour plots.
(a) Shear-wave velocity distribution for polarization along the maximum stress (x) direction.
(b) Shear-wave velocity distribution for polarization along the minimum stress (y) direction.
The velocity field in (a) or (b) is normalized by its respective far-borehole value. The numbers on the contours are the percentage difference relative to this value. Solid and dashed contours indicate positive and negative differences, respectively.

sures a higher velocity along the maximum stress direction than along the minimum stress direction.

The above analysis result is verified with numerical finite difference simulations using the technique developed by Cheng et al. (1995). Using the anisotropic stress-velocity model (equations (5.18)–(5.20)), we compute the dipole array waveforms for two dipole-shear sources pointing to the x- and y-directions, respectively. We refer to Figure 5.19 for the model characteristics. The absolute velocity difference between v_x and v_y is 16.8 % at the borehole and becomes 6.8 % far from the borehole. The center frequency of the wave in the modeling is around 3.5 kHz. Figure 5.19a shows the modeling result. The dipole-shear waves polarized in the x- (maximum stress) direction (solid curves) indeed travel faster than the waves polarized in the y- (minimum stress) direction (dashed curves), despite the complicated near-borehole variations of the anisotropic velocities (e.g., Figure 5.18). The splitting of the fast and slow dipole-shear waves in Figure 5.19a gives the measure of anisotropy, which is almost exactly the far field value of 6.8 %. The velocity difference for shear-wave

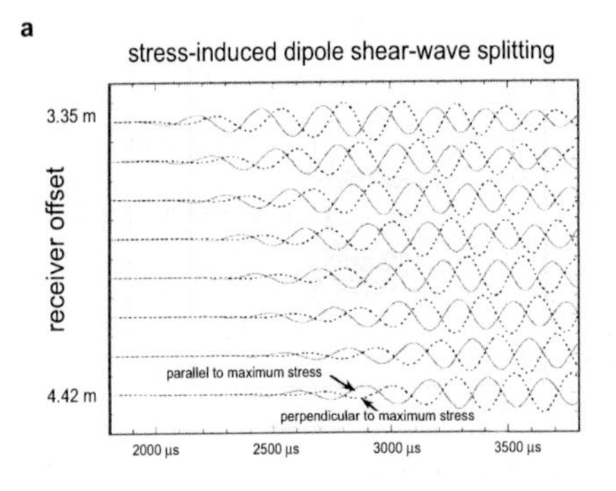

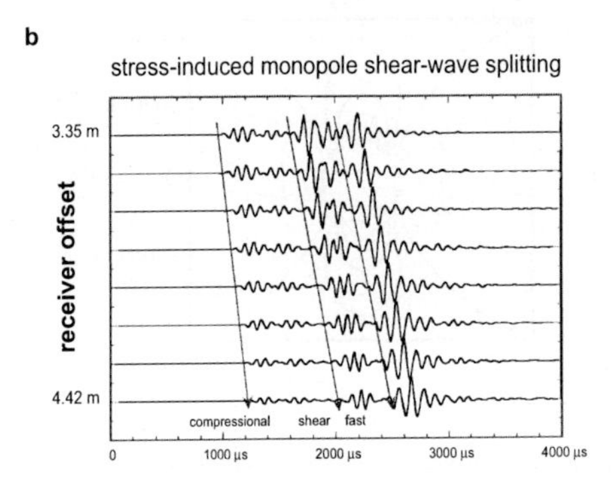

Figure 5.19. Wave propagation in a stressed borehole, simulated with a finite difference method. (a) Stress-induced dipole shear-wave splitting. The wave polarized in the maximum stress direction (solid curve) travels faster than the wave polarized perpendicular to this direction (dashed curve).
(b) Stress-induced shear-wave splitting in monopole waveforms. The splitting is caused by significant shear-wave velocity variation in near the borehole.

polarization at two orthogonal stress directions is designated as "stress-induced shear-wave anisotropy". This is an important result that forms the basis for determining stress-induced anisotropy using cross-dipole acoustic logging.

By analyzing the cross-dipole acoustic data using the previously described technique, we can reliably determine formation azimuthal anisotropy. The cross-dipole measurement gives the fast shear-wave polarization direction and fast and slow shear velocities V_x and V_y. Because the pre-stressed formation behaves like an effective HTI medium, especially in the region far from the borehole that is penetrated by a dipole wave, the applicability of the cross dipole data analysis developed for an HTI medium is justified. For a stress-induced anisotropy, the result of this study shows that the cross-dipole measurement determines the maximum stress direction as the fast shear-wave polarization direction, and finds the fast and slow shear-wave velocities along the maximum and minimum stress directions. These two velocities are given by the two equations in (5.22), respectively. These equations allow us to determine the magnitude of the difference between the maximum and minimum principal stresses perpendicular to the borehole. Taking the difference of the two equations in

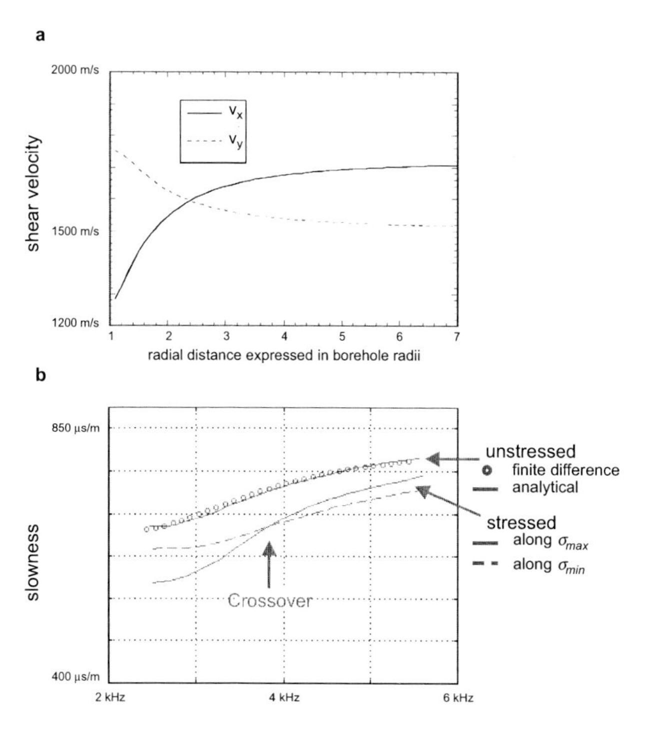

Figure 5.20. (a) Shear velocity profiles along the maximum (x) and minimum (y) stress directions. (b) Crossover phenomenon in the flexural-wave dispersion curves measured respectively along the x (solid curve) and y (dashed curve) directions.
The crossover in (b) is the result of the crossover of the velocity profiles in (a). As a check of the numerical finite difference (FD) result, the dispersion curve from the FD simulation (open circles) in the unstressed case is compared with the available analytical solution (solid curve).

(5.22), we get

$$\sigma_x - \sigma_y = \frac{V_x^2 - V_y^2}{S_{\parallel} - S_{\perp}} \ . \tag{5.23}$$

This result shows that the stress difference is directly proportional to the measured velocity difference (or anisotropy), and the proportionality constant is the inverse of the difference of the parallel and perpendicular coupling coefficients. An immediate application of equation (5.23) is to interpret the stress-induced anisotropy from cross-dipole measurement results. The azimuthal shear anisotropy, as discussed earlier in this chapter, is defined as

$$\gamma = \frac{(V_x - V_y)}{\bar{V}}, \quad \text{with} \quad \bar{V} = (V_x + V_y)/2 \quad \text{and} \quad V_x \geq V_y \ . \tag{5.24}$$

Using equations (5.23) and (5.24), we define a can formation stress indicator :

ing dipole-flexural wave senses a faster velocity in the x-direction than in the y-direction. Conversely, a high-frequency wave, being confined to the near-borehole region, senses a slower velocity in the x-direction than in the y-direction. This results in the dispersion-curve crossover shown in Figure 5.20b. At low frequencies, the slowness dispersion curve (solid) for a dipole polarization along the x-direction has lower values than the curve along the y-direction (dashed). The two curves cross each other in the medium frequency range where the dipole wavelength is approximately equal to the borehole diameter (numerical modeling also shows that the crossover frequency decreases almost linearly with increasing borehole diameter).

Based on the above results, we have two methods to detect stress-induced anisotropy measured by a cross-dipole tool. The first method is based on the shear-wave splitting phenomenon in monopole acoustic wave data. If the formation shear-wave velocity is substantially above the borehole fluid velocity (fast formation), we can inspect the shear-wave data measured by a monopole acoustic tool. If there are two shear waves and their splitting is significant such that their respective velocities or travel-times can be determined, we perform a velocity/travel-time analysis to determine the difference between the two shear velocities/travel-times. If this difference is comparable with the measured anisotropy, the anisotropy is caused by factors other than stress. However, if the difference is significantly higher than the measured anisotropy, this anisotropy is then caused by formation stress. In the second method, we analyze the cross-dipole data to look for the crossover phenomenon in the flexural wave dispersion curves. Based on the theoretical result that the fast shear polarization direction is along the maximum stress azimuth, one can use the fast shear direction, as determined from cross-dipole data analysis, to rotate the 4C cross-dipole array data into the fast and slow principal wave arrays using equations (5.3). Then, performing a dispersion analysis on the fast and slow wave arrays will give the respective dispersion curves, which can then be displayed to inspect for the crossover phenomenon.

Once a stress-induced anisotropy is detected, the orientations of the maximum and minimum principal stresses are found as the fast and slow shear-wave polarization directions, respectively. We can further determine the difference of the two principal stresses or compute the stress-indicator values by applying equations (5.23) through (5.25) to the cross-dipole measurement.

5.4.1.5 A sand-shale formation example

In this section, we demonstrate the application of the borehole stress-velocity model theory to field acoustic monopole and cross-dipole data for in-situ stress estimation. Figure 5.21 shows the cross-dipole anisotropy analysis results for a depth segment with sand/shale formations. The upper shale formation corresponds to high gamma ray (GR) values (see track 1), while the lower sandstone formation corresponds to low GR values and is interlaced with some shale streaks. Cross-dipole-determined azimuthal anisotropy is clearly identified in the sandstone formation but almost

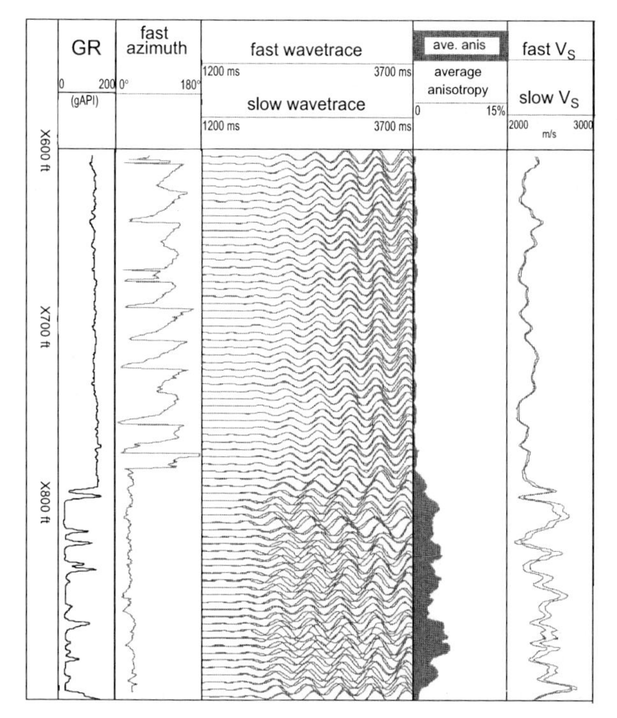

Figure 5.21. An example of stress-induced shear-wave azimuthal anisotropy obtained from cross-dipole data analysis. The stress-induced effects are negligible in the (upper) shale formation but become significant in (lower) sand formation. The stress effects are observed by the significant anisotropy values, the splitting of fast and slow dipole-shear waves, and a well-defined fast shear polarization azimuth. An additional indication using monopole shear-wave splitting is shown in Figure 5.22.

disappears in the shale formation. The anisotropy, displayed as the shaded curve in track 4, is well supported by the splitting of fast and slow dipole shear waves in track 3 (note the splitting disappears in the upper shale). It is also supported by the well-defined fast shear-wave polarization azimuth displayed in track 2. In contrast, this azimuth becomes undefined in the shale formation, showing a tendency to follow the tool azimuth (the tool was spinning during logging through the entire depth).

The monopole shear-wave data observation provides a stress indication. Figure 5.22 compares the monopole shear waves logged in the shale (a) and sand (b) formations. In the shale formation, there is only one shear wave mode in the waveform data. In contrast, there are two shear wave modes in the sand formation. This shear-wave splitting is the direct result of the stress-related wave propagation theory, as we have shown in the synthetic waveform example of Figure 5.19b. Moreover, the velocity difference between the two shear waves is about 10 %, while the cross-dipole-measured anisotropy is about 5 %. This observation agrees well with our theoretical analysis with monopole shear waves. We can therefore determine that the anisotropy is caused by formation stress. This field data example not only validates our stress-velocity theory, but also indicates the monopole shear-wave splitting as an effective

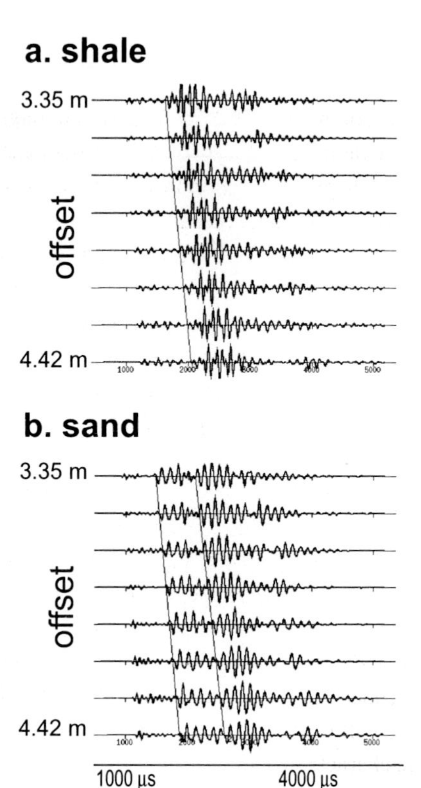

Figure 5.22. An indication of stress-induced effects from the monopole shear-wave splitting phenomenon.
(a) In the shale formation (no stress-induced anisotropy) only one shear wave is present.
(b) In the sandstone formation the stress effect splits the shear wave into fast and slow modes (compare this observation with the theoretical prediction in Figure 5.19b).

means of detecting stress-induced anisotropy in this fast formation situation.

Two additional strong indicators also suggest that the observed anisotropy be caused by stress. The first is that the same fast azimuth is also observed for a massive sand body that is about 1000 ft above the sandstone formation under study. The coincidence of the azimuth of the two sand bodies indicates that this azimuth is the direction of the maximum horizontal stress in this region. The second indicator is that the anisotropy is mainly observed in sandstone but not in shale (although shale has strong transverse isotropy (TI) rock, this TI-related anisotropy cannot be observed because the well is vertical). This agrees with our previously demonstrated laboratory result that stress-induced anisotropy is insignificant in shale. Similar phenomena have also been observed for many cross-dipole data sets from different regions around the world. Laboratory and field observations suggest that, in general, stress-induced anisotropy is insignificant in shale.

5.4.1.6 Remark on using stress indicators

Stress indicators are now commonly used in cross-dipole logging to identify stress-induced anisotropy. For example, Plona et al. (2000) demonstrated the use of the crossover characteristic to diagnose stress-induced anisotropy measured from cross-dipole logging. The crossover phenomenon can provide an effective method if the two dispersion curves are reliably measured. This depends on the reliable calculation of fast and slow array wave sets from the 4C cross-dipole data and on the accurate estimation of dispersion curves from the wave data sets. In comparison, monopole shear-wave splitting provides a much simpler approach. In practice, the monopole shear-wave splitting phenomenon is commonly observed in fast sandstone formations around the world. This wave characteristic provides an effective and straightforward diagnosis of stress effects.

A limitation of the monopole shear-wave splitting characteristic is that it can only be observed when formation shear velocity is significantly higher than the borehole fluid velocity. Another limitation, or restriction, on using stress-induced wave characteristics, including the dispersion-curve crossover, is that they cannot be observed for formations that do not exhibit significant stress-induced anisotropy effects (e.g., shales, see Figures 5.16 and 5.21). The characteristics may also be obscured or even disappear if the stress concentrations have already caused failure (e.g., breakouts) in the formation, as this will distort or destroy the near borehole velocity variation required to produce the characteristics.

5.4.1.7 Comparison with micro-fracturing results

To evaluate the cross-dipole application for stress measurement, a case study in the Eldfisk Field of the North Sea was conducted (Wade et al., 1998; Patterson et al., 1999). The determination of maximum horizontal stress orientation is crucial for a waterflood project aimed at maximizing oil recovery. The formation being evaluated was the Tor formation, which is part of the Maastrichtian-age sediments in the Late Cretaceous age. The well was drilled and the initial whole core was taken from 10 780 ft–10 795 ft and from 10 795 ft–10 824 ft. Anelastic strain recovery (ASR) measurements were performed on the cores. The maximum stress orientation results for the two intervals were 44°±6.4° and 57.9°±14.9°, respectively. The uncertainty in the magnitude was due to the low amount of physical displacement measured during the relaxation, along with the vibration effects from the production platform, since the measurements were made on location.

The initial wireline data included cross-dipole and acoustic imaging. The cross-dipole anisotropy analysis was performed on the acquired cross-dipole data to determine the fast shear-wave orientation. The processing result plot, Figure 5.23, shows the Tor formation to have low but measurable anisotropy. Track 3 shows the two shear-wave anisotropy measurements with two resolutions: array aperture and the source-to-array mid point. They show the higher resolution array measure having a maximum 4% anisotropy and the average measure having a maximum of 2 % anisotropy. Track 2 shows the fast and slow shear waveforms along with the time window that was used in the data analysis. The waveforms display some splitting, which again points to measurable anisotropy. Track 4 shows the fast-angle orientation with the plus and minus uncertainty being displayed. This shows the uncertainty of the fast angle to be small, ±6°, and the fast shear polarization direction is showing just east of northeast with an average of 51° through the entire section.

The well was then micro-fractured across two different intervals, 10 804 ft–10 812 ft and 10 778 ft–10 782 ft, and yielded the horizontal stress differentials of 1686 psi and 1838 psi, respectively. After the micro-fracturing operations, a post-fracturing acoustic image log was run to determine the orientation of the induced hydraulic fractures. These fractures were created as the hydraulic pressure worked against the minimum horizontal stress to produce a fracture along the axis of the maximum horizontal stress. The acoustic image data analysis is shown on the far right of Figure 5.23. The pre-fracturing image data shown on the left part of the image results display no visible fractures. The image also shows the borehole to be circular and in good condition with no breakout. The post-fracturing data, shown on the right, displays an induced fracture at one of the two intervals that were micro-fractured. The analysis of these two fractures yields a dip magnitude of 88° with an orientation N46°E from the bottom zone and N48°E for the top zone. The maximum stress orientation measured from the micro-fracturing test is in good agreement with that measured from the cross-dipole logging. The results demonstrate the ability of the cross-dipole to provide valuable information in the evaluation of horizontal stress orientation.

5.4.1.8 Comparison with borehole breakout from image logs

In this section we compare the cross-dipole-measured formation stress orientation with that obtained from borehole breakout observation. Borehole breakouts from image logs provide an important means for estimating formation stress orientation. This can be seen from analyzing the stress field given by equations (5.18) and from the stress distribution illustrated in Figure 5.17. The results indicate that the stress concentration effect produces a maximum (compressive) tangential (or hoop) stress $3\sigma_x - \sigma_y + p$ at the borehole wall. The maximum hoop stress occurs at the azimuth $\theta = \pm\, \pi/2$ from the maximum formation stress direction, which is the azimuth of the formation minimum stress direction. Once the maximum hoop stress value exceeds the (compressive) strength of the formation rock, the borehole wallcollapses starting from the formation minimum stress azimuth. The failed formation rock forms two stripes on the borehole wall. The centers of the stripes are respectively at $\theta = \pm\pi/2$. Thus the formation stress orientation can be obtained by determining the azimuth of the borehole breakout. The limitation of this method is that the stress field has to pose failure on the borehole wall. In other words, if the hoop stress has not reached the rock strength, no breakout can be observed to yield the stress information. The

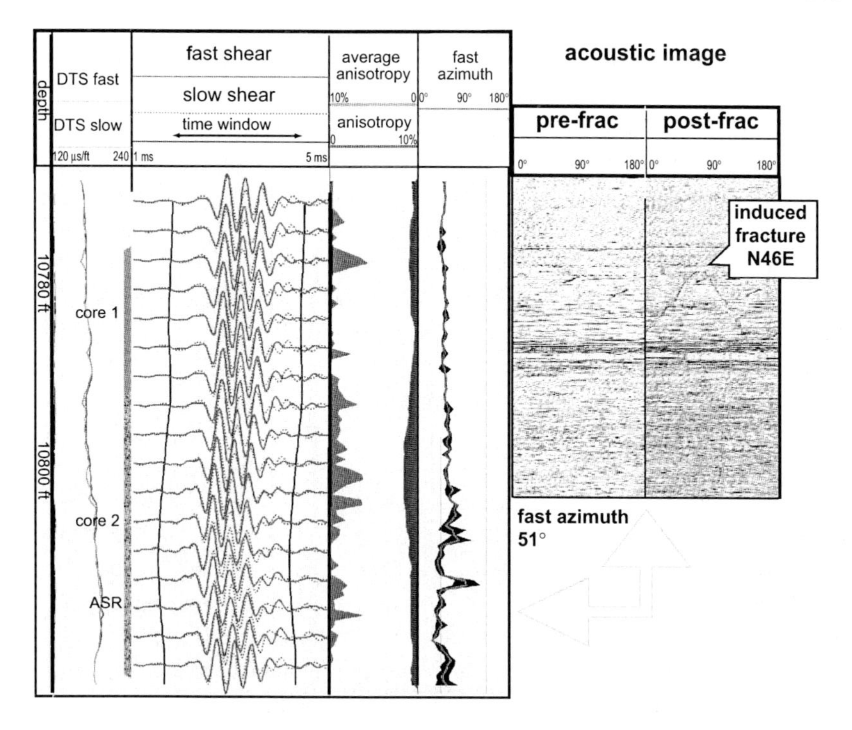

Figure 5.23. The cross-dipole analysis result plot across the Tor formation with the cored sections and ASR core analysis points being noted. The anisotropy result indicates that the fast shear azimuth through the section averages to E51°N. The acoustic imaging data obtained before (left image) and after (right image) micro-fracturing are also shown. The post-fracturing image shows an induced high angle (88°) fracture with strike direction of N46°. This direction corresponds to the maximum stress azimuth.

advantage of the cross-dipole anisotropy method is that it can be used to measure the stress information in the absence of borehole failure, due to the non-destructive nature of the method.

Figure 5.24 shows the cross-dipole and borehole acoustic imaging measurement results obtained from the upper (cross-dipole) and lower (imaging) parts of a (vertical) borehole. The image log was obtained in a sandstone formation. Stress concentrations in the lower borehole have already caused failure at the borehole wall. Breakouts can be clearly observed from the image log which shows that the minimum stress orientation is generally in the E-W direction. The upper borehole is in good shape because the caliper curve (CAL in track 1) varies smoothly and shows no abrupt changes, suggesting that the the borehole is still intact. The gamma ray curve (GR in track 1) shows that this is a sand/shale formation. Shear-wave anisotropy, shown as the array-aperture (left to right) and transmitter-to-receiver (right to left) anisotropy in track 2, is clearly observed in sand but almost disappears in shale, suggesting stress as the cause of the observed anisotropy (as shown previously in Figures 5.16 and 5.21, significant stress-induced anisotropy can be observed in sandstone but not in shale). The anisotropy azimuth, as seen from the anisotropy map in track 3,

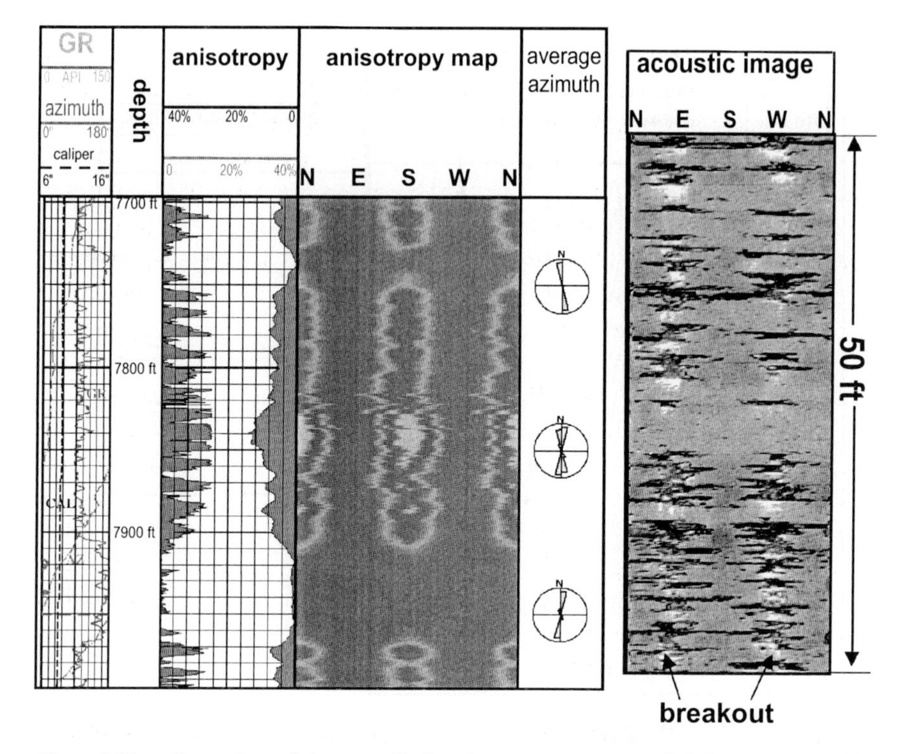

Figure 5.24. Comparison of the cross-dipole anisotropy measurement (left figure, measured in upper borehole) and borehole breakout observation from an acoustic image log (right figure, measured in lower borehole). The maximum horizontal stress orientation from the anisotropy map is in the NS direction, in good correspondence to the E- 1W direction of the minimum stress observed from the borehole breakout on the image

shows a general N-S direction, despite a significant rotation of the tool azimuth (AZ curve in track 1). This anisotropy azimuth, interpreted as the maximum stress orientation, shows an excellent agreement with that of the breakout observation from the lower borehole. This comparison also shows that the maximum regional stress orientation, whether it in the upper or lower borehole, is in the N-S direction. This example also demonstrates that the cross-dipole anisotropy and borehole breakout image methods are two complimentary techniques for down-hole stress (orientation) measurement. Combining both methods can obtain the stress information with or without failure of the borehole.

5.5 Estimating Formation Shear-Wave Transverse Isotropy

As the analysis of Chapter 2.9 demonstrates, for studying the formation VTI (i.e., borehole along the axis of TI symmetry) property using acoustic logging, Stoneley waves are the only borehole wave mode that has a significant sensitivity to the TI effects, especially when the formation is acoustically slow compared to borehole fluid. Because of the significant sensitivity of the Stoneley wave to c_{66} of the TI formation, it is desirable to estimate the TI parameter from borehole acoustic logging data, as this will yield a continuous log profile for assessing the formation anisotropy characteristic. Although the effects of TI on the Stoneley wave is well understood from theoretical analyses (see Chapter 2), application of the theory to field acoustic logging data has been hindered by two major factors. The first is lack of understanding of the effect of a compliant logging tool on Stoneley-wave propagation and an effective model to handle the effect. The second is the need for an efficient inversion method to estimate the TI effects from Stoneley wave data. This section describes a solution to these two problems. The acoustic logging tool is modeled as a cylindrical rod with an effective modulus, which can be determined using a calibration method. This model gives a good approximation when evaluated against the exact solution of a cylindrical shell model. With this approach, the effects of the tool can be accounted for, regardless of the actual structure or make-up of the tool. The inversion processing of Stoneley-wave data uses the Weighted Spectral Average Slowness theorem (equation (3.25)) which states that the Stoneley-wave slowness is a weighted average of the Stoneley wave dispersion curve over the wave's frequency band. This provides a fast and effective method for the inversion. Using this method and the estimated tool modulus and other formation and borehole parameters, V_{Sh} or c_{66} can be efficiently and reliably estimated.

5.5.1 MODELING THE EFFECT OF AN ACOUSTIC LOGGING TOOL ON STONELEY-WAVE PROPAGATION

In Chapter 2, we theoretically analyzed acoustic logging in a vertical well with a transversely isotropic formation. The main focus of this section is to further develop the analysis to incorporate the presence of an acoustic-logging tool in the borehole. Modern multipole (monopole and dipole) array-acoustic tools are quite compliant because of the requirement to isolate wave propagation along the tool during dipole logging (e.g., Cowles et al., 1994). The tool compliance will have a substantial effect on Stoneley wave propagation during monopole logging. The tool-compliance effect needs to be considered in the analysis of Stoneley-wave data acquired by a logging tool.

The model of a logging tool should provide a simple and sufficiently accurate description of the tool-compliance effect. The actual tool make-up is quite complicated (see Figure 1.7). It may be slotted or cut to allow coupling of receivers to borehole fluid. The tool may also contain sound isolation materials and electronics, etc. However, if the wavelength is long compared to the diameter of the tool, the tool can be effectively modeled as an elastic rod for the Stoneley-wave propagation. Further, in the low-frequency range corresponding to the Stoneley wave, the tool's elasticity can be adequately modeled by an effective modulus, as will be described in the following analysis.

Consider monopole wave propagation along a borehole of radius R. A logging tool of radius a is centered at the borehole. The borehole wave induces acoustic motion in the fluid annulus between the tool and formation (modeled as a TI medium). The expressions for the associated fluid radial displacement u and pressure p are given in equation (2.33), which, for the monopole case ($n = 0$), are given as

$$\left.\begin{matrix} u \\ p \end{matrix}\right\} = e^{ikz} \begin{cases} AfI_1(fr) - BfK_1(fr) \\ \rho_f\omega^2(AI_0(fr) + BK_0(fr)) \end{cases}, \quad (a < r < R) \ , \tag{5.26}$$

where A and B are the amplitude coefficient for the incoming and outgoing wavefield in the borehole fluid annulus, respectively. The other symbols have been defined previously in conjunction with equation (2.33) (note that $p = -\sigma_{rr}$).

Using a quasi-static analysis, Norris (1990) showed that borehole fluid conductance (defined as u/p) of an elastic rod concentric with the borehole is given by

$$\left(\frac{u}{p}\right)_{r=a} = -\frac{a}{M_T} \ , \tag{5.27}$$

where M_T is a tool modulus, which, in this elastic rod case, is approximately given by the rod's Young's modulus E and Poisson's ratio ν, as $M_T \approx E/(1-\nu)$. Matching fluid conductance (or impedance) at the tool surface, we extend equation (5.27) to the frequency regime assuming that 1) the quasi-static result still holds for low-frequency Stoneley waves, and 2) the compliance of a cylindrical tool can be modeled by an effective modulus M_T, regardless of the tool's internal structure. Substituting equation (5.26) into equation (5.27) gives a relation between the unknown coefficients A and B, as

$$E_{tool} = \frac{B}{A} = \frac{(M_T/a)fI_1(fa) + \rho_f\omega^2 I_0(fa)}{(M_T/a)fK_1(fa) - \rho_f\omega^2 K_0(fa)} \ , \tag{5.28}$$

where E_{tool} denotes the ratio of the Bessel function combinations related to the elastic tool. Thus only one unknown coefficient (A or B) needs to be determined from the boundary condition at the interface between the borehole and the TI formation, which is the same as equation (2.15). This leads to a dispersion equation for computing the Stoneley wave propagation in a fluid-filled borehole centered by a logging tool and surrounded by a TI formation.

$$D(k, \omega, c_{66}, c_{44}, M_T, R, a, \rho, V_f, \rho_f, c_{11}, c_{13}, c_{33}) = 0 \ . \tag{5.29}$$

The parameters in the above dispersion equation are ordered by their importance and relevance to the present problem. In fact, adding two modifications in the dispersion equation (2.52) without a tool (case $n = 0$) leads to the above dispersion equation. These modifications constitute the addition of a tool-related term to the respective Bessel function of the two matrix elements Q_{11} and Q_{21} in equation (2.51), as

$$Q_{11}: \ I_1(fR) \rightarrow I_1(fR) - E_{tool}K_1(fR) \ ,$$
$$Q_{21}: \ I_0(fR) \rightarrow I_0(fR) + E_{tool}K_0(fR) \ . \qquad (5.30)$$

Similarly, making the same modifications in the matrix elements M_{11} and M_{21} in equation (2.18) incorporates the tool effect in the isotropic analysis (equation (2.24)). The same approach can be applied to N_{11}, N_{21}, and N_{41} in equations (4.24) and (4.25) for the porous formation situation. By solving the above dispersion equation (5.29) in the frequency range of interest to find the root values of k for the Stoneley wave, as denoted by k_{ST}, the Stoneley-wave phase velocity (or slowness) dispersion curve is obtained:

$$V_{ST}(\omega) = \omega / k_{ST} \, , \text{ or } \ S_{ST}(\omega) = 1/V_{ST}(\omega) = k_{ST}/\omega \ . \qquad (5.31)$$

Two special cases can be derived by analyzing the expression in equation (5.28). The first is a fluid-filled borehole without a tool. This case is derived by setting the tool radius a to zero, leading to $E_{tool} = 0$. Equation (5.29) then reduces to equation (2.52) for the case $n = 0$. The second case is the rigid tool case studied by Tang and Cheng (1993b). This case is modeled by letting $M_T \rightarrow \infty$, resulting in $E_{tool} = I_1(fa)/K_1(fa)$, in agreement with Tang and Cheng (1993b).

5.5.2 VALIDATION OF THE SIMPLE TOOL MODEL

The validity of the simplified tool model is verified by comparing the modeling result with that of a cylindrical elastic shell model. In this example, we assume that the inner and outer radii of the shell are 0.035 m and 0.045 m, respectively. The shell's elastic parameters are $E = 48.8$ GPa and $\nu = 0.25$. The interior of the shell is filled with fluid (water). For this cylindrical shell tool model, the exact theoretical Stoneley-wave dispersion curve can be calculated using the method for modeling a multi-layered system with a fluid annulus (equations (2.33)–(2.35)). For the cylindrical shell tool model, the low-frequency limit of the Stoneley-wave velocity can be computed. This limit is then equated to that of the effective-modulus tool model to derive the modulus in equation (5.28). In fact, using a static approach, Norris (1990) derived a formula that allows the modulus to be directly computed from the Stoneley-wave velocity limit at zero frequency, as given by

$$M_T = \frac{a^2/R^2}{\left(1 - \dfrac{a^2}{R^2}\right)\left(\dfrac{1}{\rho_f V_{ST}^2(0)} - \dfrac{1}{\rho_f V_f^2}\right) - \dfrac{1}{c_{66}}} \ , \qquad (5.32)$$

where $V_{ST}(0)$ is the zero-frequency limit of velocity for the Stoneley wave in the fluid annulus between the borehole and the cylindrical shell (formation and borehole parameters are given in Table 2.3). The effective tool radius a is taken as the outer radius (0.045 m) of the cylindrical shell. Using these parameter values in equation

(5.32) gives a value of 6.73 GPa for M_T. With this value, the performance and accuracy of the approximate tool model can now be evaluated. The tool is now modeled by its effective modulus M_T regardless of its internal structure. Considering the parameters involved even in this relatively simple tool model (i.e., the shell and fluid elastic/acoustic parameters and dimensions), the effective modulus model is truly a drastic simplification, because essentially only two parameters – M_T and a – are used to model the tool effect. Using the M_T value in the dispersion relation given in equation (5.29), we obtain the Stoneley-wave phase velocity curve as a function of frequency, for the TI-formation model given in Table 2.3. Figure 5.25a shows the result (dashed curve) of the approximate model compared with the exact result (solid curve) for the fluid-filled cylindrical shell model (the dispersion curve in the absence of the tool is also shown). There is an excellent agreement between the two results at low frequencies (below 2.5 kHz), with only small differences as frequency increases. These differences are only of academic importance because in practice they are well within the errors of field Stoneley-wave measurements. The approximate model therefore provides simple but sufficiently accurate modeling for the effect of a logging tool on Stoneley-wave propagation.

The need to incorporate the logging-tool effect in the Stoneley-wave propagation and inversion problem can now be demonstrated. As shown in Figure 5.25a, the presence of the tool can substantially reduce the Stoneley-wave velocity and changes the character of the wave dispersion curve. The presence of a tool can also significantly modify the wave's sensitivity to borehole-fluid acoustic and formation elastic parameters. Using equation (5.29), the Stoneley-wave sensitivities of Figure 2.15c for the TI formation are recalculated for the presence of the logging tool (the tool is modeled using the effective modulus model with $a = 0.045$ m and $M_T = 6.73$ GPa, as derived from the fluid-filled cylindrical shell model).

Figure 5.25b shows the sensitivity analysis result (solid curves). Compared with its respective counterpart without tool (re-plotted here as dashed curves), the Stoneley-wave sensitivity to the fluid and formation parameters is reduced by the presence of the tool. The sum of the three sensitivities is appreciably less than one, as can be easily seen by adding their zero-frequency values. The reduction of the sensitivity to borehole fluid is quite significant. This is understandable because the presence of a tool replaces a portion of the borehole fluid and the wave's sensitivity is now partitioned to the tool's compliance. Fortunately, the Stoneley-wave sensitivity to V_{Sh} or c_{66} is still quite significant in the low-frequency range (below 2 kHz in this case), which ensures the estimation of this parameter from Stoneley-wave measurements. The analysis results shown in Figure 5.25 demonstrate the necessity to incorporate the effects of a logging tool in the analysis of field Stoneley-wave data.

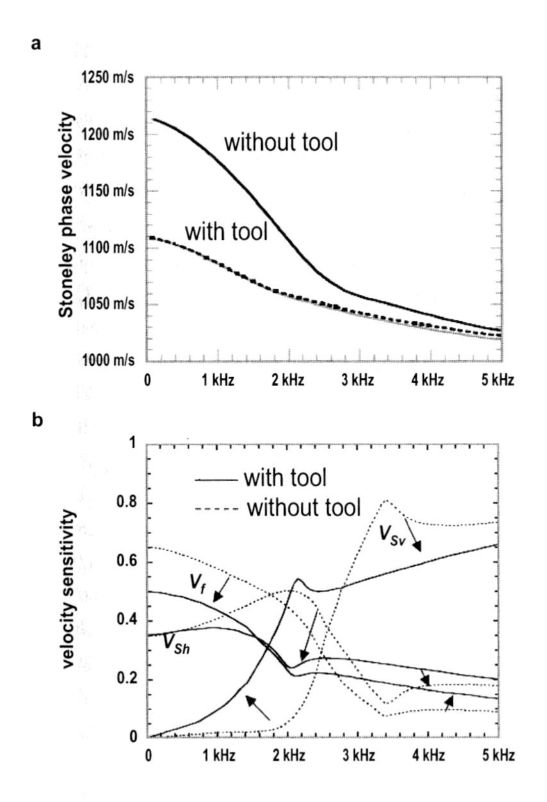

Figure 5.25. (a) Stoneley wave phase velocity for a TI formation with and without a logging tool. (b) Effect of a logging tool on Stoneley wave sensitivities to borehole fluid, formation horizontal and vertical shear velocities.

5.5.3 Inversion Formulation

This section describes a simple and effective formulation to estimate the shear-wave TI parameter from borehole Stoneley waves using dispersion equation (5.29). We also demonstrate an important application of the Weighted Spectral Average Slowness theorem, as derived in Chapter 3. The Stoneley waves acquired by an array acoustic tool are usually processed using array-coherence stacking methods (e.g., semblance or waveform inversion, as described in Chapter 3), yielding a Stoneley-wave slowness profile over the logging depth (in the following, we will use Stoneley-wave slowness, instead of velocity, to conform to the convention used in acoustic logging). These methods are non-dispersive techniques that do not consider the dispersion effect in the waveform data. However, Stoneley waves, as shown in Figure 5.25a, can exhibit a significant dispersion effect in the low-frequency range of interest. Fortunately, there is a theoretical relationship between the wave's dispersion characteristics and the wave slowness derived from a non-dispersive processing technique. This relationship is the Weighted Spectral Average Slowness theorem given in equation (3.25), which is rewritten here for the Stoneley-wave situation.

$$S^*_{ST} = \frac{\int_{-\infty}^{+\infty} S_{ST}(\omega, V_{\text{Sh}})\omega^2 A^2(\omega)d\omega}{\int_{-\infty}^{+\infty} \omega^2 A^2(\omega)d\omega}, \qquad (5.33)$$

where S^*_{ST} is the Stoneley-wave slowness, as obtained from the non-dispersive array processing; $S_{ST}(\omega,V_{\text{Sh}})$ is the Stoneley-wave phase slowness dispersion curve, as computed from equations (5.29) and (5.31). This dispersion curve is parameterized by the horizontal shear velocity V_{Sh}, assuming all other formation, borehole, and tool parameters, as needed to calculate equation (5.29), are known; $A(\omega)$ is the Stoneley-wave amplitude spectrum. The integration in equation (5.33) is over the entire wave spectrum covered by $A(\omega)$. Equation (5.33) states that the Stoneley-wave slowness resulting from a non-dispersive time-domain array processing method is a weighted spectral average of the wave's slowness dispersion curve over the frequency range of the wave spectrum. The weighting function is given by $\omega^2 A^2(\omega)$. The validity of the theoretical relationship in equation (5.33) has been numerically proved in Chapter 3.5 using synthetic array Stoneley wave data computed for a TI formation.

5.5.4 APPLICATION TO SHEAR-WAVE TI PARAMETER ESTIMATION

Equation (5.33) can be used to provide a simple and effective estimation for horizontal shear velocity V_{Sh}, or c_{66}, from the Stoneley-wave logging data. As the equation shows, the Stoneley-wave slowness, as derived by a non-dispersive array processing method (semblance or waveform inversion), is related to the wave's phase dispersion curve over the frequency range occupied by the wave's spectrum. Equation (5.33) contains only one unknown parameter, V_{Sh} or c_{66}. All other parameters, as needed to calculate the dispersion equation (5.29), are assumed known or available from log data. For example, c_{44}, c_{33}, ρ and R are available from conventional dipole shear-wave slowness, monopole P-wave slowness, density, and caliper logs, respectively. The borehole fluid parameters ρ_f and V_f can be estimated from the type of drilling fluid used. The TI parameters c_{11} and c_{33} can be treated as follows. The difference between c_{11} and c_{33} defines the P-wave TI parameter ε (Thomsen, 1986), which is generally correlated with the shear-wave TI parameter γ (Thomsen, 1986; Wang, 2001). Assuming that ε and γ are about the same order gives $c_{11}/c_{33} = c_{66}/c_{44}$. We also assume $c_{13} = c_{33} - 2c_{44}$. These treatments, although approximate, will not significantly alter the estimation for V_{Sh} or c_{66}, because the Stoneley wave is relatively insensitive to the parameters c_{11}, c_{13} and c_{33} (a complete sensitivity analysis for these parameters can be found in Ellefsen, 1990). The tool radius a is always available and the tool's effective modulus can be obtained from a calibration procedure to be described later.

By specifying all necessary parameters in the dispersion equation (5.29), the shear-wave TI parameter V_{Sh} or c_{66} can be estimated using equation (5.33) by the following procedure.

1. Process array Stoneley-wave data to obtain S^*_{ST}.
2. Compute the Stoneley-wave spectrum $A(\omega)$. Integrate $\omega^2 A^2(\omega)$ over the frequency range of $A(\omega)$ to obtain the denominator of equation (5.33).
3. For a trial horizontal shear velocity V_{Sh}, calculate the dispersion curve $S_{ST}(\omega, V_{Sh})$ using equations (5.29) and (5.31).
4. Weight $S_{ST}(\omega, V_{Sh})$ with the weighting function $\omega^2 A^2(\omega)$ and integrate the weighted slowness curve over the frequency range of $A(\omega)$. Divide the integral value with the denominator value from step 2, then equate the result to S^*_{ST}.
5. Repeat steps 3 and 4 till equation (5.33) is satisfied. Output V_{Sh} as the formation horizontal shear-wave velocity.

Repeating the above procedure for the logging depth range of interest yields a continuous formation horizontal shear-wave velocity (or slowness) profile. With the vertical and horizontal shear velocity (slowness) profiles, a continuous profile of the shear-wave TI parameter, commonly known as the Thomsen parameter γ (Thomsen, 1986), can be calculated as

$$\gamma = \frac{c_{66} - c_{44}}{2c_{44}} \approx 2\frac{V_{Sh} - V_{Sv}}{V_{Sv} + V_{Sh}} \; . \tag{5.34}$$

The anisotropy profile can then be used to assess the degree of formation anisotropy for seismic and/or formation evaluation applications.

5.5.5 Calibrating Tool Compliance

As shown in the previous modeling example (Figure 5.25), the compliance of the tool can have an important influence on Stoneley-wave propagation characteristics. A calibration procedure is used to determine the tool compliance, which is specified by the effective modulus M_T. For a formation interval of known anisotropy (e.g., zero anisotropy), the same equation (5.33) as used for TI estimation can be used to determine the tool modulus M_T. For example, for an isotropic formation the formation parameters (compressional and shear velocities and density) needed to calculate the dispersion equation (5.29) are available from log data, and the fluid parameters (V_f and ρ_f) can be estimated from the drilling fluid used. The only unknown parameter is the tool modulus M_T. The calibration procedure is essentially adjusting the value of M_T in the dispersion equation until the following equation is satisfied.

$$S^*_{ST} = \frac{\int_{-\infty}^{+\infty} S_{ST}(\omega, M_T)\,\omega^2 A^2(\omega)\,d\omega}{\int_{-\infty}^{+\infty} \omega^2 A^2(\omega)\,d\omega} \; . \tag{5.35}$$

Once the tool modulus is specified from calibration, it is used in the dispersion equation (5.29) to calculate the Stoneley-wave dispersion for the inversion processing.

5.5.6 A FIELD EXAMPLE

This method of shear-wave TI estimation has been applied to an acoustic logging data set from a well near Braggs, Wyoming. The well was drilled through the Lewis shale formation. The goal of the acoustic processing was to characterize the TI property of this shale formation. Acoustic dipole and monopole waveform logging data were acquired throughout the formation. Figure 5.26 shows detailed analysis results for the Lewis shale formation. On top of the Lewis shale, the formation is characterized by shaly sandstone and sand/shale sequences. Track 1 of the figure shows the gamma ray curve. The high gamma ray value (> 120 API) marks the beginning of the massive Lewis shale formation. Tracks 2 and 3 show, respectively, the Stoneley- and dipole-wave data. Processing the data yields the Stoneley-wave slowness in track 4 (solid curve labeled DTST) and the shear slowness in track 6 (dashed curve labeled DTSV, standing for slowness of a vertically propagating shear wave). As a quality-control indicator for the slowness curves, the curves are integrated over the transmitter-to-receiver-1 distance to give the travel time curve for the Stoneley wave (track 2) and dipole-shear wave (track 3), respectively. The travel-time curves track the respective waveform quite well, indicating the validity of the slowness results.

An interesting feature is that the travel-time curve (or waveform) for the dipole wave shows more character/variation than that of the Stoneley wave. Relative to the Stoneley wave, the dipole wave is much delayed in the shale formation compared to the formation above it. This waveform/travel-time character/delay difference between the two types of waves provides a direct indication of the TI effect. The dipole shear slowness (DTSV in track 6) and the Stoneley wave slowness (DTST in track 4) are utilized in equation (5.33) to determine the horizontal shear slowness DTSH (or $1/V_{Sh}$). The estimated horizontal shear slowness profile is shown in track 6 (solid curve labeled DTSH). This shear slowness profile shows a substantial difference (shaded area between DTSH and DTSV) compared with the vertical shear slowness DTSV. The two slowness curves are then used to calculate the shear-wave TI anisotropy parameter γ. The γ-profile is shown as the shaded curve in Track 5, scaled from 0 % to 50 %. The anisotropy shows a massive, continuous feature below the top of the Lewis shale formation, but tends to vanish in the shaly-sand interval and sandstone streaks above the formation. The anisotropy effect can also be analyzed by comparing the measured Stoneley slowness having the TI effects, with a computed slowness without the TI effects. Assuming isotropy for the formation, the vertical shear slowness curve (DTSV in track 6) can be used to compute an isotropic Stoneley-wave dispersion curve, which, after the weighted-averaging process using equation (5.33) (replacing V_{Sh} with V_{Sv}), gives the (isotropic) Stoneley slowness in track 4 (dashed curve labeled DTST (*from* DTSV)). The difference between the measured and computed (isotropic) Stoneley-wave slowness curves (shaded area

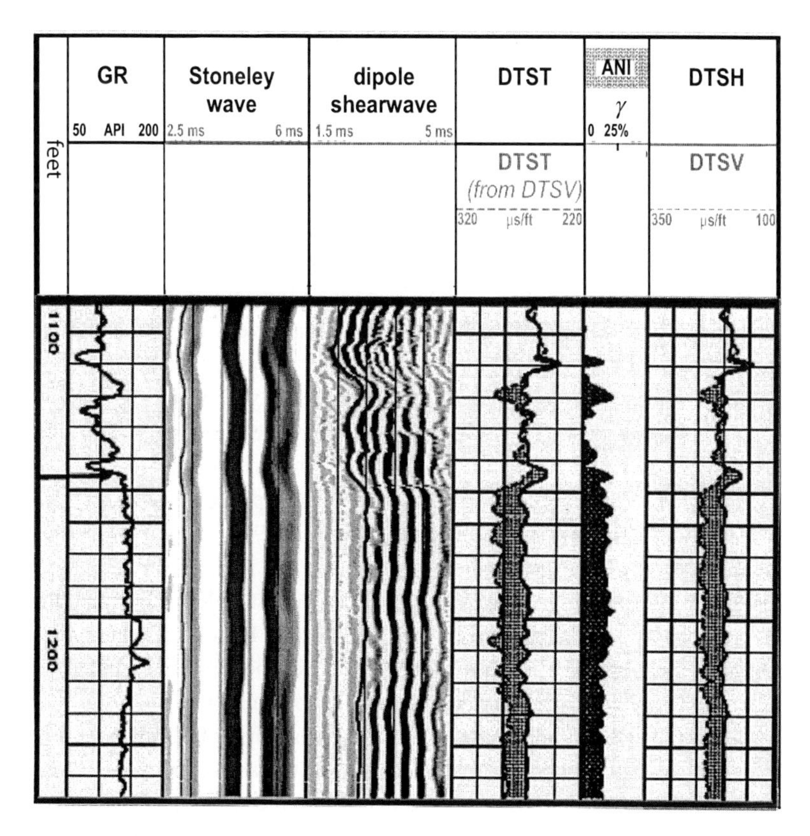

Figure 5.26. Stoneley-wave (TI) anisotropy analysis result in the top section of the Lewis shale formation (marked by high GR (>120 API) values in track 1). The significant delay of the dipole wave (track 3) relative to the Stoneley wave (track 2) in the shale indicates the TI effect. Track 6 shows the estimated horizontal (solid) versus the vertical (dashed) shear slowness curves. Track 4 shows the measured (solid) Stoneley slowness versus the (hypothetical) isotropic slowness (dashed). Track 5 is the profile for the shear-wave (TI) anisotropy parameter γ.

between the two curves) also indicates the presence of anisotropy. The anisotropy is quite significant, generally on the order of 20 %–30 %. The anisotropy estimation from the Stoneley-wave log data delineates the shear-wave (TI) anisotropy magnitude and variation for shale formation.

5.5.7 REMARKS ON THE SHEAR-WAVE TI ESTIMATION METHOD

The TI-estimation method using borehole Stoneley waves has several aspects that are worth discussing. These discussions will help clarify the strength and weakness of the method in practical applications.

In the absence of formation anisotropy (TI), the above procedure can also be used to estimate the formation shear-wave velocity (slowness) from Stoneley-wave logging data. In this case, the formation has only two elastic parameters: P- and S-wave velocities. The former is available from logging and the latter can be estimated from the above procedure. Deriving shear-wave slowness from Stoneley waves logged in slow formations was a common practice before the inception of dipole acoustic logging. However, most calculations either ignored the presence of a tool, or used inappropriate tool models (rigid or having a low impedance). With the effective elastic modulus model and the effective estimation method (equation (5.33)), as described previously, the accuracy and reliability of shear-wave velocity estimation can be enhanced.

As in the shear slowness analysis using Stoneley waves, the Stoneley-wave method is applicable mostly in slow formations where the formation shear rigidity is comparable or below the borehole fluid modulus. In this case the Stoneley wave is quite sensitive to formation shear-wave properties, isotropic or anisotropic. However, in fast formations, the sensitivity diminishes (Ellefsen, 1990) and the estimation will suffer large errors or becomes invalid.

Formation permeability significantly affects the Stoneley-wave propagation velocity, especially at low frequencies (see Chapter 4). Thus the TI-parameter estimation method using low-frequency Stoneley waves is not applicable in permeable formations unless the permeability effect can be accounted for.

In a deviated well, the borehole is tilted away from the axis of symmetry of the TI formation. In this situation, the applicability of the above analyses is reduced. However, an unpublished study indicated that for well deviation less than 30°, the sensitivity of the borehole Stoneley wave to the VTI formation is still dominated by V_{Sh} or c_{66}. Under this restriction, the above analyses can still be used, although the accuracy of the results may be degraded.

The proposed Stoneley-wave estimation method obtains only the shear-wave anisotropy information. However, in seismic migration/imaging using P and/or converted waves, the P-wave anisotropy parameter ε (Thomsen, 1986) is desired. Fortunately, in many rocks, such as shales, the P-wave anisotropy and shear-wave anisotropy are strongly correlated (Thomsen, 1986; Wang, 2001), because they are related to the same depositional processes over geological times. In this case, the shear-wave anisotropy can be correlated with the P-wave anisotropy data to delineate the magnitude and variation of the latter anisotropy. In this regard, the obtained shear-wave anisotropy still provides important information for seismic migration/imaging applications.

CHAPTER 6: SUMMARY, RELATED TOPICS, AND ROAD AHEAD

Acoustic logging has undergone significant progress over the past three decades. The technology started from being a simple acoustic transit time measurement to become an advanced, indispensable discipline for the petroleum industry. While elastic wave transit time or slowness measurements are still a routine part of acoustic logging, the advancement of the technology now allows for a broad range of measurements to obtain important formation properties. These properties, to mention just a few, are formation permeability, elastic wave attenuation, and seismic anisotropy, etc. Even for the velocity measurement alone, we can now determine formation compressional and shear velocities with much improved accuracy and resolution. With these advances, we can better detect and characterize formation fractures, estimate formation stress field, and locate/estimate petroleum reserves. The theory, methods, and applications related to common acoustic logging practice were described in the previous chapters of this book, and will be summarized in this chapter.

In this chapter, we will describe some important topics and/or new developments in acoustic logging. During decades of technical development and practice, many important applications of borehole acoustic logging have been identified or begin to emerge. Some of these applications, although important, have not yet been routinely used in common acoustic logging practice, either because they are still in the development stage, or because there are drawbacks that limit their practical use. Because of this, these applications were not elaborated or fully described in previous chapters. These applications, to mention just a few, include deviated well acoustic logging and data interpretation, near borehole acoustic tomography, single well acoustic imaging, and most recently, the development of acoustic logging while drilling. They will be described in the later part of this chapter.

6.1 Summary of Previous Chapters

The effort of this book has been on bridging the gap between borehole acoustic wave theory and field data. The emphasis was not on the fundamental theory development, but rather on using the theory to develop quantitative, practical methodologies for field data processing and interpretation. In the technical content of the book, theoretical modeling has played an important role and the insight/understanding gained from the theoretical analyses has turned out several new, important results. The major work and contents of the book are summarized below.

Chapter 1 gave a brief overview of the acoustic tool development and recent advances in acoustic logging applications. Chapter 2 systematically elaborated the basic theory for acoustic waves in boreholes. It also described the wave modeling analysis for multilayered systems, how to suppress a potential numerical instability problem in the multilayer modeling, and how to implement an acoustic ring source in the multilayer system. The results of the modeling have turned out several useful, practical applications. For example, the excessive dispersion caused by formation alteration, as demonstrated by the modeling, has been used by Tang (1996a) to diagnose formation alteration. The result of modeling dipole logging with unbonded casing has been utilized by Pampuri et al. (2003) to determine compressional-wave slowness in poorly-bonded cased holes using dipole acoustic waves. The modeling of multipole waves in the Logging-While-Drilling (LWD) configuration has been employed in the development of an LWD shear-wave tool (Tang et al., 2002a). In addition, Chapter 2 also described the theory of acoustic logging in a transversely isotropic (TI) formation. The theory provides the foundation for determining the TI anisotropy from the Stoneley wave data, as described in Chapter 5.

Chapter 3 started by elaborating several acoustic-array velocity analysis methods in both frequency and time domain. It then showed how to transform the commonly-known linear prediction theory from the frequency domain into the time domain, leading to a pair-wise waveform matching method. Combining this method with overlapping subarrays significantly enhances the resolution of the formation acoustic velocity profile. Further in Chapter 3, the analysis of dispersive waves in an array resulted in the Weighted Spectral Average Slowness theorem, which provides a simple and effective method for correcting the dispersion effect in dipole-shear waves. The theorem also provides a fast method for estimating formation shear-wave TI property from Stoneley waves (Chapter 5). The final part of Chapter 3 described a recently developed robust method for estimating elastic-wave attenuation from borehole acoustic waveform data.

Chapter 4 analyzed the role of Biot's slow wave in the interaction of formation permeability and borehole Stoneley waves, resulting in a simple and sufficiently accurate theory for forward and inverse modeling of permeability logging using Stoneley waves. Together with the permeability estimation procedure, an effective wave separation method was developed. The wave separation method suppresses noise- and scattering effects in the data and extracts reflection events from the data. A fast wave simulation method was also developed to allow for realistic modeling of Stoneley-wave propagation with an irregular borehole and an inhomogeneous formation. Besides demonstrating the validity of Stoneley-derived permeability with other measurements (e.g., core, NMR, etc.), Chapter 4 described a simple mud cake model which indicates that mud cakes, depending on their rigidity, may or may not block the Stoneley wave from interacting with formation permeability. The result of this model may have helped to resolve the decades-old mud cake mystery.

Chapter 5 was mainly concerned with borehole acoustic anisotropy measurements and applications. The very fundamental phenomenon of shear-wave anisotropy, the splitting of fast and slow shear waves, was utilized to develop a wave matching method, which improves the accuracy and robustness of the azimuthal shear-wave anisotropy estimates as compared to a conventional method. The results of the anisotropy measurement, coupled with theories of shear-wave propagation in fractured media and in pre-stressed formations, have produced very useful applications for the cross-dipole shear-wave technology. For example, the cross-dipole measurement has now been an important means for fracture characterization. The technology has even been applied to map stimulated hydraulic fractures behind casing. This chapter also described a simple phenomenological theory to model stress-induced shear-wave anisotropy around borehole, producing almost the same results as those from the nonlinear acousto-elasticity theory (e.g., Winkler et al., 1998). Consequently, we can use the cross-dipole measurement to measure the effects of the formation stress field and determine its orientation. Finally, using the theoretical foundation in Chapter 2, Chapter 5 developed a method for estimating TI anisotropy from borehole Stoneley waves. A novel part of the work is that it showed how a complicated logging tool can be accurately modeled by an elastic rod with an effective modulus, at least in the low frequency range of interest. This simplification, combined with the Weighted Spectral Average Slowness theorem described in Chapter 3, provides a fast and effective estimation method.

6.2 Related Topics and Road Ahead

In the remaining part of this chapter, we describe some interesting topics of borehole acoustic logging that were not elaborated in previous chapters. The specific topics are deviated well acoustic logging and data interpretation, near borehole acoustic tomography, single well acoustic imaging, and the development of acoustic logging while drilling. In the following, we focus on describing the concept, approach, and status/trends in these important topics that may lead to future development and applications in borehole acoustic logging.

6.2.1 Acoustic Logging and Data Interpretation in Deviated Wells

A current trend in the petroleum reservoir exploration and production is that more and more deviated (or even horizontal) wells are drilled. This is especially the case for offshore, deep water reservoirs. The need for conducting acoustic logging in deviated boreholes calls for further developing acoustic tool technology and data modeling and interpretation analyses.

6.2.1.1 *Acoustic tool development*

In a moderately or highly deviated well, the strength of a logging tool is an important issue. This is because lowering (or pushing with a conveying pipe) a tool into the borehole needs to overcome the friction generated by the weight of the tool or tool string (a suite of logging tools) acting on the lower side of the deviated well. Consequently, a tool with low (compression) strength may buckle or bend against the borehole wall and fail to reach a designated depth/formation. This situation is not in favor of a multipole acoustic tool whose acoustic isolator contains soft sound isolation materials (Cowles et al., 1994). To address the needs for deviated well logging, the logging industry has been developing acoustic tools that have significant strength to log deviated wells (e.g., Kessler and Varsamis, 2001), and at the same time maintain good tool-wave isolation to acquire quality acoustic (particularly dipole) waveform data.

6.2.1.2 *Acoustic data modeling/interpretation*

In a deviated well, the acoustic wave propagation may be complicated by tool eccentering and anisotropy. If the tool is run without a centralization device, the tool tends to lie on the lower side of the borehole, creating an eccentered tool situation. For an eccentered tool, a dipole or monopole on the tool may generate other wave modes. Understanding acoustic wave propagation in this scenario, however, needs sophisticated numerical modeling in a 3D environment.

Interpreting acoustic data acquired in deviated wells can be quite complicated, especially in the presence of strong anisotropy. The issue of anisotropy is especially important for deviated wells penetrating soft sediments of deep water reservoirs. The sedimentary formations, e.g., shales, can be highly anisotropy (for example, some shale formations have about 20 %–40 % vertical versus horizontal velocity anisotropy). We have previously shown that shear-wave anisotropy for the ideal HTI and VTI cases can be, respectively, obtained from cross-dipole and Stoneley wave measurements. However, in a deviated well, the well trajectory is neither perpendicular to, nor parallel with, the formation bedding planes. To understand acoustic logging in this environment, some studies have been carried out. For example, Wang and Hornby (2002) simulated monopole and dipole acoustic logging in the deviated well configuration with a TI formation using a 3D finite-difference technique. For a similar situation, Wang and Tang (2003) modeled the quadrupole measurement in the Logging-While-Drilling environment. The results show that the determination of the anisotropy parameters, even in the simple TI case, is a difficult problem. The measured dipole or quadrupole velocity is neither the horizontally, nor the vertically, propagating shear-wave velocity. Instead, the measured velocity is a projection, or combination, of the two shear velocities (and other anisotropy parameters) along the well deviation direction. Other auxiliary information, e.g., the Stoneley wave measurement, may be combined to resolve the TI parameters (e.g., Norris and Sinha,

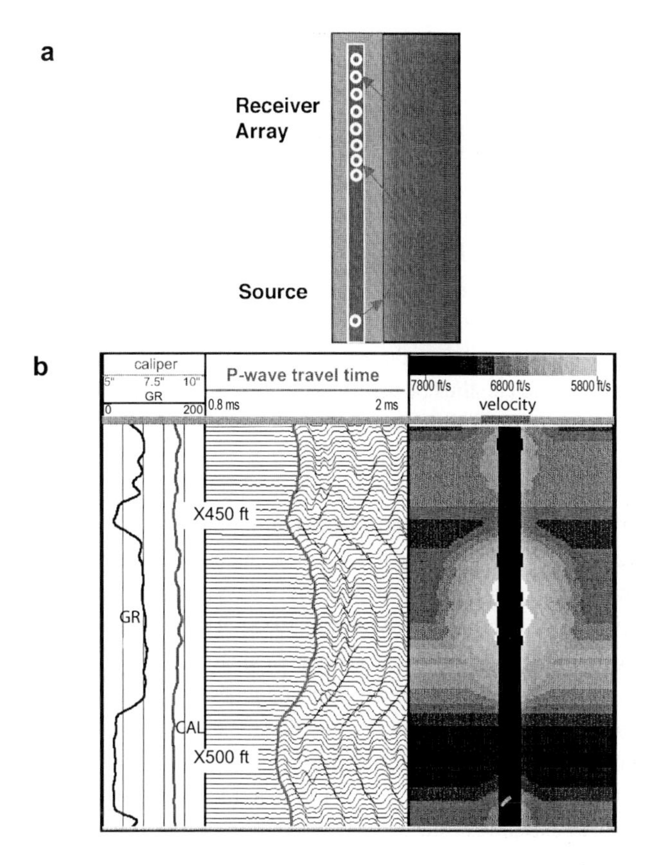

Figure 6.1 (a) Acoustic-wave ray paths in a formation with the velocity increasing away from borehole.
(b) An example demonstrating the application of acoustic travel time tomography to delineate the near-borehole alteration of a shale formation.

1993; Chi and Tang, 2003). The Stoneley-wave sensitivity to the TI formation varies with the well deviation, providing an additional measurement of the TI effect in the deviated well. More modeling and analysis studies are needed to fully understand the measurements in these situations and develop practical processing and interpretation methodologies for them.

6.2.2 Near-Well Acoustic Tomography

The formation elastic-wave velocity surrounding a borehole can vary significantly away from the borehole. This variation may be caused by a number of factors: formation damage due to drilling or stress relief, change in saturating fluid properties due to fluid invasion, and swelling-clay effects (especially in shale) caused by water take up and reaction with the drilling mud, etc. Because of the (radial) velocity variation, different acoustic tools, depending on their source-to-receiver spacing and acoustic-wave frequency content, may measure different velocity values. This situation is illustrated by the diagram in Figure 6.1a. An array acoustic tool measures the acoustic waves traveling through a formation with the velocity increasing in the radial direction. The radial velocity variation causes the individual received waves to have different radial depth of penetration, as illustrated by the first-arrival ray paths in the diagram. In this situation, the acoustic-array-derived wave velocity has some uncertainties. The velocity may or may not be that of the virgin formation, depending on the degree and radial extent of the velocity variation and the source-to-receiver spacing. A blessing in disguise is that, since the ray path samples the different radial velocity values in the formation, the wave's travel time carries the information of the velocity variation and can thus be used to determine the radial velocity changes in the formation. Acoustic travel time tomography is a technique that can be used to map the velocity distribution in the near-borehole region.

Hornby (1993) used the well-known travel time tomography technique to reconstruct a 2D velocity or slowness profile, assuming that the near-borehole formation slowness varies both axially along the borehole and radially away from the borehole. The configuration is similar to that of the diving-ray tomography used in surface seismic applications (Zhu and McMechan, 1988). The technique performs ray tracing through the 2D model to provide a forward calculation of the acoustic travel time, then minimizes the difference between the calculated and measured travel time data to find the best model slowness distribution that fits the data. Hornby's (1993) results revealed many interesting features of near-borehole slowness profiles caused by shale alteration, formation damage, and formation beds intersecting the borehole at an angle, etc. An example of the acoustic travel time tomography is shown in Figure 6.1b to demonstrate the application.

The type of formation, as shown by the gamma-ray (GR) curve in track 1, is sand (low GR value) and shale (high GR value). The P-wave signal measured across the depth interval is shown in track 2, together with a travel time curve obtained by detecting the first motion of the wave. Track 3 is the travel-time-derived slowness distribution (or tomogram) around the borehole of approximately 8.5 in diameter (the borehole caliper log curve (CAL) is given in track 1). The tomogram is displayed as a VDL image. A significant near-borehole velocity reduction is seen in the shale formation. An interpretation of this reduction is that it may be caused by *borehole damage* (because the velocity reduction is most significant in the vicinity of significant borehole changes and washouts), and by *shale alteration* (because the velocity variation penetrates about two feet into the shale formation, suggesting clay softening effects caused by fluid invasion).

The near-borehole acoustic tomography technique may have several important applications. For example, it may be used to determine the depth of investigation of an acoustic tool, obtaining the virgin formation velocity that is needed by other geophysical (e.g., seismic) applications. It may be used in the borehole stability analysis to indicate formation damage caused by drilling and stress relief. It may also be used to delineate fluid invasion in the near-borehole region, detecting the changes in the saturating fluid properties away from the borehole. A drawback that hinders the routine application of the technique is perhaps its cost in computation time for the logging situation, if the 2D tomogram needs to be calculated for the entire logging depth with a great number of source and receiver positions. An effective and efficient tomographic reconstruction technique may be the key to facilitate this useful application. Another fact that is not in favor of this application is that current generation of long space acoustic tools is designed to penetrate the shallow near borehole zones. Stephen et al. (1985) used a finite difference simulation to demonstrate this fact. Thus to effectively measure the radial velocity profile near the borehole, new tool designs, perhaps with the receiver array similar to those used in Hornby (1993), may be necessary for this application.

6.2.3 Single Well Acoustic Imaging

Acoustic imaging near a borehole can provide subsurface geological structural features at a resolution impossible to achieve with surface seismic exploration methods. For this reason, borehole acoustic measurements have recently been utilized to obtain an image of the formation structural changes away from the borehole (Hornby, 1989; Coates et al., 2000; and Li et al., 2002).

Figure 6.2a illustrates why the borehole acoustic measurement can obtain the geological structural information away from the borehole. This figure depicts the logging of a wireline acoustic tool across a dipping bed intersecting the borehole. As the acoustic source on the tool is energized, it generates acoustic waves that can be classified into two categories according their propagation direction. The waves of the first category travel directly along the borehole. These direct waves, after being recorded by an acoustic tool, give the commonly-known acoustic waveform logging data. The waves of the second category are the acoustic energy that radiates away from the borehole and reflects back to the borehole from boundaries of geological structures. These waves are called secondary arrivals in acoustic logging data because their amplitudes are generally small compared to those of the direct waves. As shown in this figure, depending on whether the tool is below or above the bed, the acoustic energy strikes the lower or upper side of the bed and reflects back to the receiver array as the secondary arrivals. Therefore, these secondary arrivals can be migrated to image the formation structural feature away from the borehole, in a way similar to migration in surface seismic processing.

The migration of the acoustic data for imaging a formation structure can use the conventional seismic processing method. Perhaps the only major difference of the borehole acoustic data, as compared to surface seismic data, is the large amplitude direct arrivals in the borehole data. These data must be removed before processing the secondary arrivals of much smaller amplitude.

As described in Li et al. (2002), a band-pass filtering technique may be used to first remove the low-frequency events, e.g., the low-frequency Stoneley waves. Afterwards, wave-separation techniques are applied to separate the secondary arrivals from the direct arrivals based on their moveout characteristics. For example, in the single-receiver data gathered for various depths (or tool positions), the direct arrivals have a small moveout because their propagation distance (source-to-receiver spacing) is fixed. In comparison, the reflection events have a large moveout because their propagation distance changes as the tool moves close to, or away from, the reflector. A number of techniques can be used for the wave separation, e.g., *f-k* filtering (Hornby, 1989), or a combination of *f-k* and median filtering (Li et al., 2002), etc. The secondary arrivals resulting from the wave separation are gathered into up-dip (reflected up-going, see Figure 6.2a) and down-dip (reflected down-going) subsets.

The secondary arrivals, or reflection events, from the wave-separation procedure are then migrated to image the formation reflector. Several migration techniques can be used, e.g., the back-projection scheme using a generalized Radon transform (Hornby, 1989), or the commonly-used Kirchoff depth migration method (Li et al., 2002). The migration procedure needs a velocity model to correctly map the reflection events to the position of a formation reflector. The acoustic wave velocity obtained from the acoustic logging measurement is conveniently used to build the velocity model. (Hornby, 1989; Li et al., 2002).

After migration, the acoustic component data are mapped into a two-dimensional (2D) domain. One dimension is the radial distance away from the borehole axis; the other is the logging depth, or the tool position along the borehole axis. Figure 6.2b shows an example of acoustic imaging of near-borehole formation geological structures. If the boundaries of a geological structure have large acoustic and/or other petrophysical property contrast, they can be detected from conventional well log curves. As in track 1, the sharp changes in the compressional slowness (DTP) and gamma-ray (GR) curves indicate the location of the boundaries of geological structures at the borehole. However, these well log curves do not depict the structural features in the formation. In contrast, acoustic imaging determines the trends of the geological structures deep into the formation. This scenario is demonstrated in track 2 of the figure. The image in this track comprises the up-dip image projected into the left column, and the down-dip image projected into the right column, of the borehole. The imaging result clearly shows several geological structures crossing the borehole at an angle about 50°. Moreover, some structural features can still be identified at a radial distance of about 45 ft in this example.

The single well acoustic imaging can be applied to several important scenarios. As we have demonstrated, the technique can be used to map geological boundaries intersecting a well bore. It can also be used to image fractures/faults away from the borehole. It can even be applied in a horizontal well to trace the boundary of a reservoir (Esmersoy et al., 1998). The technique also has the potential to obtain the geological structure information ahead of the drill bit during the drilling operation. The acoustic imaging technology, at its present stage, has several drawbacks, as described in the following.

A major drawback of the technique is that it requires that the secondary arrivals reflected from formation boundaries have significant, measurable signal amplitude.

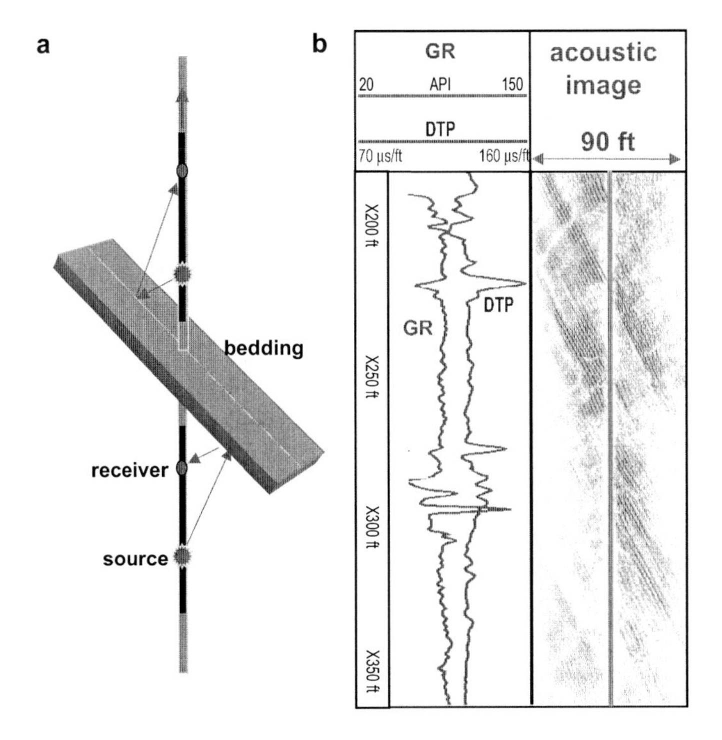

Figure 6.2. (a) Imaging near-borehole bedding boundary using acoustic reflections from the lower (tool below bed) and upper (tool above bed) side of the boundary.
(b) An example of imaging geological structures crossing the borehole. The structural boundaries are detected by sharp changes in the gamma-ray and P-slowness log curves in track 1. The structural features in the formations are mapped by the image in track 2.

The reflection amplitude, as recorded by an acoustic tool in the borehole, is strongly influenced by several factors such as impedance contrast between borehole and formation, formation type (fast or slow), source radiation (Meredith, 1990) and receiver reception patterns (Peng, 1994) in connection with a borehole, wave frequency, reflectance of the formation structure to be imaged, formation anelasticity, and acoustic tool configuration (e.g., source-receiver spacing), etc. The reflection events, as measured in connection with the acoustic logging data, are usually quite small. The events are often contaminated by the bed-boundary-caused near-borehole reflection and mode conversion. In the presence of these influences, or data noise, the signal-to-noise ratio of the reflection waves is crucially important. Our experience is that the technique performs better in slow formations (e.g., the case of Figure 6.2b) than in fast formations. The reason may be that the lower borehole-formation impedance contrast of a slow formation, as compared to a fast formation, allows for a larger portion of the acoustic energy to radiate into the formation. Acoustic tool development may optimize the acquisition of the reflection events. As with the near borehole tomography application, the current generation of full waveform acoustic tools is not designed optimally for single well imaging. Esmersoy et al. (1998) described an acoustic tool which is slightly modified from the conventional tool (it still has a conventional receiver array) to improve the separation of the direct and reflected arrivals. Another drawback of the current acoustic imaging result is that the technique, because it obtains only a 2D image of the formation reflectors, cannot resolve the azimuth of an imaged formation structure. A recent study (Tang, 2003) uses acoustic waves from a directional acoustic tool (i.e., a tool with dipole source(s) and/or receivers) to image a formation structure and determine its azimuth. Further development in acoustic tool technology, theoretical modeling, and data processing and imaging algorithms is needed to fully realize the potential of the single well acoustic imaging application.

6.2.4 Acoustic Logging-While-Drilling

The Logging-While-Drilling (LWD) acoustic technology was developed in recent years to address the need for rig time saving and real-time applications such as seismic tie and/or pore pressure determination, and as a replacement of wireline logging. The goal of LWD acoustic measurement is to determine formation compressional and shear velocities even while the well is being drilled. The two elastic wave velocities provide information important for the exploration and production of oil and gas. The LWD acoustic technology has been successful in the measurement of compressional-wave velocity of earth formations (Minear et al., 1995; Leggett et al., 2001) since the 1990s. The next challenge to the LWD technology has been the shear-wave velocity measurement, especially in slow formations. Shear velocities are increasingly utilized in converted-wave imaging, so that updating a shear velocity model while drilling will become important in the near future. Shear velocities are now being introduced for pore pressure prediction, with an abnormally large V_P/V_S ratio being indicative of overpressure. In some wells that need to be drilled with water-based muds, severe shale alteration and/or large borehole damage can make it difficult to determine shale properties from later wireline logging. In this situation, high quality LWD shear and other data may be the only chance to estimate these shale properties. In all these cases, the availability of high quality LWD shear velocity data will significantly help to calibrate pre-drill earth models. Because of the needs in measuring formation shear-wave velocity, a series of developments have taken place in recent years.

6.2.4.1 LWD acoustic tool development

Figure 6.3a is a schematic view of an LWD acoustic system built into a drill collar. A distinctive difference between the LWD and wireline (see Figure 1.1) acoustic tools is that the LWD tool occupies a large portion of the borehole, which is also the obstacle facing LWD compressional and shear measurements.

The first generation of LWD acoustic tools focused on the compressional-wave measurement. A commonly used design was an acoustic transmitter and receiver

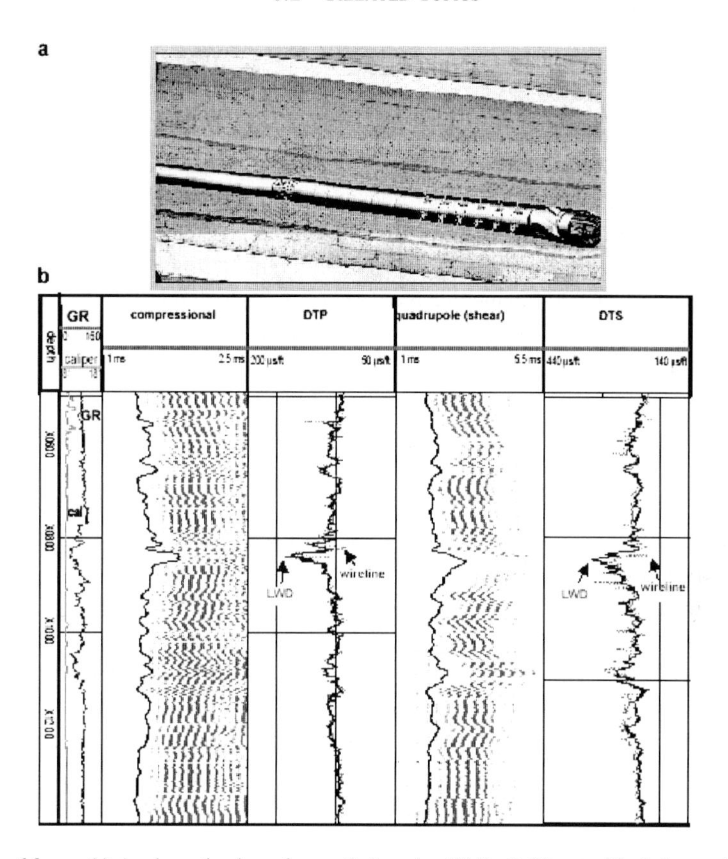

Figure 6.3. (a) A schematic view of acoustic Logging-While-Drilling tool built into a drill collar.
(b) Comparison of the LWD and wireline compressional and shear slowness measurements. Tracks 2 and 4 show, respectively, the LWD compressional and quadrupole-shear wave data. A transit time derived from respective slowness is also shown. Tracks 3 and 4 show, respectively, the overlay of the LWD and wireline compressional and shear slowness profiles. The LWD results agree well with those of the wireline measurements.

system that is mounted transversely to the longitudinal axis of the drill collar (Minear et al. 1995; Kostek et al., 1994; Kostek et al., 1998). In this system, the transmitter and receiver transducers are aligned along the collar and pointed away from the collar axis. The benefit of this design is its small, compact size suitable for sustaining the drilling vibration. In one single measurement, the transversely-mounted system measures the formation property only for a small azimuthal sector of the borehole. The full azimuthal coverage is obtained as the system rotates during drilling. Later designs include a dual-mode system (Varsamis et al, 1999) and an omni-directional system (Joyce et al., 2001) to improve both the azimuthal coverage and acoustic signal amplitude of the measurement.

Acoustic isolation is crucial for the compressional-wave measurement. In such a measurement, a dominant tool mode, the collar extensional wave, is excited and propagates along the collar. This tool wave has a much larger amplitude than that of the formation arrivals and therefore must be suppressed in order to measure the later arrivals. Fortunately, the tool extensional wave has a natural stop band (of about 2 kHz–3 kHz width) in the frequency range of 10 kHz–20 kHz, the actual location of the stop band depending on the collar thickness. In this band, the tool wave has the highest attenuation and smallest amplitude. Interestingly, this stop band coincides with the Airy phase frequency (i.e., group velocity minimum) of the tool extensional wave, suggesting that the attenuation is caused by the radiation loss of the Airy phase into the borehole fluid and surrounding formation. By taking advantage of the tool-wave stop band, the isolator can be made by cutting grooves on the collar to further enhance the attenuation. The grooved structure, like any periodic or quasi-periodic structure, also has a stop band, which, coupled with the tool wave's natural stop band, creates attenuation sufficiently large to suppress the tool wave (a 40 dB attenuation was reported in Leggett et al., 2001). From the above described operation principle of a LWD compressional-wave isolator, one can understand why the LWD compressional-wave operating frequency (> 10 kHz) is usually higher than its wireline counterpart (< 10 kHz).

Later generations of LWD tools began to address the shear-wave measurement in slow formations. The first attempt to measure the shear wave in LWD was using the dipole source and receiver system, because the dipole technology has been very successful and widely accepted in wireline logging. However, the application of the dipole measurement to LWD is seriously limited by the drill collar. The drill collar, under the dipole excitation, produces a strong collar flexural wave that contaminates the formation arrival. In addition, there is a large velocity difference between formation shear and flexural waves, as have been demonstrated by the modeling results in Chapter 2. In comparison, the LWD quadrupole wave is a better candidate for the shear measurement. As the modeling results of Chapter 2 show, the advantages of using LWD quadrupole waves are (i) the tool quadrupole wave is absent when operating in the low-frequency range, and (ii), the quadrupole wave in a slow formation travels at formation shear velocity at low frequencies. Based on these advantages, a multipole (monopole and quadrupole) LWD acoustic tool has been built (see Tang et al., 2002a). A benefit of using the quadrupole design is that no acoustic isolation is needed for the quadrupole wave because the tool quadrupole wave does not exist when operating in the low-frequency (< 10 kHz) range.

6.2.4.2 Field example

Several field measurements have been made with the LWD multipole acoustic tool, which showed the promising results of compressional and, particularly, shear velocity measurement using LWD acoustic technology (Tang et al. 2002a). Figure 6.3b shows a field data comparison example for a deviated well in a slow sand-and-shale formation (see gamma-ray and caliper curves shown in track 1). Track 2 shows good quality LWD compressional-wave data (shown as a VDL image). The LWD (solid curve) and wireline (dotted curve) compressional slowness profiles agree well, as shown in track 3. Track 4 shows the measured LWD quadrupole-wave data. The quadrupole-shear wave shows almost no significant tool-wave contamination. This confirms the theoretical prediction of Figure 2.11. However, a small amplitude wave with invariant arrival time can still be seen ahead of the quadrupole wave. This wave is interpreted as the collar flexural/extensional wave caused by tool eccentering in this deviated well (theoretically, an off-centered quadrupole source can generate monopole and dipole waves). The slowness-derived compressional- and quadrupole-wave transit time curves track the respective waveform data in tracks 2 and 4 quite well, validating the slowness results. The quadrupole-wave slowness (solid curve in track 5), compared with the shear slowness profile (dotted curve) obtained from a later wireline dipole measurement, shows a good agreement between the two measurements. The small differences between the two curves are attributed to effects such as resolution difference, anisotropy, and data noise, etc., in the respective measurements. This example validates the compressional and shear slowness data measured using the LWD acoustic technology. Particularly, it demonstrates the practicability of measuring formation shear velocity using LWD quadrupole waves. However, the application of the quadrupole technology for the full range of LWD environments remains to be tested because there are still issues and concerns about this technology, as discussed below.

6.2.4.3 Issues on LWD acoustic measurement

In the actual LWD environment, several influences on the acoustic measurements are always present and can cause difficulties. One major influence is the noise caused by drilling and drilling mud circulation. The drilling environment is a very tough one. Acoustic noises can be generated by various vibrations of the drill string in its axial, radial, lateral, and azimuthal directions. Besides, the impact of the drill string on the borehole and the impact of drill bit on the formation generate strong drilling noise. Field measurements (Joyce et. al., 2001) have shown that the frequency range for typical drilling noise is 0—3.5 kHz. The circulation noise spectrum is focused in the 1 kHz–2.3 kHz range. These noises inevitably influence the frequency range for the measurement of shear wave velocities in slow formations. Therefore, enhancing the signal-to-noise ratio is the key to a successful low-frequency shear-wave velocity measurement. The measurement is also influenced by the tool waves traveling along the collar. In the quadrupole measurement, the tool wave is absent if the tool is perfectly centered in the borehole (see Figure 2.11). However, in the real world of LWD, tool waves may still exist during quadrupole logging. Because of the complexity of collar movement during drilling, perfect tool centralization is unlikely. An off-center quadrupole source inevitably generates some monopole and dipole components to excite tool waves along the collar. Fortunately, modeling (Wang and Tang, 2003; Huang, 2003) and measurement (e.g., Tang et al., 2003; see also Figure 6.3b) show that the resulting acoustic signals are still dominated by the quadrupole component, with some minor contamination from the non-quadrupole tool wave component.

Another concern is the dispersion effect on the quadrupole measurement. As shown by the quadrupole dispersion curve in Figure 2.11, the LWD quadrupole waves, like wireline dipole waves, can be quite dispersive when the measurement frequency is high. The dispersion effect is aggravated if heavy drilling mud is used. Practice experience with wireline dipole logging suggests that the LWD quadrupole dispersion effects can be dealt with by two approaches. One is improving the LWD tool engineering and data acquisition to enhance the signal-to-noise ratio in the low frequency range. The other is developing an effective dispersion correction method to compensate for the dispersion effects in the measured data.

A last, but certainly not the least, issue is the interpretation of LWD acoustic data, including the quadrupole, in a deviated well with strong anisotropy, as presently more and more deviated wells are drilled into highly anisotropic soft sediments of deep water reservoirs. With the development of the LWD acoustic technology, some efforts are already made to investigate this issue (Wang and Tang, 2003; Tang et al., 2003). In the coming years, we expect to see more developments in the LWD acoustic technology to address the needs of the petroleum industry. An integrated effort combining theoretical analysis/modeling, tool engineering and development, and data interpretation is the key to a more fruitful outcome and applications of the LWD acoustic technology.

REFERENCES

Aki, K., and Richards, P.G., 1980, QUANTITATIVE SEISMOLOGY: THEORY AND METHODS: W. H. Freeman and Co.

Alford, R.M., 1986, Shear data in the presence of azimuthal anisotropy: 56th Ann. Internat. Mtg., Soc. Explor. Geophys.: Expanded Abstracts, 476–479.

Auriault, J.L., Borne, L., and Chambon, R., 1985, Dynamics of porous saturated media, checking the generalized law of Darcy: J. Acoust. Soc. Am., 77, 1641–1650.

Behrmann, L.A., and Elbel, J.L., 1991, Effects of perforations on fracture initiation, J. Petr. Tech. (May 1991), 608–615.

Biot, M.A., 1956a, Theory of propagation of elastic waves in a fluid-saturated porous solid, I: Low frequency range: J. Acoust. Soc. Am. 28, 168–178.

Biot, M.A., 1956b, Theory of propagation of elastic waves in a fluid-saturated porous solid, II: Higher frequency range: J. Acoust. Soc. Am. 28, 179–191.

Biot, M.A., 1962, Mechanics of deformation and acoustic wave propagation in porous media: J. Appl. Phys. 33, 1482–1498.

Bracewell, R., 1965, THE FOURIER TRANSFORM AND ITS APPLICATIONS, McGraw-Hill, New York.

Brie, A., Pampuri, F., Marsala, A.F., and Meazza, O., 1995, Shear sonic interpretation in gas-bearing sands: SPE 30595, 1995 SPE annual technical conference and exhibition.

Chen, X.,F., Quan, Y., and Harris, J.M., 1994, Seismogram synthesis for radially multi-layered media using the generalized reflection/transmission coefficients method: 64th Ann. Internat. Mtg., Soc. Explor. Geophys., Expanded Abstracts, 4–7.

Cheng, C.H., and Johnston, D.H., 1981, Dynamic and static moduli, Geophys. Res. Lett. 8, 39–42.

Cheng, C.H., and Toksöz, M.N., 1981, Elastic wave propagation in a fluid-filled borehole and synthetic acoustic logs: Geophysics 56, 1603–1613.

Cheng, C.H., Toksöz, M.N., and Willis, M.E., 1982, Determination of in situ attenuation from full waveform acoustic logs: J. Geophys. Res. 87, 5477–5484.

Cheng, C. H., Zhang, J., and Burns, D.R., 1987, Effects of in-situ permeability on the propagation of Stoneley waves in a borehole: Geophysics 52, 1279–1289.

Cheng, N.Y., 1994, Borehole wave propagation in isotropic and anisotropic media: Three dimensional finite difference approach, Ph.D. Thesis, Massachusetts Institute of Technology, Cambridge, Massachusetts, USA.

Cheng, N.Y., and Cheng, C.H., 1995, Decomposition and particle motion of acoustic dipole log in anisotropic formation: 65th Ann. Internat. Mtg., Soc. Explor. Geophys., Expanded Abstracts, 1–4.

Cheng, N.Y., Cheng, C.H., and Toksöz, M.N., 1995, Borehole wave propagation in three dimensions: J. Acoust. Soc. Am. 84, 2215–2229.

Chi, S. H., and Tang, X. M., 2003, Accurate approximations to qsv and qp wave speeds in TIV media and Stoneley wave speed in general anisotropic media, in 44th Society of Processional Well Log Analysts Annual Logging Symposium, paper OO.

Chunduru, R.K., and Tang, X.M., 1998, Method for determining acoustic velocity of earth formations by simulating receiver waveforms for an acoustic array logging instrument, US Patent 5,740,124.

Chunduru, R.K., Sen, M.K., and Stoffa, P.L., 1996, 2-d resistivity inversion using spline parameterization and simulated annealing: Geophysics 61, 151–161.

Coates, R., Kane, M., Esmersoy, C., Fukuhara, C., and Yamamoto, H., 2000, Single-well sonic imaging: High-definition reservoir cross-sections from horizontal wells, SPE #65457.

Cowles, C.S., Leveille, J.P., Hatchell, P.J., 1994, Acoustic multi-mode wide-band logging device: U.S. Patent No. 5,289,433.

Dvorkin, J., and Nur, A., 1993, Dynamic poroelasticity: A unified model with the squirt and Biot mechanisms: Geophysics 58, 524–533.

Ellefsen, K.J, 1990, Elastic wave propagation along a borehole in an anisotropic medium, Sc.D. Thesis, Massachusetts Institute of Technology, Cambridge, Massachusetts, USA.

Ellefsen, K.J., Cheng, C.H., and Toksöz, M.N., 1991, Effects of anisotropy upon the normal modes in a borehole: J. Acoust. Soc. Am. 89, 2597–2616.

Esmersoy, C., Boyd, A., Kane, M., and Denoo, S., 1995, Fracture and stress-evaluation using dipole-shear anisotropy logs, paper J, in 39th Annual Meeting Transactions: Society of Professional Well Log Analysts.

Esmersoy, C., Chang, C., Kane, M., Coates, R., Tichelaar, B. and Quint, E., 1998, Acoustic imaging of reservoir structure from a horizontal well: The Leading Edge 17, no. 07, 940–946.

Gelinsky, S. and Tang, X. M., 1997, Fast forward modeling of Stoneley waves for irregular boreholes and heterogeneous formations, 67th Ann. Internat. Mtg: Soc. of Expl. Geophys., 317–320.

Hornby, B.E., 1989, Imaging near-borehole of formation structure using full-waveform sonic data, Geophysics, 54, 747–757.

Hornby, B.E., 1989: Method for determining formation permeability by comparing measured tube waves with formation and borehole parameters, U.S. Patent No. 4,797,859.

Hornby, B.E., 1993, Tomographic reconstruction of near-borehole slowness using refracted borehole sonic arrivals: Geophysics 58, 1726–1738.

Hornby, B.E., Johnson, D.L., Winkler, K.H., and Plumb, R.A., 1989, Fracture evaluation using reflected Stoneley-wave arrivals: Geophysics 54, 1274–1288.

Hsu, K., and Chang, S. K., 1987, Multiple-shot processing of array sonic waveforms: Geophysics 52, 1376–1390.

Huang, X., 2003, Effects of tool positions on borehole acoustic measurements: a stretched grid finite difference approach, Ph.D. thesis, Massachusetts Institute of Technology, Cambridge, Massachusetts, USA

Hudson, J., 1981, The overall property of cracked solid: Math. Proc. Cambridge Phil. Soc. 88, 371–384.

Ingber, L., 1989, Very fast simulated reannealing: Journal of Math. Computation 12, 967–973.

Jackson, J.D., 1962, CLASSICAL ELECTRODYNMICS, John Wiley & Sons, New York.

Jaeger, J.C., and Cook, N.G.W., 1977, FUNDAMENTALS OF ROCH MECHANICS, Halsted Press.

Johnson, D.L., Koplic, J., and Dashen, R., 1987, Theory of dynamic permeability and tortuosity in fluid-saturated porous media: Journal of Fluid Mechanics 176, 379–400.

Joyce, R., Patterson, D., and Thomas, J., 1998, Advanced Interpretation of Fractured Carbonate Reservoirs Using Four-component Cross-dipole Analysis, Paper R, in 39th Annual Logging Symposium Transactions: Society of Professional Well Log Analysts.

Joyce, B., Patterson, D., Leggett, J.V., III, and Dubinsky, V., 2001, Introduction of a new omni-directional acoustic system for improved real-time LWD sonic logging-Tool design and field test results, paper G, in 42nd Annual Logging Symposium Transactions, Society of Professional Well Log Analysts.

Kenyon, W.E., Day, P.I., Straley, C., and Willemsen, J.F., 1986, Compact and consistent representation of rock NMR data for permeability estimation: Trans., 61st Ann. Tech. Conf. Soc. Petr. Eng., 22.

Kessler, C., and Varsamis, G., 2001, A new generation crossed dipole logging tool: Design and case histories: SPE Annual Conference and Exhibition, paper 71740, New Orleans, Louisiana.

Kimball, C.V., 1998, Shear slowness measurement by dispersive processing of borehole flexural mode: Geophysics 63, 337–344.

Kimball, C.V., and Marzetta, T.L., 1986, Semblance processing of borehole acoustic array data: Geophysics 49, 274–281.

Kitsunezaki, C., 1980, A new method for shear wave logging, Geophysics, 45, 1489–1495.

Klimentos, T., 1995, Attenuation of P- and S-waves as a method of distinguishing gas and condensate from oil and water: Geophysics 60, 447–458.

Klimentos, T., and McCann, C., 1990, Relationships among compressional wave attenuation, porosity, clay content, and permeability in sandstones: Geophysics 55, 998–1014.

Kostek, S., Plona, T.J., and Chang, S.K., 1994, Receiver apparatus for use in logging-while-drilling: U.S. Patent No. 5,309,404.

Kostek, S., Chang, S.K., McDaniel, G., Plona, T., Randall, C., Masson, J-P., Mayes, J.C., and Hsu, K., 1998, Method of logging while drilling a borehole traversing an earth formation: U.S. Patent No. 5,796,677.

Leggett, J.V. III, Dubinsky, V., Patterson, D., and Bolshakov, A., 2001, Field test results demonstrating improved real-time data quality in an advanced LWD acoustic system, SPE Annual Conference and Exhibition, paper 71732, New Orleans, Louisiana.

Li, Y., Zhou, R., Tang, X., Jackson J.C., and Patterson, D., 2002, Single-well imaging with acoustic reflection survey at Mounds, Oklahoma, USA, paper P141, 64th EAGE Conference & Exhibition, Florence, Italy. (May 27–30).

Mao, N.-H., 1987, Shear-wave transducers for stress measurements in boreholes: U.S. Patent No. 4,641,520.

Marquardt, D.W., 1963, An algorithm for least-squares estimation: J. Soc. Ind. Appl. Math. 11, 431–441.

Marrauld, J., Arnaud, Jblanco, J., and Paugue, L., 2001, How can log and core velocity anisotropy measurements improve seismic processing: Paper B-18, in 63rd Conference and Technical Exhibition, European Association of Geoscientists and Engineers.

McFadden, P.L., Drummond, B.J., and Kravis, S., 1986, The n-th root stack: Theory, application, and examples: Geophysics 51, 1879–1892.

Meredith, J.A., 1990, Numerical and analytical modeling of downhole seismic sources: The near and far field, Ph.D. thesis, Massachusetts Institute of Technology, Cambridge, Massachusetts, USA.

Minear, J., Birchak, R., Robbins, C., Linyaev, E., Mackie, B., Young, D., and Malloy, R., 1995, Compressional slowness measurements while drilling, paper VV, in 36th Annual Logging Symposium Transactions, Society of Professional Well Log Analysts.

Morse, P.M., and Feshbach, H., 1953, METHODS OF THEORETICAL PHYSICS, McGraw-Hill Book Company, New York.

Mueller, M.C., Boyd, A.J., and Esmersoy, C., 1994, Case studies of the dipole shear anisotropy log: 64th Ann. Internat. Mtg., Soc. Explor. Geophys., Expanded Abstracts, 1143–1146.

Nolte, B., and Huang, X.J., 1997, Dispersion analysis of split flexural waves: ANNUAL REPORT OF BOREHOLE ACOUSTICS AND LOGGING AND RESERVOIR DELINEATION CONSORTIA, Massachusetts Institute of Technology.

Norris, A.N., 1990, The speed of a tube wave: J. Acoust. Soc. Am., 87, 414–417.

Norris, A.N., and Sinha, B.K., 1993, Weak elastic anisotropy and the tube waves:

Geophysics 58, 1091–1098.

Nur, A., and Simmons, G., 1969, Stress-induced velocity anisotropy in rock: An experimental study: Jour. Geoph. Res. 72, 6667–6674.

Paillet, F.L., and Cheng, C.H., 1991, ACOUSTIC WAVES IN BOREHOLES, CRC Press, Inc.

Pampuri, F., Borghi, M., Deias, S., Giorgioni, M., Brambillla, F., Tang, X.M., and Patterson, D., 2003, Compressional slowness determination behind poorly-cemented casing (case history), OMC 2003, 6th Offshore Mediterranean Conference and Exhibition, Ravenna, Italy, March 26–28, 2003.

Patterson, D., and Skjong, G., and Wade, J. M., 1999, Horizontal Stress Orientation Analysis Using the Cross-dipole Acoustic log in the Eldfisk Field, Paper SS, in 40th Annual Logging Symposium Transactions: Society of Professional Well Log Analysts.

Peng, C., 1994, Borehole effects on downhole seismic measurements, Ph.D. thesis, Massachusetts Institute of Technology, Cambridge, Massachusetts, USA.

Plona, T., Kane, M., Sinha, B., Walsh, J., and Viloria, O., 2000, Using acoustic anisotropy, Paper H, in 41st Annual Logging Symposium Transactions: Society of Professional Well Log Analysts.

Press, W.H., Flannery, B.P., Teukosky, S.A., and Vetterling W.T., 1989, NUMERICAL RECIPES, Cambridge University Press, Cambridge, UK.

Prony, R., 1795, Essai experimental et analytique: L'ecole Polytech.1, 24–76.

Qobi, L., Kuijper A., Tang, X.M., and Strauss, J., 2001, Permeability determination from Stoneley waves in the Ara Group carbonates, Oman, GeoArabia 6, 649–666.

Rai, C.S., and Hanson, K.E., 1988, Shear-wave velocity anisotropy in sedimentary rocks: A laboratory study: Geophysics 53, 800–806.

Rosenbaum, J.H., 1974, Synthetic microseismograms: logging in porous formations: Geophysics 39, 14–32.

Schmitt, D.P., 1988a, Effect of radial layering when logging in saturated porous formations: J. Acoust. Soc. Am. 84, 2200–2214.

Schmitt, D.P., 1988b, Shear-wave logging in elastic formations: J. Acoust. Soc. Am. 84, 2215–2229.

Schmitt, D.P., 1989, Acoustic multipole logging in transversely isotropic poroelastic formations: J. Acoust. Soc. Am. 86, 2397–2421.

Schmitt, D.P., 1993, Dipole logging in cased boreholes: J. Acoust. Soc. Am. 93, 640–657.

Schmitt, D.P., Bouchon, M., and Bonnet, G., 1988, Full-waveform synthetic acoustic logs, in radially semi-infinite saturated porous formations: Geophysics 53, 807–823.

Schoenberg, M., and Sayers, C.M., 1995, Seismic anisotropy of fractured rock: Geophysics 60 204–211.

Sinha, B.K., and Kostek, S., 1995, Identification of stress induced anisotropy in formations: U.S. Patent No. 5,398,215.

Sinha, B.K., Norris, A.N., and Chang, S.K., 1994, Borehole flexural modes in anisotropic formations: Geophysics 59, 1037–1052.

Smith, M.L., Sondergeld, C.H., and Norris, J.O., 1991, The Amoco array sonic logger: The Log Analysts 32, 201–214.

Stephen, R.A., Pardo-Casas, F. and Cheng, C.H., 1985, Finite-difference synthetic acoustic logs : Geophysics 50, 1588–1609.

Sun, X., Tang, X.M., Cheng, C.H., and Frazer, N., 2000, P- and S- wave attenuation logs from monopole sonic data, Geophysics 65, 755–765.

Tang, X.M., 1996a, Processing dipole waveform logs for formation alteration identification: 65th Ann. Internat. Mtg., Soc. Expl. Geophys., Expanded Abstracts.

Tang, X.M., 1996b, Fracture hydraulic conductivity estimation from borehole Stoneley wave transmission and reflection data, paper HH, in 37th Annual Logging Symposium Transactions: Society of Professional Well Log Analysts.

Tang, X.M., 2003, Imaging near-borehole structure using directional acoustic-wave measurement: 73rd Ann. Internat. Mtg., Soc. Explor. Geophys., Expanded Abstracts.

Tang, X.M., and Cheng, C.H., 1993a, Borehole Stoneley wave propagation across permeable structures: Geophysical Prospecting 41, 165–187.

Tang, X.M., and Cheng, C.H., 1993b, The effects of a logging tool on the Stoneley wave propagation in elastic and porous formations: The Log Analysts 34, 46–56.

Tang, X.M., and Cheng, C.H., 1996, Fast inversion of formation permeability from Stoneley wave logs using a simplified Biot-Rosenbaum model: Geophysics 61, 639–645.

Tang, X.M., and Martin, R.J., 1994, High-resolution evaluation of formation flow properties from a borehole acoustic imaging tool: Research Report under Gas Research Institute Contract No. 5093-260-2753.

Tang, X.M., and Patterson, D., 2000, Fracture measurements in open/cased holes using cross-dipole logging: Theory and field results: 70th Ann. Internat. Mtg., Soc. Explor. Geophys., Expanded Abstracts.

Tang, X.M., and Patterson, D., 2001, Detecting thin-gas beds in formations using Stoneley-wave reflection and high-resolution slowness measurements, paper OO, in 42nd Annual Meeting Transactions: Society of Professional Well Log Analysts.

Tang, X.M., Cheng, C.H., and Toksöz, M.N., 1991a, Dynamic permeability and borehole Stoneley waves: A simplified Biot-Rosenbaum model: J. Acoust. Soc. Am. 90, 1632–1646.

Tang, X.M., Cheng, C.H., and Toksöz, M.N., 1991b, Stoneley wave propagation in a fluid-filled borehole with a vertical fracture: Geophysics 56, 447–460.

Tang, X.M., Cheng, N., and Cheng, A., 1999, Identifying and estimating formation stress from borehole monopole and cross-dipole acoustic measurements, paper QQ, in 40th Annual Meeting Transactions: Society of Professional Well Log Analysts.

Tang, X.M., Patterson, D., and Melvin, H., 2001, Evaluating hydraulic fracture stimulation with cross-dipole acoustic logging technology, Paper 72500, SPE Journal of Reservoir Engineering and Formation Evaluation.

Tang, X.M., Dubinsky, V., Wang, T., Bolshakov, A., and Patterson, D., 2002a, Shear-velocity measurement in the Logging-While-Drilling environment: Modeling and field evaluations, Paper RR, in 43rd Annual Meeting Transactions: Society of Professional Well Log Analysts.

Tang, X.M., Wang, T., and Patterson, D., 2002b, Multipole Acoustic Logging-While-Drilling: 72nd Ann. Internat. Mtg., Soc. Explor. Geophys., Expanded Abstracts, 364–368.

Tang, X.M., Dubinsky, V., Harrison, C.W., Bolshakov, A., and Patterson, D., 2003, Logging-While-Drilling shear and compressional measurements: case histories, 44th Annual Logging Symposium Transactions, Society of Professional Well Log Analysts, paper II.

Tezuka, K., Cheng, C.H., and Tang, X.M., 1997, Modeling of low-frequency Stoneley-wave propagation in an irregular borehole: Geophysics 62, 1047–1058.

Thomsen, L., 1986, Weak elastic anisotropy: Geophysics 51, 1954–1966.

Toksöz, M.N., and Johnston, D., 1981, SEISMIC WAVE ATTENUATION, Geophysics reprint series No. 2., Society of Exploration Geophysicists.

Tongtaow, C., 1982, Wave propagation along a cylindrical borehole in a transversely isotropic formation, Ph.D. thesis, Colorado School of Mines.

Tsang L., and Radar, D., 1979, Numerical evaluation of the transient acoustic waveform due to a point source in a fluid-filled borehole: Geophysics 44, 1706–1720.

Tubman, K., Cheng, C.H., and Toksöz, M.N., 1986, Synthetic full waveform acoustic logs in cased boreholes II: Poorly bonded casing, Geophysics 51, 902–913.

Varsamis, G.L., Wisniewski, L.T., Arain, A., and Althoff, G., 1999, A new MWD full wave dual mode sonic tool design and case histories, paper F, in 40th Annual Logging Symposium Transactions, Society of Professional Well Log Analysts.

Wade, J.M., Hough, E.V., Pedersen, S.H., 1998, Practical methods employed in determining permeability anisotropy for optimization of a planned water flood, SPE 48961, 1998 SPE Annual Technical Conference and Exhibition.

Walsh, J.B., 1965, The effects of cracks on the compressibility of rock: J. Geophys., Res. 70, 831–842.

Wang, T., and Tang, X.M., 2003, Investigation of LWD quadrupole wave propagation in real environments, 44th Annual Logging Symposium Transactions, Society of Professional Well Log Analysts, paper KK.

Wang, X., and Hornby B.E., 2002, Dipole sonic response in deviated boreholes penetrating an anisotropic formation: 72nd Ann. Internat. Mtg., Soc. Explor. Geophys., Expanded Abstracts, 360–363.

Wang, Z., 2001, Seismic anisotropy in sedimentary rocks, 71st Ann. Internat. Mtg: Soc. of Expl. Geophys., 1740–1743.

Watson, G.A., 1944, A TREATISE ON THE THEORY OF BESSEL FUNCTIONS, Cambridge University press, second edition.

White, J.E., 1967, The Hula log, a proposed acoustic tool: Paper I, in 8th Annual Logging Symposium Transactions: Society of Professional Well Log Analysts.

White, J.E., 1983, UNDERGROUND SOUND, Elsevier Science Publishing Company, Inc.

White, J.E., and Tongtaow, C., 1981, Cylindrical waves in transversely isotropic media: J. Acoust. Soc. Am. 70, 1147–1155.

Williams, D.M., 1990, The acoustic log hydrocarbon indicator: Paper W, in 31st Annual Logging Symposium Transactions: Society of Professional Well Log Analysts.

Williams, D.M., Zemanek, J., Angona, F.A., Dennis, C.L., and Caldwell, R.L., 1984, The long space acoustic logging tool, Paper T, in 25th Annual Logging Symposium Transactions: Society of Professional Well Log Analysts.

Winbow, G.A., Chen, S.T., Rice, J.A., 1991, Acoustic quadrupole shear wave logging device, US Patent No. 5,027,331.

Winkler, K.W., Liu, H.L., and Johnson, D.L., 1989, Permeability and borehole Stoneley waves: Comparison between experiment and theory: Geophysics 54, 66–75.

Winkler, K.W., Sinha, B.K., and Plona, T.J., 1998, Effects of borehole stress concentrations on dipole anisotropy measurements: Geophysics 63, 11–17.

Xu, P.-C. and Parra, J., 1999, Effects of a vertical fluid-filled fracture on borehole dipole logs, 69th Ann. Internat. Mtg: Soc. of Expl. Geophys., 41–44.

Zemanek, J., Angona, F.A., Williams, D.M., and Caldwell, R.L., 1984, Continuous shear wave logging, Paper U., in 25th Annual Logging Symposium Transactions: Society of Professional Well Log Analysts.

Zhang, T., Tang, X.M., and Patterson, D.L., 2000, Evaluating laminated thin beds in formations using high-resolution slowness logs, Paper XX., in 41st Annual Logging Symposium Transactions: Society of Professional Well Log Analysts.

Zhao, X., Toksöz, M.N. and Cheng, C.H., 1993, Stoneley wave propagation in heterogeneous permeable porous formation, 63rd Ann. Internat. Mtg: Soc. of Expl. Geophys., 76–79.

Zhao, X.M., 1994, Effects of heterogeneity on fluid flow and borehole permeability measurement, Ph.D. Thesis, Massachusetts Institute of Technology, Cambridge, Massachusetts, USA.

Zhu X., and McMechan, G.A., 1988, Estimation of near-surface velocities by tomography: 58th Ann. Internat. Mtg., Soc. Expl. Geophys., Expanded Abstracts, 1236–1238.

Names and Places

K

L

M

N

P

Q

R

S

T

Index

A

D

E

F

Q

R

T

U

V

Y

Z